Education

Presentations at the 2010 AIChE Annual Meeting

Salt Lake City, Utah, USA
7-12 November 2010

ISBN: 978-1-61782-150-9

Printed from e-media with permission by:

Curran Associates, Inc.
57 Morehouse Lane
Red Hook, NY 12571

Printed by Curran Associates, Inc. (2011)

For permission requests, please contact AIChE
at the address below.

AIChE
3 Park Avenue
New York, NY 10016-5991

Phone: (203) 702-7660
Fax: (203) 775-5177

www.aiche.org

Additional copies of this publication are available from:

Curran Associates, Inc.
57 Morehouse Lane
Red Hook, NY 12571 USA
Phone: 845-758-0400
Fax: 845-758-2634
Email: curran@proceedings.com
Web: www.proceedings.com

TABLE OF CONTENTS

(4a) Surface Functionalization of Nanomaterials to Elicit "Smart" Properties

Allan E. David and Victor C. Yang, Department of Pharmaceutical Sciences, The University of Michigan, Ann Arbor, MI

Rapid advances in the field of nanotechnology have brought enormous potential for improving the quality of life. The effects of this growth are now being felt in such varied industries as drug delivery, catalysis, tissue engineering, renewable energy, and optical coatings, to name just a few. While the control of length scales at the nanometer level alone yields some exciting properties, further functionalization of the surface with "smart" ligands that provide environment-responsive performance is generating tremendous interest. My research goals are towards the development of novel, response nanomaterials for a broad field of applications. During this poster session, I will present some of the work done in developing functional nanoparticles and nanocomposites for applications in biocatalysis and drug delivery, and discuss the future potential of this technology.

My graduate research was focused on the development of a nanoporous organic-inorganic composite matrix, using sol-gel techniques, for enzyme immobilization. With proper management of surface energies, we developed an organic-inorganic hybrid material with the following desirable properties: 1) high surface area (~650 m2/g) and uniform surface functionalization which together yielded an extremely high loading capacity (>700 mg protein/g matrix); 2) complete retention of enzyme's catalytic efficiency; 3) good pH and thermal stability; and 4) open pore morphology to facilitate substrate/ production transport. The extremely high enzymatic activity of this material allowed for reactor miniaturization, which was validated with large-scale, GLP production of a therapeutic peptide (currently undergoing preclinical evaluation for Investigational New Drug approval under a $2.4 million NIH SBIR Phase II Continuation grant).

This material technology was further enhanced during my postdoctoral research with the inclusion of microenvironment-responsive biopolymers within the inorganic matrix for drug delivery applications. Particles formed with this nanocomposite displayed pH-dependent mucoadhesive and swelling properties – allowing for targeted and controlled delivery of drugs to the oral cavity, stomach, or intestinal tract. We have specifically pursued the utilization of this material for oral delivery of insulin and antibiotics. Additionally it was found that this nanocomposite could be utilized to modulate water activity close to the surface, which in turn would provide control of immobilized enzyme activity – yielding control of both reaction extent and side-product generation. There is also potential for this material in wound healing and tissue engineering, as well as in separation processes. I also served as group leader for the MION sub-group, of Dr. Victor Yang's lab (Department of Pharmaceutical Sciences, University of Michigan, Ann Arbor MI), which focused on the utilization of magnetic nanoparticles for simultaneous imaging and therapy of brain tumors. In this capacity, I mentored and trained an undergraduate student, five PhD students, and four postdoctoral fellows, a group which has already produced 10 peer-reviewed papers with more in preparation.

Overall, I have an extensive background, and had success, in the development of novel nanomaterials – both functional nanoparticles and nanocomposites (yielding 4 patent applications). I have displayed leadership abilities by maximize the productivity of a diverse group of students/researchers. Additionally, I am very familiar with the grant funding mechanism - having served as ad hoc member for six different NIH SBIR grant review committees, and successfully obtained more than $3 million in NIH grant funding for the various projects. The common thread I have going through all my research interests is the surface modification of nanomaterials to elicit "smart" properties. I expect to initially continue the development of novel, environment-responsive nanoparticles for cancer therapy and nanocomposite materials for oral drug delivery, while I evaluate the other applications for this technology.

See more of this Session: Meet the Faculty Candidate Poster Session
See more of this Group/Topical: Education

(4aa) Chemical Looping Strategy and Its Commercial Potential for Carbon Negative Energy Conversions

Fanxing Li and L. - S. Fan, Chemical and Biomolecular Engineering, The Ohio State University, Columbus, OH

Energy and global warming are two intertwined issues of critical importance in the modern era. With the pressing needs for clean, efficient, and cost effective energy sources, the chemical looping strategy has evolved as a promising alternative to the traditional carbonaceous fuel conversion processes. With aid of oxygen carrier particles, the chemical looping process efficiently converts carbonaceous fuel while capturing CO_2 in an effective manner by looping reaction design. Two unique chemical looping gasification processes, i.e. the Syngas Chemical Looping (SCL) and the Coal/Biomass Direct Chemical Looping (CDCL) processes were developed and investigated. Unlike other chemical looping processes that are combustion based, the SCL and CDCL processes convert a variety of fuels including coal and biomass into a combination of hydrogen, liquid fuel, and/or electricity. Over the past 6 years, the processes have been advanced from a novel concept to successful sub-pilot (25 kW_{th}) scale demonstrations. A 250 kW_{th} SCL pilot plant is being designed and will be operated at the National Carbon Capture Center (NCCC). The various aspects of the looping process development including the formulation and performance of the oxygen carrier particles, the bench and sub-pilot scale process testing results, and the energy integration optimization and techno-economic analyses of the processes will be presented. The scale-up strategy of the processes for commercial use will also be discussed. The ionic transfer mechanism that underlines the oxygen carrier reaction along with several energetically optimized novel applications of the chemical looping for gasification process systems will be highlighted.

See more of this Session: Meet the Faculty Candidate Poster Session
See more of this Group/Topical: Education

(4ab) Fabrication and Evaluation of Omniphobic Surfaces for Liquid Repellency and Reduced Ice Adhesion

Adam J. Meuler, Chemical Engineering, Massachusetts Institute of Technology, Cambridge, MA

Currently I am a postdoctoral associate advised by Dr. Joseph M. Mabry (Edwards Air Force Base), Professor Gareth H. McKinley (MIT Mechanical Engineering), and Professor Robert E. Cohen (MIT Chemical Engineering). Following the conclusion of my postdoctoral tenure, I hope to begin a tenure track faculty position in a Chemical Engineering or Materials Science and Engineering Department. My research program as a faculty member will focus on the fabrication of polymeric materials, including block polymers, nanocomposites, and coatings, and the evaluation of their structure-property relationships. Core experimental competencies of my laboratory will include polymer synthesis, structural characterization (e.g., small-angle X-ray scattering), surface characterization (e.g., contact angle measurements, electron microscopy), and dynamic measurements (e.g., rheology).

Our research team investigates the fundamentals of solid-liquid and solid-solid interactions and uses the principles elucidated from these studies (and those of other groups) to design and fabricate surfaces with desired wettability and adhesion characteristics. My research has largely involved omniphobic surfaces that resist wetting by all liquids and adhere poorly to other solids. In this poster I will highlight two of these specific projects:

Relationships between Water Wettability and Ice Adhesion

Reports describing superhydrophobic surfaces that resist wetting by water droplets have appeared with increasing frequency in the literature over the past twenty years. Typically this liquid repellency is imparted using a combination of surface texture and low-energy organic coatings. The fundamentals of ice adhesion have, relative to their water wettability counterparts, received relatively little attention in the literature, and it is not widely understood which attributes must be tuned to design icephobic surfaces. Here we probe the relationships between advancing/receding water contact angles and the strength of ice adhesion to more than 20 nominally smooth (Wenzel roughness < 1.01) test coatings with a broad range of wettabilities. Contact angles are measured using a commercial goniometer while the shear strengths of ice adhesion are evaluated with a home-built laboratory-scale apparatus. We find that high receding water contact angles correlate strongly with reduced ice adhesion. We believe that these results allow us to estimate the maximum reduction in ice adhesion strength that is attainable by coating smooth surfaces with known low energy coatings. Current efforts are focused on examining the adhesion of ice and other freezing/vitrifying liquids to surfaces with a variety of surface textures and liquid wettabilities.

Towards Robust Hydrophobic and Oleophobic Surfaces

Sustenance of metastable composite (solid-liquid-air) interfaces is critical to achieving the high contact angles and low contact angle hystereses that are characteristic of a strong repellency of low surface tensions liquids. Unfortunately many such surfaces described in the literature are of limited utility in applications due to poor mechanical properties, optical opacity, expensive fabrication procedures, and/or non-robust repellency against low surface tension liquids (e.g., oils). Here we describe the development of a replica molding process used to prepare mechanically robust crosslinked elastomeric substrates comprised of perfluoropolyethers or polysiloxanes that are characterized by submicron re-entrant surface texture. Some of these crosslinkable materials are available commercially, while others are prepared in our laboratory. Essential steps in this protocol include the preparation of a master mold using photolithographic techniques, filling the mold and subsequently curing the crosslinkable oligomers, and releasing the elastomeric replica from the template. Substrates with a range of reentrant texture feature sizes and spacings are prepared and the wettabilities of these surfaces are probed using a variety of high (e.g., water) and low (e.g., hexane) surface tension liquids.

See more of this Session: Meet the Faculty Candidate Poster Session
See more of this Group/Topical: Education

(4ac) Atmospheric Organic Particulate Matter: Measurements, Models and Mitigation

Lea Hildebrandt, Chemical Engineering, Carnegie Mellon University, Pittsburgh, PA

Submicron atmospheric particles adversely affect human health, and they have a highly uncertain effect on climate. Organic aerosol globally comprises a significant fraction (20-90%) of the submicron particle mass (PM1). While the formation of inorganic aerosol is relatively well understood, organic aerosol is not. In contrast to inorganic aerosol, which is mostly composed of a few well-characterized components such as sulfates, nitrates and ammonium, organic aerosol is composed of thousands of species, many of them unidentified, and has a myriad sources – both anthropogenic and biogenic, particle-phase and gas-phase. Even though much progress has been made in recent years, these studies have revealed that atmospheric organic aerosol is much more complex than originally thought, complicating the analysis of policy options aimed at reducing organic aerosol concentrations.

The goal of my research is to improve our understanding of the processes governing the properties and concentrations of organic aerosol in the atmosphere through a series of laboratory experiments and ambient measurements. The findings from these studies are implemented in our three-dimensional chemical transport model PMCAMx, and the model is then tested against observations. Finally, the improved model can be used to evaluate different emission control options for reducing organic aerosol concentrations and their adverse effects.

My experimental work focuses on investigating the initial formation, mixing and aging of organic aerosol. Laboratory experiments using environmental chambers are complicated by issues such as losses of particles and vapors to the chamber walls. Even though much work remains to be done to truly understand wall losses, we have treated these issues thoroughly (Pierce et al., 2008; Hildebrandt et al., 2009), and we are continuing to work on them. In the aerosol formation experiments, we found that the aerosol mass yields from anthropogenic organic aerosol precursors such as toluene are much higher than previously reported values (Hildebrandt et al., 2009). We suggested new model parameters based on our studies, and their use is significantly improving the ability of chemical transport models to reproduce observations.

Most chemical transport models such as PMCAMx use a pseudo-ideal mixing assumption for modeling organic aerosol. This is a critical assumption which greatly affects modeled organic aerosol concentrations, but the assumption has not been confirmed experimentally. A main experimental challenge is that organic aerosol from different sources exhibits very similar mass spectra in an aerosol mass spectrometer (AMS), especially after the aerosol has undergone further reactions. The AMS from Aerodyne, Inc. is currently the most powerful and the most widely used instrument to obtain online information on aerosol composition at high time resolution. In order to be able to distinguish and track organic aerosol from different sources using an AMS, I conducted novel experiments using a High

Resolution Time-of-Flight AMS (HR-ToF-AMS) and isotopically labeled compounds. Analyzing these experiments included identifying all isotopically labeled organic fragments based on their exact mass, and subsequently separating the labeled and non-labeled organic fragments using mathematical methods such as positive matrix factorization and chemical mass balance. I was able to confirm that anthropogenic and biogenic secondary organic aerosol components mix ideally. In the future, we will use this experimental approach to also probe the ideal mixing assumption between primary and secondary organic aerosol components.

Chemical transformation (aging) of organic aerosol is highly complex and currently poorly understood. In aging experiments, we have found that the concentration and degree of oxidation of organic aerosol change with aging, but that the extent of these effects depends on organic aerosol type, as well as on experimental (or ambient) conditions. In general, studying the aging of organic aerosol in the laboratory is challenging due to the limited time scale of laboratory experiments. Ambient measurements in a remote location have allowed us to study highly aged organic aerosol. We have found that highly aged organic aerosol has very similar characteristics (oxidative state, vapor pressure), regardless of its original source (Hildebrandt et al., 2010a; Lee et al., 2010; Pikridas et al., 2010). In the winter when oxidative conditions are milder, the organic aerosol is less oxidized and its characteristics are more variable than under the harsher oxidative conditions in the summer (Hildebrandt et al., 2010b). Overall, the variability between different organic aerosol types decreases significantly with aging. Thus, the photochemical age of organic aerosol may be just as important as the aerosol source in understanding organic aerosol concentrations and characteristics.

A complete understanding of the formation and evolution of atmospheric organic aerosol is still far away. Nevertheless, our studies have allowed us to develop new parameterizations for our chemical transport model PMCAMx, and to improve existing ones. The updated PMCAMx performs well in predicting measured organic aerosol concentrations and approximate oxidative states in urban and rural environments, and in different seasons. It represents organic aerosol processes and concentrations more accurately than any other available model, making it suitable for evaluating different policy options aimed at reducing atmospheric aerosol concentrations.

References:

Hildebrandt, L., et al. (2009), High formation of secondary organic aerosol from the photo-oxdiation of toluene, Atmospheric Chemistry and Physics, 9, 693-733.

Hildebrandt, L., et al. (2010a), Aged organic aerosol in the Eastern Mediterranean: The Finokalia Aerosol Measurement Experiment – 2008, Atmospheric Chemistry and Physics, 10, accepted.

Hildebrandt, L., et al. (2010b), Formation of low-volatility oxygenated organic aerosol in the atmosphere: Insights from the Finokalia Aerosol Measurement Experiments -2008/2009, Geophysical Research Letters, in preparation.

Lee, B.-H., et al. (2010), Volatility of organic aerosol sampled during FAME-2008, Atmospheric Chemistry and Physics Discussions, in preparation.

Pierce, J. R., et al. (2008), Constraining particle evolution from wall losses, coagulation, and condensation-evaporation in smog-chamber experiments: optimal estimation based on size distribution measurements, Aerosol Science and Technology, 42, 1001-1015 .

Pikridas, M., et al. (2010), The Finokalia Aerosol Measurement Experiments - 2008 (FAME-08): An Overview, Atmospheric Chemistry and Physics Discussions, 10, 6641–6679.

See more of this Session: Meet the Faculty Candidate Poster Session
See more of this Group/Topical: Education

(4ad) One- and Two-Probe Nonlinear Microrheology: Normal Stress Differences, Osmotic Pressure, and Nonequilibrium Depletion Flocculation

Roseanna N. Zia, Mechanical Engineering, California Institute of Technology, Pasadena, CA

The effect of single- and two-particle motion in nonlinear microrheology is studied, with a view toward connecting microstructural mechanics to normal stress and osmotic pressure. In active microrheology, the motion of a microscale probe (or set of probes) is tracked while being driven by an external force through a colloidal dispersion. Recent work on microrheology has focused on the average motion of a single probe, with the microviscosity defined theoretically and determined experimentally by application of Stokes' drag law. Fluctuations in probe motion are also of interest; collisions between probe and bath particles cause velocity fluctuations, scattering the probe from its mean path. We determined in our recent work that the (long-time) probe scattering is diffusive. The microdiffusivity, $\boldsymbol{D}^{\text{micro}}$, is transversely anisotropic, scaling linearly in the volume fraction of bath particles ϕ (for small ϕ) for all Peclet numbers, *Pe*, which gives the strength of probe forcing compared to thermal forces: $Pe = F^{ext}/(kT/b)$, where kT is the thermal energy and b the bath particle size. In light of this anisotropy, the notion that self-diffusion is driven by gradients in the particle or osmotic pressure prompts investigation of normal stresses in active microrheology — the anisotropy of the microdiffusivity indicates the presence of normal stress differences. We take two approaches to determine normal stresses: First, we derive the stresses directly from the deformed microstructure. Second, via the relationship of the microdiffusivity to particle (osmotic) pressure gradients, &partΣ/&partϕ, where Σ is the suspension stress. Owing to the axisymmetry of the motion about a spherical probe, $N_2 = 0$, while N_1 is linear in *Pe* for $Pe >> 1$ and vanishes when $Pe << 1$. The two approaches agree, suggesting that normal stress differences can be measured in active microrheological experiments if both the mean and mean-square motion of the probe are monitored. Next we study anisotropy arising due to external forcing of *two* probes through the bath — and the corresponding normal stresses. Much work in two-probe microrheology has focused on the linear response regime; here, we investigate the nonlinear response of a colloidal dispersion to the motion of two probe particles, both of size a, kept at constant separation R (scaled by a) and driven at constant velocity $\boldsymbol{U}^{ext}$ (perpendicular to their line of centres) through the suspension. Brownian dynamics simulations were conducted for concentrated and dilute suspensions, where the probe and bath particles are all of equal size; the bath microstructure and interactive force between probes were measured. The probes attract each other for separations less than a critical value R_c due to the depletion of bath particles between them and repel each other for separations of $R_c \leq R < 4.25$ as bath particles are able to accumulate between them. For separations greater than 4.5 radii, the probes no longer interact.

(4ae) Regulation of Stem Cell Behavior by Biomimetic Microenvironment Created with Polymer Nanoengineering

Yong Yang, Biomedical Engineering, Duke University, Durham, NC

Stem cell therapy holds great promise for diseases that are untreatable at present, such as various gene and neurological disorders, heart and blood diseases, as well as diabetes. A key challenge is to expand and derive these cells into a specific lineage, which requires a thorough understanding of how the stem cell microenvironment controls their fate decision. My research goals are to develop a well-defined biomimetic microenvironment for stem cell research and regenerative medicine through a solid understanding of the nanoscale properties of polymers and application of environmentally and biologically benign nanoengineering techniques.

The microenvironment in living tissues comprises numerous cells, extracellular matrix (ECM) proteins and a variety of soluble and ECM-bound factors distributed in three-dimensional (3-D) space. The ECM serves as structural support and provides, in concert with the spatio-temporally arranged signaling molecules and external stimuli, cues for cell adhesion, migration, proliferation and differentiation. Faithfully mimicking the in vivo cell microenvironment is a priority for basic cell biology and stem cell therapy. My proposed research in the near term will focus on creating a well-defined biomimetic microenvironment from biomaterials, micro-/nanofabrication and microfluidic approaches and investigate how these external stimuli, in particular, physical cues influence stem cell behavior. Through the mechanistic study with the biomimetic microenvironment, the merits of the physical cues in regulation of stem cell fate will be clarified, thus leading to the development of new biomaterials and biomedical devices. My proposed research in the long term will focus on applying this fundamental understanding, transport phenomena principles, and bioreactor approach to scale up the expansion and differentiation of stem cells for clinical transplantation.

Creation of the biomimetic microenvironment demands advances in polymer nanoengineering and polymer nanomateirals. Our severely limited understanding of structure-property relations in polymers at the nanoscale poses a major obstacle toward rational development of novel materials and processes for fabrication of nanodevices. Only by improving our fundamental understanding of the nanoscale properties of polymers, will we be able to elucidate scientific and engineering principles related to nanoscale structure and processes. My proposed research will focus on developing nanoengineering technologies to generate polymeric nanostructures and advancing experimental tools to probe the thermal, rheological and mechanical responses of the nanoconfined polymers. With the understanding of the properties of nanoconfined polymers, I will manipulate the nanoparticles of desired size, shape, and surface chemistry at three levels, namely particle aggregation/dispersion, 2-D and 3-D particle orientation and interconnection, and functionalization in the matrix for targeted properties. By utilizing a combination of controlled chemistry, controlled morphology, and polymer compounding techniques, a cost-effective, scalable procedure is expected to be developed to engineer functional polymer nanomaterials for biomedical applications as well industrial applications.

See more of this Session: Meet the Faculty Candidate Poster Session
See more of this Group/Topical: Education

(4af) Biosynthesis of Fungal Resorcylic Acid Lactones

Hui Zhou, Chemical and Biomolecular Engineering, University of California, at Los Angeles, Los Angeles, CA and Yi Tang, Chemical and Biomolecular Engineering, Chemistry and Biochemistry, University of California, at Los Angeles, Los Angeles, CA

Resorcylic acid lactones (RALs) are macrocyclic polyketides that display diverse biological activities. The biosynthesis of these compounds is proposed to involve two fungal polyketide synthases (PKS) that function collaboratively to yield a 14-membered macrolactone with a resorcylate core. This tandem enzymatic working mode was investigated based on hypothemycin biosynthetic system. Using purified Hpm8 and Hpm3 recombinantly expressed from *Saccharomyces cerevisiae*, we successfully reconstituted the >30 steps required for the synthesis of DHZ from malonyl-CoA in vitro. Kinetic analyses of the tandem IPKS system revealed that synthesis of the reduced hexaketide by Hpm8 is the rate-limiting step, and Hpm3 is only required at substoichiometric concentrations. The SAT domain of Hpm3 was also confirmed to mediate the communication between the Hpm8 and Hpm3 and is critical for the transacylation of the hexaketide intermediate between the two IPKSs. Biochemical study on PKS13, a non-reducing PKS (NRPKS) from zearalenone biosynthetic pathway revealed that it displayed broad starter-unit specificities towards fatty acyl-CoAs ranging in sizes between C6 and C16. Interestingly, PKS13 can also synthesize numerous alkyl pyrones and alkyl resorcylic esters in *Escherichia coli*, which is demonstrated to be interacting with *E. coli* fatty acid biosynthetic machinery. Heterologous reconstitution of Rdc1 and Rdc5 in *S. cerevisiae* led to the production of monocillin II, the precursor to radicicol. The generation of monocillin II via precursor feeding into Rdc1 culture further suggested that the chain length control by Rdc1 and Rdc5 differed from the former two systems (Hpm8 and Hpm3, PKS4 and PKS13).

See more of this Session: Meet the Faculty Candidate Poster Session
See more of this Group/Topical: Education

(4ag) Rational Design of the Heterogeneous Catalysts and Their Opportunities in Renewable Fuels

Wenqin Shen[1], Gerald Huffman[2], Goeff A. Tompsett[1], W. Curtis Conner Jr.[1] and George W. Huber[1], (1)Chemical Engineering, University of Massachusetts-Amherst, Amherst, MA, (2)Consortium for Fossil Fuel Science, University of Kentucky, Lexington, KY

Abstract:

A sustainable economy requires the development of alternative energy to replace the energy based on fossil fuels, thus the hydrogen production and the biomass conversion attract a lot of research attentions recently to make the world sustainable. The objective of the research is to develop novel catalysts and processes for the hydrogen production and the upgrading biomass-derived oxygenated compounds to useful chemicals and fuels. The nanoparticle impregnation method was developed to prepare supported nanopartilce (np) catalysts with controlled particle sizes and composition. And these np catalysts were applied for the non-oxidative dehydrogenation of methane for the production of CO_x-free hydrogen and sequestering carbon as potentially value-added by-product carbon nanotubes (CNTs). Monosized $Fe_{0.65}Ni_{0.35}$ oxide nanoparticles with an average particle size of 9 nm were prepared by thermal decomposition of an Fe, Ni oleate-surfactant complex in octadecene under reflux; these nanoparticles were dispersed onto a Mg(Al)O support to form a Fe-Ni np/Mg(Al)O catalyst. Compared with the Fe-Ni IW/Mg(Al)O catalyst, the nanoparticle catalyst was more easily reduced and exhibited enhanced methane dehydrogenation and 5 times longer life-times. Each reduced Fe-Ni nanoparticle functioned as an active site for the growth of CNTs. An invar-like Fe-Ni-C alloy phase is believed to be more active phase for methane dehydrogenation than Fe-Ni austenitic phase, Fe metal and metastable Fe-C carbides. Another effort was the combination of computational and experimental approaches to design shape selective solid base catalysts for the liquid phase furaldehydes condensations with ketones, where the zeolites with desired pore structures and sizes were nitrided in ammonia at the elevated temperatures to replace the bridging oxygens to lower electronegativity nitrogens for the generation of the strong basic sites. In one of these researches, the nitrogen-substituted NaY showed comparable catalytic activity to $MgO\text{-}ZrO_2$, but higher selectivity to furadehyde acetone monomer product, indicating the shape selective significance of nitrogen-substituted NaY. The combination of advanced material synthesis with computation, characterization, temperature programmed surface reactions, and kinetic studies provide a useful tool to rational design heterogeneous catalysts.

Keywords: Non-oxidative methane dehydrogenation; monodispersed nanoparticles; nanoparticle catalysts; solid base catalysts; nitrogen-substituted zeolites

See more of this Session: Meet the Faculty Candidate Poster Session
See more of this Group/Topical: Education

(4ah) Process and Catalysis Development for Sustainable Fuels Production

Andrew A. Peterson[1], Jefferson W. Tester[2] and Jens K. Nørskov[1], (1)Department of Chemical Engineering, Stanford University, Stanford, CA, (2)Chemical and Biomolecular Engineering, Cornell University, Ithaca, NY

The field of chemical engineering was built around the processing of petroleum into fuels and chemicals, and it is perhaps the field's biggest challenge to transition the industry's feedstocks to non-fossil sources of carbon. The only possible non-fossil carbon feedstocks are carbon dioxide and biomass. In this poster, I will report on two processes that I have worked on that offer innovative solutions to this challenge.

No material has been shown to (photo-)electrochemically produce hydrocarbon fuels from CO_2 both selectively and efficiently. Producing fuels and chemicals from CO_2 in a such a process would produce an ideal storage medium for intermittent renewable energy sources such as solar and wind. In my postdoctoral work at Stanford and at DTU, we are unraveling mechanistic details of how the selective catalysts (primarily copper and its alloys) and the efficient catalysts (primarily the active centers of enzymes) operate. Through these insights, we are designing future catalysts for efficient and selective CO_2 fixation to hydrocarbon fuels and chemicals.

In the second part of the poster, I will report on a novel biomass process that we developed while I was in my graduate studies at MIT. Current biofuels processes suffer from low efficiencies and produce fuels that are chemically distinct from their conventional analogs. In collaboration with co-workers in the metabolic engineering laboratory, we developed processes by which efficient fermentations could be employed to produce partially deoxygenated intermediate compounds. In a subsequent step, these intermediate compounds are hydrothermally reformed in sub- or supercritical water. This results in the selective production of either propane or propylene, depending on the pathway chosen. Since hydrothermal reforming is capable of being undertaken directly on the fermentation broth and the products (propane and propylene) are immiscible with the aqueous effluent, no costly distillation step is necessary to separate them. This is expected to result in an efficient, selective process towards producing conventional fuels from biomass feedstocks.

See more of this Session: Meet the Faculty Candidate Poster Session
See more of this Group/Topical: Education

(4ai) Statistical Process Inference, Control and Engineering (SPICE)

Kris Villez, School of Chemical Engineering, Purdue University, West Lafayette, IN

The field of systems engineering incorporates a broad range of techniques for advanced process inference, control and design. These techniques are commonly deployed to obtain better understanding, tracking and/or manipulation of complex processes in view of economical and energy efficiency, improved safety and reduction of environmental impacts. Historically, first-principle models have been most popular for the tasks of process state estimation, inference and control. Data-driven, statistical techniques have been developed more recently and require less fundamental process knowledge. As such, they have become increasingly popular since the nineties for ill-understood processes. Today, a tremendous amount of techniques is available within the Statistical Process Control (SPC) field.

Despite the boom in SPC research, adoption of these techniques in the industries is limited. This can at least in part be accounted to the lack of oversight in the field. Often, different techniques are aimed to tackle the same problem yet without elaborate comparison or real-life testing. In other cases, the absence of guidelines for practical implementation is a significant barrier for technology transfer. Therefore, I expect to devote a large part of my research career to bringing structure and guidelines to the SPC field.

Even if the amount of SPC techniques is vast, Fault Detection and Identification is largely based on the on-line analysis of data streams without direct inference with the process. As such, they can be labelled as Passive FDI techniques. However, it is technically possible to manipulate process variables in such a way that the produced data stream is more informative about the status of the monitored process. By doing so, the outcomes of on-line inference procedures can be made more accurate. To search for optimal manipulations of process with respect to on-line process inference accuracy, Optimal Experimental Design (OED) techniques will be adopted, refined and integrated into FDI theory and applications. As the active manipulation of a process for improved FDI is a new concept, I expect to take a lead in the emerging field of Active Fault Detection and Identification (Active FDI).

From a theoretical angle, it can also be said that there is no consensus on the theory and application of non-linear data-driven techniques for process monitoring and diagnosis. At this point, the leading research groups in this niche each have their own technique and a status quo is observed on the theoretical level. A better analysis of existing techniques and further theoretical development is essential to make further advances.

Based on the above analysis, a research plan is set out along three tracks. Each of these tracks relates to the field of applied statistics in process systems modelling, inference and control. The three tracks are listed as follows, from the most theoretical track to the most application-oriented track:

- Track I will result in the establishment of proper techniques for non-linear process monitoring. It will result in the specification of new model structures and algorithms.

- Track II will deliver techniques for improved on-line fault detection and identification (FDI) by means of the development and application of Active Fault Detection and Isolation methods. This will require the integration of Optimal Experimental Design (OED) techniques within FDI theory as well as specific adjustments to enable real-time computation. Extensive testing in both simulation and laboratory will be important.

- Track III is oriented at bridging the gap between theory and practice in Statistical Process Control (SPC). This track relies on comparative studies, validation and integration into real-life systems. Industrial partnerships are considered important to ensure impact on the field.

Together, the three tracks provide a balanced plan for research in systems engineering. The first track is largely theoretical. The second and third are more practice-oriented with the third focusing the removal of the gap between statistical theory and industrial process systems engineering.

See more of this Session: Meet the Faculty Candidate Poster Session
See more of this Group/Topical: Education

(4aj) Synthetic Biomaterials for Enhancing the Delivery and Effectiveness of DNA-Based Therapeutics

Christopher M. Jewell, Departments of Bioengineering, Materials Science and Engineering, and The Ragon Institute of MGH, MIT, and Harvard, Massachusetts Institute of Technology, Cambridge, MA

Over the past several decades, synthetic materials have contributed dramatically to new prophylactic and therapeutic treatments for a range of illness and disease, but drug delivery – the time, concentration, location, and specificity with which biological cargo is delivered – continues to be a major obstacle for medicine. My research interests are focused on designing and exploiting synthetic materials to achieve increased levels of control over the delivery of biological cargo. During this poster session I will highlight examples from my graduate and postdoctoral research which demonstrate the utility of synthetic biomaterials for enhancing the delivery of drugs, proteins, and nucleic acids *in vitro* and *in vivo*, and more recently, in the context of translational research focused on vaccination and immunotherapy [1-9]. Building on this past work, I will describe my plans to nucleate a distinct, independent research program at the interface of immunology and materials engineering.

During my graduate work, I took advantage of the properties of synthetic polymers and lipids to achieve high levels of spatial and temporal control over the delivery of nucleic acids and proteins. One major research area centered on the use of degradable multilayered polyelectrolyte coatings incorporating synthetic polyamines and DNA or proteins [4-6]. By incubating cells with substrates coated with ultrathin multilayered assemblies containing DNA encoding green fluorescent protein, we achieved localized gene expression (transfection) in up to 40% of cells without the aid of additional transfection agents. DNA-containing films fabricated on the surfaces of intravascular stents were conformal and robust, and able to withstand the mechanical forces associated with arterial deployment in vivo. Further, these coatings mediated significant levels of reporter gene expression in an in vivo rabbit model. Like cationic polymers, cationic lipids have also been widely investigated as gene delivery agents owing to their ability to aggregate with DNA to form lipid/DNA complexes which efficiently transport DNA into cells. Thus, a second research area focused on a cationic, ferrocene-containing lipid which has a charge density that can be reversibly transformed – using chemical or electrochemical methods – from +1 (reduced) to +3 (oxidized) [7-9]. We discovered that complexes prepared from reduced lipid yield high levels of transfection, while lipoplexes prepared using oxidized lipid yield very low levels of transfection, and we identified corresponding differences in the physical properties of these assemblies. Conventional lipids are typically active immediately upon formation and this presents a challenge for strategies aimed at achieving spatial or temporal control over the delivery of DNA to specific cells or tissue. Thus, our approach provides a means to address this challenge by permitting activation of DNA delivery through modulation of lipid redox state.

More recently, my research has focused on developing biomaterial-mediated strategies for improving DNA vaccination and immunotherapy. Our approach is based on enhancing vaccination through controlled delivery of antigen-encoding DNA and a novel class of immunomodulatory small molecules. We synthesized micro and nanoparticles with a core of poly(lactide-co-glycolide) or other bioresorbable polymers enveloped by a conformal phospholipid bilayer shell using a emulsion/solvent evaporation process. Inclusion of charged or reactive lipids enabled reversible association of plasmid DNA with the surface of these materials and facilitated the prolonged expression of DNA-encoded antigens *in vivo*. To enhance the function of these particles in vaccination, we encapsulated small-molecule immunomodulators in the particle core designed to amplify or tune the quality of immune responses generated following vaccination. Using optimized particle formulations, the ability of these materials to modulate immune responses following immunization with model antigens and HIV-relevant antigens was also investigated. Biomaterials permit levels of control over drug delivery that are not currently achievable using conventional vaccination regiments, thus this approach could serve as a route for enhancing existing DNA vaccination strategies for HIV, and hold relevance for the broader vaccine field.

Selected References

1. **C. M. Jewell** and D. M. Lynn, "Multilayered Polyelectrolyte Assemblies as Platforms for the Delivery of DNA and Other Nucleic Acid-Based Therapeutics." *Advanced Drug Delivery Reviews* **2008**, *60*, 979-999.

2. **C. M. Jewell** and D. M. Lynn, "Surface-Mediated Delivery of DNA: Cationic Polymers Take Charge." *Current Opinion in Colloid and Interface Science* **2008**, *13*, 395-402.

3. D. J. Irvine and **C. M. Jewell**, "Vaccine and Immunotherapy Delivery", *Comprehensive Biomaterials*, Ed. P. Ducheyne, Elsevier Press, **2010** – in press.

4. **C.M. Jewell**, J. Zhang, N. J. Fredin, D.M. Lynn, "Multilayered Polyelectrolyte Films Promote the Direct and Localized Delivery of DNA to Cells." *Journal of Controlled Release* **2005**, *106*, 214-223.

5. **C. M. Jewell**, J. Zhang, N. J. Fredin, M. R. Wolff, T. A. Hacker, D. M. Lynn, "Release of Plasmid DNA from Intravascular Stents Coated with Ultrathin Multilayered Polyelectrolyte Films." *Biomacromolecules* **2006**, *7*, 2483-2491.

6. **C. M. Jewell***, S. M. Fuchs*, R. M. Flessner, R. T. Raines, D. M. Lynn, "Multilayered Films Fabricated from an Oligoarginine-Conjugated Protein Promote Efficient Surface-Mediated Protein Transduction." *Biomacromolecules* **2007**, *8*, 857-863.

7. N. L. Abbott, **C. M. Jewell**, M. E. Hayes, Y. Kondo, D. M. Lynn, "Ferrocene-Containing Cationic Lipids: Influence of Redox States on Cell Transfection." *Journal of the American Chemical Society* **2005**, *127*, 11576-11577.

8. **C. M. Jewell**, M. E. Hayes, Y. Kondo, N. L. Abbott, D. M. Lynn, "Ferrocene-Containing Cationic Lipids for the Delivery of DNA: Oxidation State Determines Transfection Activity." *Journal of Controlled Release* **2006**, *112*, 129-138.

9. **C. M. Jewell**, M. E. Hays, Y. Kondo, N. L. Abbott, and D. M. Lynn, "Chemical Activation of Lipoplexes Formed from DNA and a Redox-Active, Ferrocene-Containing Cationic Lipid." *Bioconjugate Chemistry* **2008**, *19*, 2120-2128.

See more of this Session: Meet the Faculty Candidate Poster Session
See more of this Group/Topical: Education

(4ak) A Full-Chain Stochastic Tube Model for Entangled Polymeric Liquids: Improvement for Extended Applications

Joontaek Park, David W. Mead and Morton M. Denn, Benjamin Levich Institute for Physico-Chemical Hydrodynamics, The City College of City University of New York, New York, NY

A full-chain stochastic tube model developed by Xu et al. [J. Rheology (2006) 50:477-494] was modified to simulate entangled melts and solutions of linear polymers at high shear rates. This model, which has only one adjustable parameter equivalent to the Rouse time, successfully describes shear and normal stress data at low to moderate rates, but the model deviates systematically from data at high rates. The algorithm for re-entanglement has been modified to reduce the deviation in the calculated viscosity at high rates. The discussion focuses on the structural changes in terms of the constraint renewal mechanism, the effect of a non-uniform drag coefficient of chains, and the entanglement density.

See more of this Session: Meet the Faculty Candidate Poster Session
See more of this Group/Topical: Education

(4al) Rational, Model-Guided Design and Experimental Evaluation of Targeted Drug Delivery Vehicles

Derek W. Bartlett, Abraxis BioScience, Costa Mesa, CA

The success of a therapeutic drug is directly related to its ability to be safely and effectively delivered to its intended site of action within the body. Given the complexity of drug delivery systems and the steps required to achieve their desired activity in a biological organism, mathematical modeling is becoming an increasingly important tool to complement experimental approaches in the drug development process. Here, specific examples from my research will be presented demonstrating how this model-based approach, accompanied by *in vitro* and *in vivo* experimental studies, has been applied toward the design of nanoparticle delivery systems for nucleic acid therapeutics and chemotherapeutics.

During my time as a graduate student in the laboratory of Professor Mark E. Davis at the California Institute of Technology, we employed a novel experimental technique using noninvasive bioluminescence imaging and positron emission tomography (PET) to quantify the pharmacokinetic and pharmacodynamic parameters of targeted siRNA nanoparticles. Mathematical models were developed to guide and interpret the experimental studies performed *in vitro* and *in vivo*. This analysis revealed how cell doubling time profoundly impacts the kinetics of siRNA-mediated gene silencing, providing important guidelines for the design of optimized dosing regimens using siRNA. Furthermore, these studies demonstrated that targeting ligands may have a minimal impact on the overall tumor localization of nanoparticles, but they can be critical for achieving intracellular delivery of a therapeutic payload. More recently, a similar model-based approach is being applied to my current research at Abraxis BioScience on nanoparticle delivery systems for water-insoluble chemotherapeutics. These studies aim to understand how a drug's formulation impacts its interaction with biological systems.

Technological advances, such as noninvasive molecular imaging, are making biological systems increasingly amenable to quantitative engineering analysis. This will enable a paradigm shift in which novel drug delivery systems will be rationally and efficiently designed through the integration of quantitative mathematical models with experimental approaches.

REFERENCES

Bartlett DW, Su H, Hildebrandt IJ, Weber WA, Davis ME. Impact of tumor-specific targeting on the biodistribution and efficacy of siRNA nanoparticles measured by multimodality *in vivo* imaging. Proceedings of the National Academy of Sciences USA. 2007;104(39):15549-54.

Bartlett DW, Davis ME. Insights into the kinetics of siRNA-mediated gene silencing from live-cell and live-animal bioluminescent imaging. Nucleic Acids Research. 2006;34(1):322-33.

See more of this Session: Meet the Faculty Candidate Poster Session
See more of this Group/Topical: Education

(4am) De-Oxygenation Catalysis On Titania for Renewable Fuel Applications

Prashant Reuben Daggolu, Dave C. Swalm School of Chemical Engineering, Mississippi State University, Mississippi State, MS

Background My PhD Project concerns with converting short chain organic oxygenates to fuel range hydrocarbon. This project is part of a campus wide effort to develop sustainable energy at Mississippi State University. Under this program a collaborative research group has been established called Sustainable Energy Research Center (SERC), www.serc.msstate.edu, comprising of researchers in the departments of Chemical Engineering, Forest Products, Chemistry, Agricultural and Biological Engineering to name a few. As part of this endeavor, our group under my advisor, Prof. Mark White has been developing non-Fisher Tropsch technology for biomass derived syngas to be converted to gasoline. One approach to this challenge is to first convert syngas to alcohols and then convert alcohols to gasoline. This approach while effective yields oxygenate by-products. My Project has been to study the use of titania as a catalyst to convert these by products to fuel range hydrocarbon to increase the viability of the biomass-to-gasoline technology.

Accomplishments The main oxygenate by-products formed during conversion of syngas to alcohols are acetic acid, acetaldehyde, methyl acetate and acetone. Hence initially, I performed batch reactor studies using both acid catalysts like H+/ZSM-5 and silica alumina, and base catalysts like titania and magnesium oxide. It was concluded that base catalysts yield almost no coke when reacted with these oxygenates. Then I studied interactions between these oxygenates by reacting pairs of these compounds on titania in a batch reactor. The results were documented and showed no prohibitive interactions. Then I made mechanistic studies of each of the compounds using quantum mechanics. Molecular modeling software called Spartan from Wavefunction Inc was utilized for this purpose. Hartree – Fock method was used to predict thermodynamic energies.

Acetic acid has been shown in literature as one of the most coke-causing compounds in both syngas derived products as well as in bio-mass pyrolysis products when tried to convert to hydrocarbon. Hence titania as a catalyst has all the more significance as it can convert acetic acid to acetone and subsequently acetone to mesitylene. I decided to study the acetone condensation to mesitylene much more closely as it forms an important link in converting acetic acid to hydrocarbon. Also, Acetone itself can be converted to liquid hydrocarbon. I studied this reaction in high pressure conditions similar to those employed in syngas conversion to alcohols. This was done keeping in mind the possibility of using a single bed to convert syngas to gasoline including all by-products. I had excellent success in this endeavor and reported high yield and selectivity of mesitylene from acetone. The data was presented in national meetings and I am presently writing a paper on the results. As alcohols can be easily converted to gasoline range hydrocarbon on zeolite catalyst, I wanted to show that alcohols were essentially inert on titania. This would ensure that only by-products are converted to hydrocarbon on titania and alcohols are retained as such. In course of making this study, when ethanol was reacted on titania, I found that tiny amounts of 1-butanol was found. I immediately saw an opportunity to upgrade alcohols in carbon number. This is of significance because higher alcohols (>C2) have higher energy content than lower ones. This would make them better fuel additives than ethanol. Also, previous research in our lab showed that higher alcohols can give better yields of gasoline when reacted on zeolites than methanol or ethanol. Hence I decided to pursue making 1-butanol from ethanol as well. Pacific Northwest National Laboratories (PNNL) has been working on direct conversion of syngas to higher alcohols but was successful only to make ethanol and not substantial yields of any higher ones. Hence our group collaborated with PNNL and I travelled to Richland, WA to work as an intern at PNNL for 10 weeks to show that titania can be used to convert ethanol to 1-butanol. I was successful in doing so. Also, my interaction with scientists at PNNL was a rich experience for me. After returning to Mississippi State University, I continued to work on converting ethanol to 1-butanol not only on titania but also on catalysts with similar known surface chemistry. I found that Zirconium dioxide (Zirconia) could convert ethanol to 1-butanol with much better yields than titania. Also, while titania made 1-butanol as the highest alcohol from ethanol, zirconia could convert ethanol to not only 1-butanol but also various pentanols, hexanols and even octanol. The data will be soon documented in a publication.

For the past few weeks I have been finishing up my research by studying the surface chemistry of titania. This includes Surface area analysis, Fourier Transform Infra red Spectroscopy, Transmission Electron Microscopy, Scanning Electron Microscopy, X-ray diffraction Microscopy and Thermogravimetric Analysis techniques. Also, I have been studying the surface active sites of titania using acetone condensation to mesitylene as a probe reaction. This is being performed by selectively poisoning the catalyst and running the reaction to study the loss in rate of reaction.

From the work that I performed for the last 31 months, I gave 4 presentations, received 1 provisional patent and was selected for 1 award.

See more of this Session: Meet the Faculty Candidate Poster Session
See more of this Group/Topical: Education

(4an) Engineering the Yeast Saccharomyces Cerevisiae for Drug Discovery and Bioenergy Applications

Michelle A. O'Malley[1], Anne Skaja Robinson[2] and Chris A. Kaiser[1], (1)Department of Biology, Massachusetts Institute of Technology, Cambridge, MA, (2)Department of Chemical Engineering, University of Delaware, Newark, DE

Molecular engineering has enabled the economical production of several valuable products with the aid of microbial systems, from vaccines and therapeutics to biofuels. Yeast systems, such as *S. cerevisiae*, are ideal for protein production and other industrial applications because they are easy to grow and scale-up, inexpensive to work with, and straightforward to genetically manipulate. Since yeast have a eukaryotic secretory pathway, they also are equipped with the cellular machinery to facilitate protein folding and critical post-translational processing steps that often are necessary for the production of authentic proteins.

My doctoral research involved the expression, purification, and biophysical characterization of human G-protein coupled receptors (GPCRs) from *S. cerevisiae*. GPCRs are membrane proteins that mediate responses to extracellular stimuli, represent over 50% of all drug targets, and have been directly linked to several diseases including HIV infection, cancer, and diabetes. However, the development of improved, structure-based therapeutics to target these proteins requires milligram amounts of properly folded, purified receptors, which are generally not achievable from native tissues. To structurally characterize this important class of proteins, GPCRs were over-expressed in *S. cerevisae* to circumvent the problem of low natural abundance, and their trafficking, folding, and activity were evaluated. Although all GPCRs were successfully expressed at the mg/L scale, we discovered that most receptors trigger cellular stress responses when expressed in yeast and exhibit compromised activity. These observations were closely linked to improper processing of the N-terminal leader sequence, which suggests that translocation is a critical bottleneck for receptor over-expression in yeast. Of all the receptors analyzed, only the human adenosine A_2a receptor (hA_2aR) was found in large quantities at the plasma membrane, where it was able to bind to adenosine analogs. The hA_2aR was purified to homogeneity in milligram amounts and reconstituted with mixed surfactant micelles, which allowed for extensive structural characterization using biophysical techniques.

Building on my expertise with *S. cerevisiae*, my postdoctoral research is focused on engineering *S. cerevisiae* as a tool for biofuel production from diverse cellulose-rich feedstocks. Our approach involves the characterization of fungal cellulosomes, or large multisubunit complexes of carbohydrate hydrolyzing enzymes and cellulose binding domains, from anaerobic fungi that are native to the rumen of large herbivorous animals. The multicomponent cellulosome complexes produced by these fungi are superior to similar complexes that have been widely studied from bacteria since they are much more efficient at degrading cellulose and have novel cellulolytic properties. As the components of the fungal cellulosome have not been completely identified, we are using powerful genomic tools to identify genes within the anaerobic fungi *Piromyces equi* that are upregulated when grown in the presence of cellulose-rich feedstocks. We are currently working towards expressing and reconstituting components of the *Piromyces* cellulosome within *S. cerevisiae* to facilitate consolidated bioprocessing of cellulose to ethanol. Furthermore, we seek to understand the structural basis for efficient cellulose degredation, and optimize the cellulosome through genetic engineering to maximize *S. cerevisiae*'s ability to degrade complex forms of cellulose.

See more of this Session: Meet the Faculty Candidate Poster Session
See more of this Group/Topical: Education

(4ao) Machine Learning: Extracting Research Pathways From Nanoscale Phenomena

Andres F. Hernandez Moreno, Georgia Institute of Technology, Atlanta, GA

Every day many research papers come out to the scientific community, summarizing its conclusions as experimental data in important fields like cell biology, medicine and nanotechnology. With the amounts of information available, there is a necessity to extract significant research trends. For over 50 years, machine learning has developed algorithms in computer science and mathematics to automatically recognize complex patterns and make intelligent decisions from information. Our research philosophy looks for a synergistic effect between first principal models and machine learning tools to reveal insights in new research areas. The application of machine learning techniques provides architecture to simultaneously integrate experimental data with our traditional chemical engineering modeling. Machine learning also offers a different perspective to understand and model systems with a scalable computational platform.

Here, we present the results of this philosophy describing the growth of platinum nanoparticles under supercritical conditions using a mathematical tool called Gaussian process regression. The results describes how the understanding of the nanoscale phenomena leads to the dynamic description of the process at the manufacturing level. We conclude this presentation with future applications of machine learning algorithms for biological and medical problems.

See more of this Session: Meet the Faculty Candidate Poster Session
See more of this Group/Topical: Education

(4ap) Targeted Nanoparticles for Systemic Delivery of Therapeutic Agents to Solid Tumors in Animals and in Humans

Chung Hang J. Choi and Mark E. Davis, Chemical Engineering, California Institute of Technology, Pasadena, CA

Targeted, nanoparticle delivery of therapeutic molecules to solid tumors has the potential to provide safer and more effective therapies for cancer patients. Our research focuses on understanding how targeted nanoparticles can reach inside cancer cells of solid tumors from systemic administrations and how to optimize their ability to do so via their proper design.

PEGylated gold nanoparticles (PEG-AuNPs) are used as model nanoparticles to investigate the effects of physicochemical parameters (e.g., size, surface charge, and ligand density) on their pharmacokinetics and in vivo distribution, allowing for study at the whole body, tissue, and sub-cellular levels all from the same animal (1). When covalently attaching human transferrin (Tf), an endogenous protein commonly used as a ligand in active targeting therapies, on PEG-AuNPs with varying amounts spanning two orders of magnitude (while holding all other parameters constant), their intravenous injection into mice bearing Neuro2A tumors (mouse neuroblastoma cells over-expressing Tf receptors) reveals the key mechanism of active targeting in solid tumors. The pharmacokinetics and in vivo distribution (including particle content in the solid tumor) of the targeted nanoparticles are primarily a function of the colloidal properties of the nanoparticle and not the amount of Tf. However, the intracellular localization within cancer cells of the tumor is strongly a function of the Tf amount: It becomes more pronounced with increasing Tf amount only above the threshold Tf amount.

Results from mechanistic studies can provide insights for designing effective targeted nanoparticles for use in the clinic. In a recent Phase I clinical trial, melanoma patients received infusions of Tf-targeted, siRNA-containing nanoparticles. Using imaging techniques developed and verified in animal models, tumor biopsies obtained from several patients reveal intracellular accumulation of Tf-targeted nanoparticles within the melanoma cells (and not in adjacent tissues) in a dose dependent fashion. That is, more targeted nanoparticles reside inside the melanocytes within the tumor from patients who received higher doses of targeted nanoparticles (first reported for any targeted nanoparticles) (2). Thus, the patient data are largely consistent with the proposed mechanism of active targeting in solid tumors based on results from animal models.

References: (1) Choi, C.H.J., Alabi, C.A., Webster, P., Davis, M.E. Mechanism of active targeting in solid tumors with transferrin-containing gold nanoparticles. Proc. Nat. Acad. Sci. USA, 107, 3, 1235-1240 (2010). (2) Davis, M.E., Zuckerman, J.E., Choi, C.H.J., Seligson, D., Tolcher, A., Alabi, C.A., Yen, Y., Heidel, J.D., Ribas, A. Evidence of RNAi in humans from systemically adminstered siRNA via targeted nanoparticles. Nature, 464, 7291, 1067-1070 (2010).

See more of this Session: Meet the Faculty Candidate Poster Session
See more of this Group/Topical: Education

(4aq) A Property Based Approach for Simultaneous Process and Molecular Design

Nishanth Chemmangattuvalappil and Mario Richard Eden, Department of Chemical Engineering, Auburn University, Auburn University, AL

The selection of product/product mixtures that give the optimum performance of a process is a critical issue for a design engineer. The process performance is usually understood in terms of physical properties and on many occasions, the physical properties of the product rather than their chemical structure determine the suitability of a specific product as the input to the process. Molecular design algorithms generally require target properties to design the molecules. Therefore, for obtaining the optimal solution for this type of problem, it is necessary to have a methodology to represent the product performance in terms of measurable physical properties and identify the molecule/mixture that gives the property targets corresponding to the optimum process performance.

In my dissertation work, systematic algorithms have been developed for the design of molecules corresponding to the optimum performance of a process. The concept of property clustering has been extended to the formation of molecular clusters based on second and third order group contribution methods. An algebraic approach has been developed utilizing higher order groups built from first order groups subject to the constraints of overlapping. The advantage of using an algebraic approach is that it can handle any number of molecular groups or properties and can generate all possible compounds within the required range of properties. The most significant aspect of the aforementioned method is that both the application range and reliability of the molecular property clustering technique are considerably increased by incorporating second and third order estimation. A methodology has been developed for incorporating the property contribution predicted using combined group contribution and connectivity indices into the design framework in case the property contributions of any of the molecular groups of interest are not available in literature. For the design of simple monofunctional molecules, a modified visual approach has been used whereas for the design of more complicated structures and/or for treating more than three properties at a time, an algebraic method has been used.

The developed algorithm can be used as a tool for the solution of integrated process and product design problems. Since an algebraic approach for the solution of process design problems is already available, it is now possible to solve the integrated problem irrespective of the number of properties involved. The process design problem can be solved in terms of constitutive variables without committing to any component beforehand. The property targets estimated in the process design step along with the molecular groups will form the input to the molecular design algorithm. The algebraic equations formed using the molecular operators are solved simultaneously to generate the possible molecular structures that match the optimum process performance defined by the property targets. However, apart from the targets set by the process design, the molecule has to satisfy a number of other environmental and safety constraints in order to be used in an actual process. In addition, there are many properties which cannot be estimated through group contribution methods, because it is not always possible to find a correlation between the molecular groups and the properties. Similarly, not all possible atomic arrangements are represented in group contribution methods. So, there is a need for an efficient methodology for the design of molecules with more diverse property targets. In my dissertation, I have developed an algorithm that uses the concept of molecular signature descriptors for molecular design. The signature is a systematic coding system of atom types and the signature of a molecule can be obtained as a linear combination of its atomic signatures. It has been proven that any topological indices of molecules can be represented in terms of molecular signatures and it is possible to correlate the topological indices to the actual properties and biological activities. Here, the new algorithm utilizes molecular property operators based on signatures for solving the reverse problem of obtaining the molecular structures that satisfy the property targets estimated in the process design step. A new set of equations is employed to ensure that the molecule meets the safety and environmental constraints as well. The principles in graph theory are incorporated to avoid the generation of infeasible structures. The property models required to describe the target properties may be based on QSPR/QSAR (quantitative structure-property/activity relationship) models or in the form of group contribution methods. The accuracy of this method depends only on two factors, i.e. how well the actual property-topological index relationships are estimated and the height of atomic signatures used to describe the topological indices. Since, many topological indices can be used to describe each property, this algorithm generally provides reliable results. Finally, a framework has been developed to incorporate the design of process flow sheets into the molecular design problems. The recently developed computer aided flowsheet design techniques are utilized along with the molecular design techniques to obtain the required process performance.

As a future faculty candidate, my primary interest is in the area of process and product design. However, different tools used and developed in the research I have conducted so far have the ability to be applied in many related areas as well. Similarly, many of the methods I have developed in my dissertation can be applied in different area of chemical engineering and a number of multi-disciplinary fields.

The initial projects I will be working on are:

1. Extend the applicability of the developed algorithms into more rigorous industrial problems by including the effects of varying process conditions and incorporating more flowsheet properties. This approach will be helpful for the design of materials with good accuracy.

2. The iterative nature of the molecular design problems will be reduced into a systematic generation process by introducing three dimensional descriptors in the molecular design step to expand the application range of reverse problem formulations and by including constraints in the form of signatures that ensure the stability of the designed structures

My long term projects are:

1. The extension of the molecular signature concept into the exploration of biochemical reaction pathways

2. A property based approach to designing optimum molecules by designing the reaction networks using the concept of molecular signatures in process design

See more of this Session: Meet the Faculty Candidate Poster Session
See more of this Group/Topical: Education

(4ar) Design of Novel Artificial Allosteric Proteins and Their Applications

Jingjing Li, Department of Chemical and Biomolecular Engineering, the Ohio State University, Columbus, OH and David W. Wood, Chemical and Biomolecular Engineering, the Ohio State University, Columbus, OH

Allosteric regulation of protein function is one of essential pathways for cellular processes. The basic feature for natural allosteric proteins is that ligand binding will induce a significant conformational alteration and subsequently influence their activity. Inspired by this mechanism, artificial allosteric proteins might be developed from various receptor proteins as artificial molecular switches. While nuclear receptors, with their inherent molecular switch behavior, are suitable for directly coupling ligand binding with a reporter protein or enzyme, other natural receptors may require a synthetic allosteric domain. With the advancement of directed-evolution technology, robust artificial binding proteins originating from engineered scaffolds with high stability and affinity can be easily generated. These artificial affinity proteins might serve as an abundant source for artificial allosteric proteins, either through coupling with natural allosteric domains or synthetic switches. These artificial allosteric proteins might then be used to control metabolic pathways for target optimization, or be used to interfere with the transcription of target genes for therapeutic applications. They may even be able to directly act as controllable therapeutics for a series of severe diseases such as cancer and cardiovascular disorders. In this context, binding of the allosteric protein to the target receptor or target protein will activate or release a functional domain. On the other hand, the linkage of affinity proteins with reporter enzyme can be used to develop highly sensitive biosensors for drug discovery. In this poster, the basic strategy for the design of artificial allosteric proteins and the potential applications of these proteins will be introduced and discussed.

See more of this Session: Meet the Faculty Candidate Poster Session
See more of this Group/Topical: Education

(4as) Analysis and Control of Self-Assembling Nucleic Acid Systems

Victor A. Beck, Bioengineering, California Institute of Technology, Pasadena, CA

The specificity and predictability of Watson-Crick base pairing has been exploited to direct the assembly of oligonucleotides into myriad structures, chemical logic circuits, walkers and machines. However, we have only recently begun to develop and test models capable of describing the self-assembly processes. We previously introduced hybridization chain reactions (HCR), in which metastable DNA hairpins undergo conditional self-assembly to form long nicked double-stranded 'polymers' in the presence of a DNA initiator molecule (PNAS 2004, 10, 15275). Here, we model the equilibrium and kinetic properties of HCR, revealing sources of non-ideal behavior and methods for controlling system performance. Our results demonstrate that upon initiation, HCR is accurately modeled as a living alternating copolymerization.

See more of this Session: Meet the Faculty Candidate Poster Session
See more of this Group/Topical: Education

(4at) Synthesis of Multifunctional Nanoparticles for Cancer Imaging, Diagnostics and Therapy

Yun Wu, Nanotechnology Science and Engineering Center, The Ohio State University, Columbus, OH and L.James Lee, Chemical and Biomolecular, the Ohio State University, Columbus

Synthesis of multifunctional nanoparticles with combined imaging, diagnostic and therapeutic functions is a rapidly developing field that has attracted a great deal of research interest in recent years. The composition and structure uniformity of the multifunctional nanoparticles play an extremely important role in the success and efficacy of their applications in vitro and in vivo.

Although the conventional nanoparticle preparation method, bulk mixing, is straightforward, the formation of nanoparticles is not well controlled. We developed a novel one-step micromixer-electrospraying method to produce multifunctional lipoplex nanoparticles. In this method, a micromixer was used to bring the therapeutic drugs/genes and imaging agents in a well controlled manner and the mixture is dispersed into tiny particles via electrospraying. In our study, cy5 labeled oligonucleotide (Cy5-G3139) was chosen as the model therapeutic agent and quantum dots (QD605-Amine) were chosen as model imaging or diagnostic agents. Cy5-G3139 and QD605-Amine form a fluorescence resonance energy transfer (FRET) pair that provides the ability to study the intracellular fate of the multifunctional lipoplex nanoparticles.

We investigated the effects of synthesis methods, micromixer-electrospraying vs. bulk mixing, on the physical properties (size, surface charge and structure) of multifunctional lipoplex nanoparticles using dynamic light scattering, zeta potential and CryoTEM. The biological properties, i.e. cellular uptake, intracellular trafficking and antisense activity against Bcl-2 protein, of multifunctional lipoplex nanoparticles produced by micromixer-electrospraying and bulk mixing were also examined in KB cells using confocal microscopy, flow cytometry, real time PCR and Western blot.

See more of this Session: Meet the Faculty Candidate Poster Session
See more of this Group/Topical: Education

(4au) Multiscale Chemical Product Design Using the Reverse Problem Formulation

Charles C. Solvason and Mario Richard Eden, Department of Chemical Engineering, Auburn University, Auburn University, AL

The National Research Council (NRC) has recently recognized the importance of developing integrated formulated product design tools. In response, it created the Committee on Integrated Computational Materials Engineering (CICME) which developed and published a road map for what it termed a "Grand Challenge" [1]. In the report, the committee noted that in order to alleviate strain put on U.S. manufacturers from the swiftly changing and increasingly global marketplace, integrated design closely coupling computational models with manufacturing processes would be required. The term 'integrated' recognizes that the properties of products are controlled by a multitude of separate and often competing mechanisms that operate over a wide range of length and time scales. It is the linkage of the scales that remains the 'Grand Challenge' [1]. In response to this call from CICME, the design research community has shifted focus from developing only the physical form, function, and aesthetics of assembled products to the design of chemically formulated products [2]. Chemically formulated products are products designed at the molecular level to deliver a specific desired attribute that may exist at multiples scales such as the quantum-, nano-, meso-, macro-, and mega-scales. Examples of formulated products include pharmaceuticals [3], proteins [4], organic semi-conductors [5], nano-structured materials [6] and many more product types.

Using today's technology, designing these optimized molecules for a specific end-use has never more attainable. Recent achievements owe much of their successes to increased computational speeds and the advent of novel algorithms that improved the transfer of information between the formulated scales [6]. One type of promising new algorithm utilizes the reverse problem formulation which has been shown to significantly reduce the complexity and required computational time of product design problems. For example, Eden et al. [7] and Eljack et al. [8] demonstrated the ability of the reverse problem formulation to decouple product design problems from constitutive equations so that molecules can be designed prior to verification at the macro scale process simulation, effectively reducing the computational complexity. Furthermore, Satyanarayana et al. [9] recently demonstrated the use of the reverse problem formulation combined with grid technology for polymer design with long range order approaching the meso-scale. While each of these applications of the reverse problem formulation has incorporated elements of multi-scale product design, the integration of the reverse problem formulation into the complete multi-scale framework has yet to be achieved.

The main objective for this research was to demonstrate how the reverse problem formulation algorithm could include aspects of each of the length scales, thereby creating a framework where product design could be achieved with significantly reduced computational time. The plan to achieve the objective entailed the following: (1) the development of a centralized framework, (2) the development of an algorithm that chooses the appropriate properties to bridge each of the length scales, (3) the successful implementation of a property clustering algorithm to load the property descriptions into the framework. Independent validation steps were built in to the method and verified using published experimental results because of the "profound importance of experimental results to calibrate and validate computational methods and fill gaps in the theoretical understanding" [1]. Due to the size of this project, proof-of-concept case studies initially focused on experimentally derived parameters. Information from the molecular scale on short range order, such as group structure, conformation, and stereoregularity, have been combined with information from the microscale on long range order, such as the structure of the lattice, polymorph form, and particle size, in the design of excipients for direct compressed acetaminophen tablets.

The developed method's success in combining characterization data from the micro-scale with the molecular scale suggested it could be applied to other characterization based chemical product designs, such as those with nanoscale architectures. Based on this inference, I authored and received a grant from the Alabama EPSCoR Graduate Research Scholars Program (GRSP) for a proof-of-concept application example regarding the design of green functional groups that improve dispersion of single walled carbon nanotubes (SWNTs). In this work, systems techniques for utilizing IR/NIR spectroscopic based combinatorial methods were used to identify suitable environmentally benign functional groups, sizes, and orientations that improved the dispersion of SWNTs in polypropylene in order to quantify experimental results found by the Davis Laboratory at Auburn University [10].

It is my belief that based on the preliminary results of this research, a new multi-scale design method can be developed that offers many advantages over existing approaches. One of the most beneficial advantages is a reduction in the computational expense needed to solve multi-scale design problems. For example the traditional method of solving a design problem has been to compute information at smaller length scales and pass it to models at larger length scales by removing degrees of freedom (coarse-graining) with the objective being to predict macroscopic properties from molecular information [6]. While often the most accurate method for predicting properties, simulation has two limitations: (1) it has an immense computational cost due to hierarchical nesting and (2) it utilizes a priori knowledge of the molecular architecture (i.e. the number and types of atoms or electrons present). The large computational cost typically prevents an accurate modeling of mesoscopic structure such as the morphology of polymers without the use of constraints that significantly limit the degrees of freedom in the simulation (e.g. it may take a supercomputer six weeks to generate a single data point) [6]. Furthermore, when these models are integrated within a product-process design framework, such as those presented in the senior-design course, the computational intensiveness exponentially increases because each projected molecular architecture must now be simulated to determine its physico-chemical properties [2]. To minimize the computational cost in these types of problems, the property prediction simulations are typically approximated with constitutive equations based on structure descriptor models such as property based group contribution (GCM), quantitative structure property/activity relationships (QSPR/QSAR) using topological indices (T.I.), or chemometric based models. The introduction of structure descriptor models improves computational efficiency by avoiding nested simulation at the cost of prediction accuracy. However, these structure descriptor models are still too computationally expensive when utilized in design algorithms because of their often highly nonlinear nature. It has been shown that the reverse problem formulation can be used to minimize this issue by decoupling the constitutive equations from the constraint and balance equations to identify the property solution domain without committing to any components a priori [7]. Making use of this significant find in addition to parallel programming could dramatically improve the computational efficiency of multi-scale product and process design problems, especially those with significant architecture influences at the nano-, micro-, and meso-scales. Hence, I believe we can begin to transform chemical product and process design from a candidate selection problem to a true candidate DESIGN problem that identifies the optimum molecular architecture from ALL possibilities. To achieve this goal, my research plan consists of the development of cutting edge computation reduction approaches in two main research thrusts: (1) characterization studies using empirical structure models to establish information work flows between scales using the reverse problem formulation and (2) simulation studies using coarse-grained beads to investigate parallel computation algorithmic efficiency, accuracy, and applicability. The domains studied will include both physically relevant properties, such as cohesive energy density, and mathematical constructs, such as latent properties in decomposition techniques. Applications of this research include the development of SWNT-polymer structured products, drug-delivery mechanisms for pharmaceuticals and therapeutics, and reaction network flow sheet designs for the conversion of biomass to structured chemical products beyond biofuels.

References

[1] Committee on Integrated Computational Materials Engineering, National Research Council (2008). Integrated Computational Materials Engineering: A Transformational Discipline for Improved Competitiveness and National Security. The National Academies Press, USA.

[2] Hill, M. (2004) Product and Process Design for Structured Products. AIChE Journal, 50(8), pp. 1656-1661.

[3] Solvason C.C., Chemmangattuvalappil N.G., Eden M.R. (2009) Decomposition Techniques for Multi-Scale Structured Product Design: Molecular Synthesis. Computer Aided Chemical Engineering (In Press).

[4] Floudas C.A., Fung H.K., McAllister S.R., Monningmann, M., Rajgaria R. (2006) Advances in protein structure prediction and de novo protein design: A review. Chemical Engineering Science, 61(3), pp.966-988.

[5] Farhang A.R and Deeter T. (1999) Theoretical characterization of phenylene-based oligomers, polymers, and dendrimers. Synthetic Metals, 100(1), pp. 141-162.

[6] Fermeglia M. and Pricl S. (2009) Multiscale molecular modeling in nanostructured material design and process system engineering. Computers & Chemical Engineering (In Press).

[7] Eden M.R., Jergensen S.B., Gani R., and El-Halwagi M.M. (2003) Reverse problem formulation

based techniques for process and product synthesis and design. Computer Aided Chemical Engineering, 15(1), pp. 451-456.

[8] Eljack F.T., Eden M.R., Vasiliki K., Kazantzi Q., and El-Hawagi M.M. (2007) Simultaneous process and molecular design - A property based approach. AIChE Journal, 53(5), pp. 1232-1239.

[9] Satyanarayana K.C., Abildskov J., and Gani, R. (2009) Computer-aided polymer design using group contribution plus property models. Computers & Chemical Engineering, 33, pp. 1004-1013.

[10] Radhakrishnan V.K., Davis E.W., Davis V.A. (2010). Influence of Initial Mixing Methods on Melt Extruded Single-Walled Carbon Nanotube-Polypropylene Nanocomposites. Polymer Engineering and Science, (accepted for publication 12/8/09).

See more of this Session: Meet the Faculty Candidate Poster Session
See more of this Group/Topical: Education

(4av) Microfluidic Three-Dimensional in Vitro System Enabling An Information-Rich Assay to Investigate Breast Cancer Progression

Kyung Eun Sung[1], Ning Yang[2], Carolyn Pehlke[1], Patricia J. Keely[3], Kevin W. Eliceiri[4], Andreas Friedl[2] and David J. Beebe[1], (1)Department of Biomedical Engineering, University of Wisconsin, Madison, WI, (2)Department of Pathology and Laboratory Medicine, University of Wisconsin, Madison, WI, (3)Department of Pharmacology, University of Wisconsin, Madison, WI, (4)Laboratory for Optical and Computational Instrumentation, University of Wisconsin, Madison, WI

The earliest stage at which breast cancer can be diagnosed is known as ductal carcinoma in situ (DCIS), which is uncontrolled growth of epithelial cells within breast ducts. This cancer is non-invasive as carcinoma cells have not yet spread into surrounding stromal tissues, but it can proceed to invasive ductal carcinoma (IDC) and stromal alterations have been implicated in initiating this transition. Studies show that one out of every five new breast cancer diagnoses each year is DCIS, and one out of every three women with untreated DCIS will later develop invasive breast cancer. Despite the fact that DCIS is one of the most common breast cancers and it may be life-threatening if it progresses to IDC, the understanding of the mechanisms involved and the identification of potential therapeutic targets are still vague. One reason for this is the lack of a reliable in vitro experimental model. While the transition from DCIS to IDC has been observed clinically and has been recapitulated in mouse models, we lack a good in vitro model that allows screening for inhibitors and furthers our ability to elucidate the underlying molecular mechanisms that govern the transition.

We introduce a simple microfluidic 3D compartmentalized co-culture system that supports the DCIS to IDC transition in vitro. The platform employs microfluidic channels with two input ports to load MCF10-DCIS cells adjacent to stromal fibroblasts, recapitulating the general spatial relationship observed in vivo. The compartmentalized model enables both spatial (distance-dependent transition, i.e., more transition at interface) and temporal (transition from larger clusters) control of the microenvironment. In addition, we present the use of second harmonic generation (SHG) intensity information to measure quantitatively the degree of invasive transition of MCF10-DCIS clusters in the compartmentalized co-culture model. Uniquely, the compartmentalized culture system enables imaging collagen structures that are altered by MCF10-DCIS cells alone because each cell type is in a separate compartment. The arrayed microchannel-based model is compatible with the existing infrastructure and provides a cost effective approach (~ 40 times less sample required) to test for inhibitors of pathways involved in DCIS progression to IDC, allowing a screening approach to the identification of potential therapeutic targets. It is important to note that the model can be easily adapted and generalized to a variety of cell-cell signaling studies.

See more of this Session: Meet the Faculty Candidate Poster Session

(4aw) Development of a FRET-Based Tension Sensor for Measuring Forces Across Proteins in Living Cells

Brenton D. Hoffman, Cardiovascular Research Center, University of Virginia, Charlottesville, VA

Force and other mechanical variables are important regulators in many biological processes. For instance, forces from blood flow are key factors in the development of disease states, like atherosclerosis, and the mechanical environment cells experience can determine the differentiation state of stem cells and the metastatic potential of cancer cells. But how cells detect and interpret these forces, a process called mechanotransduction, is not well understood.

Most research on mechanotransduction has focused on cataloging cellular responses and identifying proteins that change activity or localization in response to applied force. This is mainly due to a lack of tools to study force at a molecular level in a living cell. We therefore developed a calibrated, genetically encoded Forster Resonance Energy Transfer (FRET) based tension sensor with pico-Newton sensitivity consisting of a spring-like peptide sequence between two fluorescent proteins. When tension is applied to the sensor the spring extends and the energy transfer between the fluorescent proteins decreases. The sensor has been put into the protein vinculin, as it is a major component of the mechanical link between integrins, which mediate cell adhesion to the extra-cellular matrix, and the actin cytoskeleton, the primary force generation system of the cell. The sensor detects changes in response to inhibition of cellular contractility, in different areas of a migrating cell, and as focal adhesions assemble or disassemble.

With this new tool, we can visualize the spatio-temporal dynamics protein deformation in live cells to study the relationships between forces and cell signaling in ways that were previously impossible. Initial efforts will focus on kinases that have been strongly implicated in mechanotransduction. Additional efforts include development of a polymer physics based model of the sensor to further optimize this sensor and aid in the development of new sensors for other proteins. Future projects include developing a microscopy-based screening system to detect proteins that localize to areas of high force in focal adhesions and are therefore likely key players in mechanotransduction.

See more of this Session: Meet the Faculty Candidate Poster Session
See more of this Group/Topical: Education

(4ax) Exploiting Immune Function and Response for Biopharmaceutical Engineering

Sai T. Reddy, Chemical Engineering, University of Texas Austin, Austin, TX

Biopharmaceutical drugs are defined as therapeutics and in vivo diagnostics engineered from a biological source. Nearly half of all clinically approved biopharmaceuticals are comprised of subunit vaccine and antibody-based therapeutics; of which, their development has been heavily reliant on applied immunology (i.e., hybridomas and recombinant vaccine antigens). Here, I will describe novel approaches that exploit immune function and response for vaccine and antibody engineering.

Subunit vaccine development is often focused on antigen targeting and adjuvancy schemes that respectively facilitate delivery of antigen to dendritic cells and elicit their activation. Here I will present a novel polymer nanoparticle-based vaccine platform, which targets lymph node– residing dendritic cells via interstitial flow and activates these cells by in situ complement activation. After intradermal injection, interstitial flow transported ultra-small nanoparticles (25 nm) highly efficiently into lymphatic capillaries and their draining lymph nodes. The surface chemistry of these nanoparticles activated the complement cascade, generating a danger signal in situ and potently activating dendritic cells. Using nanoparticles conjugated to the model antigen ovalbumin, the generation of humoral and cellular immunity was shown in mice in a size- and complement- dependent manner. Such a vaccine technology platform is both effective and inexpensive, and therefore allows it to be explored in global health applications.

The ability of humoral B cell immunity to generate a vastly diverse antibody repertoire as a response to stimuli (i.e., immunization) has been substantially exploited for monoclonal antibody discovery. Here, I will present a novel technology for monoclonal antibody discovery that exploits high-throughput DNA sequencing of B cell antibody repertoires from immunized mice. It was found that antigen-specificity of antibodies could be identified based on sequence analysis of the immune repertoire; subsequently we utilized automated synthetic gene construction for the production of antigen-specific antibodies in recombinant expression systems. This work illustrates the potential to apply high-throughput DNA sequencing for molecular and protein engineering applications. Taken within the context of the rapid advances and declining costs for DNA sequencing and gene synthesis, this technology offers a powerful alternative to conventional methodologies for monoclonal antibody discovery (e.g., hybridoma and phage/microbial display screening). In addition to antibody discovery technology, we have also engineered antibody effector functions. We have used a bacterial display screening system to isolate aglycosylated IgG antibody variants that bind with high affinity and selectivity to activating Fcγ receptors on the surface of immune cells. Importantly, anti-Her2 aglycosylated IgG variants (targeting breast cancer specific receptors) bound to FcγRI and exhibited a unique ability to activate human monocyte-derived dendritic cells (mDCs) to kill target breast cancer cells. The antibody-mediated activation of mDC cytotoxicity may be of clinical value in antibody-mediated immunotherapy.

1. **S.T. Reddy**, X. Ge, A. Miklos, R. Hughes, S.H. Kang, K.H. Hoi, C. Chrysostomou, S.P. Hunicke-Smith, B.L. Iverson, P.W. Tucker, A.D. Ellington, G. Georgiou, Isolation of monoclonal antibodies without screening by mining the variable gene repertoire of plasma cells, **(Nature Biotechnology, In Press)**

2. S.T. Reddy and G. Georgiou, Frequency analysis of the antigen-specific antibody repertoire from plasma cells of immunized mice, (*In preparation*).

3. S.T. Jung, **S.T. Reddy**, T.H. Kang, M.J. Borrok, I. Sandie, P.W. Tucker, G. Georgiou, Aglycosylated IgG variants expressed in E. coli that selectively bind FcγR1 potentiate tumor cell killing by monocyte-dendritic cells, **Proc. Nat. Acad. Sci. USA**, 107(2): 604-609, (2010). *Featured by: -Nature Biotech Research Highlights, 28(1): 45, (2010) -Nature Science-Business Exchange, 3(3): 1-5, (2010)*

4. A. Mercanzini, S.T. Reddy, D. Velluto, P. Colin, A. Maillard, J.C. Bensadoun, J.A. Hubbell, P. Renaud, Controlled release nanoparticle embedded coatings reduce the tissue reaction to neuroprostheses, **J. Cont. Rel.**, 145(3): 196-202, (2010).

5. H.Y. Lim, J.M. Rutkowski, J. Helft, **S.T. Reddy**, M.A. Swartz, G.J. Randolph, V. Angeli, Hypercholesterolemic mice exhibit lymphatic vessel dysfunction and degeneration. **Am. J. Pathol.**, 175 (3): 1328-37, (2009).

6. M.A. Swartz, J.A. Hubbell, and **S.T. Reddy**. Lymphatic drainage function and its immunological implications: From dendritic cell homing to vaccine design. **Seminars Immunol**, 20(2): 147-56, (2008).

7. **S.T. Reddy**, A.J van der Vlies, E. Simeoni, V. Angeli, G.J. Randolph, L.K. Lee, M.A. Swartz, and J. A. Hubbell. Exploiting lymphatic transport and complement activation in nanoparticle vaccines. **Nature Biotechnology**, 25(10): 1159-64, (2007). *Featured by: - Nature Nanotechnology Research Highlights, 10(1): (2007) - Faculty of Biology 6.0 – Must Read*

8. **S.T. Reddy**, M.A. Swartz, and J.A. Hubbell. Targeting dendritic cells with biomaterials: developing the next generation of vaccines. **Trends Immunol**. 27(12): 573-579, (2006).

9. **S.T. Reddy**, A. Rehor, H.G. Schmoekel, J.A. Hubbell, and M.A. Swartz. In vivo targeting of dendritic cells in lymph nodes with poly(propylene sulfide) nanoparticles. **J. Control. Rel.** 112(1): 26-34, (2006). *- Top 10 most cited article - J. Controlled Release 2007*

10. **S.T. Reddy**, D.A. Berk, R.K. Jain, and M.A. Swartz. A sensitive in vivo model for quantifying interstitial convective transport of injected macromolecules and nanoparticles. **J. Appl. Physiol,** 101(4): 1162-1169, (2006).

(4ay) A Fresh Look at Nanomaterials, Energy, and the Environment

Ludovico Cademartiri, Chemistry and Chemical Biology, Harvard University, Cambridge, MA

I will outline my past and present research and teaching efforts, as well as my future plans. I will detail how my research progressed from (i) developing novel gram-scale and green nanosynthetic strategies, to (ii) inventing a nanomaterial processing platform, to (iii) discovering polymer-like nanowires, to (iv) finding new principles in CO2 capture and in the electrical manipulation of combustion, to (v) my future plans of using (nano)materials strategies for the solution of environmental emergencies. I will also outline my past and present efforts as a textbook author and my teaching philosophy. Behind the diverse range of research topics I explored is the determination to follow the most unconventional ideas about the creation, manipulation, and use of materials, with the final objective of contributing to the solution of the most difficult and important scientific questions and problems.

See more of this Session: Meet the Faculty Candidate Poster Session
See more of this Group/Topical: Education

(4az) Computational Methods and Research in Chemical Engineering

Eric L. Maase, Chemical and Biomolecular Engineering Department, Lafayette College, Easton, PA

Analytical and computational modeling represents, I believe, an area in chemical engineering that has and will continue to become a fundamental aspect of our work. Students at all levels, faculty and industrial partners ever increasingly use and rely upon computational tools in all aspects of their work and my strength and focus in the field is toward the creation, investigation and use of new analytical and computation models and tools appropriate and valuable in our endeavors. My recent research work including the computational modeling of biological systems for tissue engineering as well as other simulation based experimentation including the design and development of a first course in molecular modeling and computational simulation for undergraduates (an elective) represent some of the specific projects in which I have been involved. With my research and strength and focus as a computational analysis / modeler I intend to build a practical (and nationally recognized) research program focused on dynamic data analysis and modeling (which include presently systems analysis for complex processes, and biochemical & thermodynamics modeling) . In these research strengths it is my objective to form mutually beneficial collaborations with colleagues in the department and other departments across campus on grant proposals and research goals.

See more of this Session: Meet the Faculty Candidate Poster Session
See more of this Group/Topical: Education

(4b) Cybernetic Modeling for Metabolic Engineering and Reactor Optimization

Hyun-Seob Song, School of Chemical Engineering, Purdue University, West Lafayette, IN

The cybernetic approach offers a unique framework for dynamic modeling of metabolic systems by describing metabolic regulation as optimal response of the organism to changes in its environment by conserving its resources. The cybernetic framework, developed by Ramkrishna and his coworkers, has evolved over the last two decades and continues a remarkable growth. Classical cybernetic approaches which are based on a lumped view of metabolism and thus called lumped cybernetic models (LCMs) have been successful in predicting various uptake patterns of mixed substrates (e.g., Kompala et al., 1986). The current state of cybernetic modeling, presented by Young et al. (2008), has the significantly augmented capacity of dealing with large metabolic networks. Successful application of Young's model to mutant as well as wild-type strains is an example showing a great value of cybernetic models as a promising tool for metabolic engineering. Formulation of such an elegant model would, of course, require extensive measurements for the identification of multitudes of kinetic parameters in the model. Meanwhile, hybrid cybernetic modeling (HCM) has been introduced as an interesting alternative to Young's framework due to its simpler computational structure (Kim, 2005). HCM is able to handle larger networks by invoking the quasi-steady-state approximation for intracellular metabolites under which network is decomposed into a set of subnetworks called elementary modes (EMs). It is viewed in HCM that the total uptake flux is divided into individual fluxes through EMs. Taking into account of the whole set of EMs into HCM is, however, hindered by overparameterization.

This poster presents various innovative developments made by the author to make cybernetic models practically more useful not only for reactor optimization but also for metabolic engineering applications.

1. EM reduction for systematic formulation of HCM (Song and Ramkrishna, 2009; Song et al., 2009): Overparameterization is an intrinsic problem of HCM as it takes dynamic measurements of only extracellular metabolites for parameter identification while the number of EMs (and thus parameters) exponentially increases with the network size. To resolve this problem, we propose 'yield analysis,' a new method of extracting a subset of EMs essential for describing metabolic behaviors. Yield analysis can be defined as the analysis of metabolic pathways in yield space where the solution space is a bounded convex hull. A consistent way of choosing the unique, minimal active EMs among a number of possible candidates is presented. The proposed idea is tested in a case study of a metabolic network of recombinant yeasts fermenting both glucose and xylose.

2. Development of lumped HCM (L-HCM) for prediction of metabolic function from limited data (Song and Ramkrishna, 2010): Motivated by the need for a quick quantitative assessment of metabolic function without extensive data, we present an adaptation of the cybernetic framework, denoted as L-HCM, which combines the attributes of the classical LCM and the recently developed HCM. The basic tenet of L-HCM and HCM is the same, that is, they both view the uptake flux as being split among diverse pathways in an optimal way as a result of cellular regulation such that some chosen metabolic objective is realized. The L-HCM, however, portrays this flux distribution to occur in a hierarchical way, that is, first among lumped pathways, and next among individual EMs in each lumped pathway. The distribution of uptake flux at the first split is dynamically regulated according to environmental conditions, while the subsequent split is based purely on the stoichiometry of EMs. The resulting model is conveniently represented in terms of L-EMs which are fully identified with respect to yield coefficients of all products. These characteristics enable the model to account for the complete set of EMs for arbitrarily large metabolic networks despite containing only a small number of parameters which can be identified using minimal data. In a case study involving aerobic batch growth of Saccharomyces cerevisiae, LHCM is compared with LCM.

3. Prediction of "dynamic" behavior of knockout strains: In silico prediction of cellular response to gene knockouts is a challenging task as mutated strains often exhibit the significant change in uptake rate as well as in (relative) internal flux distribution. While constraint-based approaches including Flux Balance Analysis (FBA), Minimization Of Metabolic Adjustment (MOMA), and Regulatory On/Off Minimization (ROOM) have been applied to the simulation of knockout strains, they have no machinery to address the change of "uptake rate" and thus their prediction is only about re-routing of intracellular fluxes for a given uptake rate. Consequently, these approaches might be useful with respect to "yield," but not reliable with respect to actual "rate." We present here a powerful methodology to predict the "dynamic" behaviors of knockout mutants from limited metabolic data of wild-type strains using the L-HCM recently proposed by Song and Ramkrishna (2010). L-HCM makes a satisfactory prediction on dynamic behaviors of knockout strains as shown in various case studies. This proves the universal validity of the cybernetic regulatory mechanisms which work not only for wild-type strains as having been shown so far in various cases in the literature, but also for mutated strains as shown in this work.

4. Development of an automated software for dynamic metabolic modeling: An automated software for developing various dynamic metabolic models is developed using MATLAB and named as AUMIC which stands for **AU**tomated tool for **M**etabolic modeling **I**ntegrated with the **C**ybernetic regulatory mechanisms. While AUMIC is designed mainly for constructing cybernetic models including LCM, HCM, and L-HCM , it can also handle various other approaches to dynamic metabolic modeling such as lumped kinetic model (LKM), macroscopic bioreaction model (MBM) (Provost and Bastin, 2004), and dynamic flux balance analysis (dFBA) (Mahadevan et al., 2002) for comparative studies. LKM, MBM and dFBA can be considered as respective counterparts of LCM, HCM and L-HCM because of analogous features in each pair. In AUMIC, all of the foregoing approaches are subsumed as quasi-steady-state models which are formulated through consistent procedures within a unifying framework. Moreover, AUMIC provides the parameter identification routine so that one can develop their own models using specific experimental data. Finally, the results are quickly analyzed using the post-processing module. Conveniently, all tasks in AUMIC are performed using graphical user interface. Continuing efforts are being made towards developing AUMIC as an even more effective and advanced tool and its first version will be released soon for public use.

References :

Kim JI, Varner JD, Ramkrishna D. 2008. A Hybrid Model of Anaerobic E. coli GJT001: Combination of Elementary Flux Modes and Cybernetic Variables. Biotechnology Progress 24(5):993-1006

Kompala DS, Ramkrishna D, Jansen NB, Tsao GT. 1986. Investigation of Bacterial-Growth on Mixed Substrates - Experimental Evaluation of Cybernetic Models. Biotechnology and Bioengineering 28 (7):1044-1055

Mahadevan R, Edwards JS, Doyle FJ. 2002. Dynamic flux balance analysis of diauxic growth in Escherichia coli. Biophysical Journal 83(3):1331-1340

Provost A, Bastin G. 2004. Dynamic metabolic modelling under the balanced growth condition. Journal of Process Control 14(7):717-728.

Song HS, Morgan JA, Ramkrishna D. 2009. Systematic Development of Hybrid Cybernetic Models: Application to Recombinant Yeast Co-consuming Glucose and Xylose, Biotechnology and Bioengineering 103(5): 984-102

Song HS, Ramkrishna D. 2009. Reduction of a Set of Elementary Modes Using Yield Analysis. Biotechnology and Bioengineering 102(2): 554-568

Song HS, Ramkrishna D. 2010. Prediction of Metabolic Function from Limited Data: Lumped Hybrid Cybernetic Modeling (L-HCM). Biotechnology and Bioengineering 106(2): 271-284

Young JD, Henne KL, Morgan JA, Konopka AE, Ramkrishna D. 2008. Integrating cybernetic modeling with pathway analysis provides a dynamic, systems-level description of metabolic control. Biotechnology and Bioengineering 100(3):542–559.

See more of this Session: Meet the Faculty Candidate Poster Session
See more of this Group/Topical: Education

(4ba) Molecular Mobility in Non-Linear Optical Glassy Chromophores

Daniel B. Knorr Jr., Chemical Engineering, University of Washington, Seattle, WA

Organic second-order nonlinear optical (NLO) materials are being actively pursued for applications in photonic devices such as high-speed electro-optic (EO) modulators, optical switches, and frequency converters. For practical applications, NLO materials must have both high macroscopic EO activity and thermal stability within the operating temperature range. High macroscopic EO activity can be achieved by acentrically ordering a system containing a high density of high dipole chromophores via electric field poling at elevated temperatures. Thermal stability requires the system to have internal constraints to prevent collapse of the acentric order at operating temperatures. Recent efforts for achieving both requirements have focused on dendrons capable of self-assembly through various 'soft' non-covalent interactions within self-assembling molecular glassy chromophores (SAMGCs). An example of these 'soft' interactions are arene-perfluoroarene (e.g. phenyl-pentafluorophenyl) quadrupolar interactions, which provide excellent EO activity and good thermal stability. In this study, experimental (novel scanning probe microscopy techniques) and theoretical techniques (molecular dynamics simulation) were used to determine the energetics of the underlying molecular mobility within SAMGCs incorporating phenyl-, naphthyl- or anthryl-pentafluorophenyl interactions. Further, the energetics of these mobilities are tied to their thermal relaxation behavior. Insights gained are being used to guide synthesis of next-generation organic NLO materials.

See more of this Session: Meet the Faculty Candidate Poster Session
See more of this Group/Topical: Education

(4bb) Multiscale Simulation to Advance Micellar Drug Delivery

Sharon M. Loverde, Chemical & Biomolecular Engineering, University of Pennsylvania, Philadelphia, PA

Biodegradable block copolymer self-assemblies are rapidly emerging for drug delivery and other 'green' applications. An overall theme of my research during my postdoctoral studies is to address packing and interfacial effects in micellar assemblies, using multiscale simulation techniques. Atomistically accurate, coarse-grained (CG) polymer models are being developed and studied, with the latest example being poly(ethylene oxide) - poly(caprolactone) (PEO-PCL) with simulations comparing well with experimental phase behavior. The anti-cancer drug paclitaxel is also being CG'd with intramolecular interactions again obtained from all-atomistic molecular dynamics. Solubility of CG paclitaxel in the worm micelle morphology at experimentally relevant loading concentrations addresses several issues such as: partitioning within the micelle, aggregation of drug, interfacial effects, and possible mechanisms of drug release.

See more of this Session: Meet the Faculty Candidate Poster Session
See more of this Group/Topical: Education

(4bc) Methods for Assessing Biocompatibility and the Foreign Body Response of Polymers and Drug Delivery Systems

Kaitlin M. Bratlie, Chemical Engineering, MIT, Cambridge, MA

In vivo fluorescence imaging of cathepsin and macrophage activity and genomic profiling have been developed as powerful tools for understanding biocompatibility. Traditionally, biocompatibility of materials has been determined via histology which has the severe limitation in that only one time point can be examined, making kinetic observations of the interaction between the immune system and the implant improbable. Genomic and molecular level understanding of what influences the foreign body response has remained largely unknown. Fluorescence imaging circumvents this issue by allowing observations of immune system responses in vivo in conjunction with genomic profiling to identify potential cytokines influential in eliciting the foreign body response. In vivo imaging holds the advantage that multiple time points can be analyzed while genomic profiling can lead to mechanistic understanding of the foreign body response process. Together, we are able to have a more detailed understanding of the biocompatibility of a certain material. These techniques have been used to study the biocompatibility of polymers subcutaneously injected in an array format in mice. Validation via histology through the use of positive and negative controls established these methods as a means of observing immunological responses to biomaterials. Through surface functionalization, the phagocytitic response to different chemical groups can be ranked. A library of chemically modified alginates were synthesized and assayed for biocompatibility through this technique which allowed for rapid and complete analysis of the library. Drug delivery systems were also interrogated for their ability to curtail foreign body responses.

See more of this Session: Meet the Faculty Candidate Poster Session
See more of this Group/Topical: Education

(4bd) Environmental Perspectives On the Interactions of Nanomaterials and Microorganisms

Teresa L. Kirschling, Department of Chemical Engineering, Carnegie Mellon University, Pittsburgh, PA

The number of nanomaterials in commercial production is rapidly rising. Much of this production is for new generations of nanomaterial enhanced consumer products. As a result of escalating production, release of engineered nanomaterials to the environment will increase significantly. In some cases, release may be intentional as when reactive iron nanoparticles are deployed for environmental remediation. In most cases release will be unintentional, as when nanoparticles are leached from consumer products into wastewater and concentrate in sludges that are ultimately applied as fertilizer to agricultural fields. Microorganisms will normally be the first organisms to interact with engineered nanomaterials in the environment, but interactions between microorganisms and engineered nanomaterials are poorly understood. This poster summarizes work that demonstrates both the impact of nanomaterials on microbial communities and the impact of bacteria on nanomaterials themselves. For example, nanomaterials currently being designed for in situ remediation of groundwater contamination by chlorinated organic compounds are added at high concentrations directly into aquifers. Microcosm studies carried out with contaminated aquifer materials show that these can dramatically change the composition of microbial communities, despite not showing broad biocidal effects. Investigations are currently underway to develop strategies to combine nanoparticle-based remediation treatments with dechlorinating microorganisms to achieve synergism and more rapid contaminant degradation. This poster also presents the results of an investigation that demonstrates the ability of bacteria to directly degrade polymeric coatings on nanoparticles. The implications of such microbial processing on the environmental fate, partitioning and potential ecotoxicity of engineered nanoparticles will be discussed.

See more of this Session: Meet the Faculty Candidate Poster Session
See more of this Group/Topical: Education

(4be) Molecular Motions of the Beta Relaxation and Glassy Biomolecular Preservation

David S. Simmons, Marcus T. Cicerone and Jack F. Douglas, Polymers Division, Materials Science & Engineering Laboratory, National Institute of Standards and Technology, Gaithersburg, MD

It has long been known that sugar glasses can stabilize a variety of macromolecules. This effect has been observed in the natural protection of lipid bilayers against freezing and drying, and it is employed artificially in the preservation of otherwise unstable dehydrated protein drugs. Recent experimental results employing antiplasticizer additives have suggested that it is the fast dynamics, rather than structural relaxation, of an encapsulating sugar glass that is most coupled to biomolecular preservation. This conclusion could have important implications in the leveraging of this preservation phenomenon, and it exemplifies a broader need to better understand the molecular origins of fast relaxation processes in glasses and supercooled liquids.

A central issue both in understanding the fast dynamics of glasses and in clarifying the coupling of these processes to biomolecule preservation is the nature of the fast beta relaxation. Although the behavior of this relaxation has been proposed to relate to dynamic heterogeneities characteristic of glass forming liquids, the exact nature of the molecular motions corresponding to this relaxation are not known. Such short-time dynamics have been characterized by the Debye-Waller factor – a measure of caging size – and yet the exact physical relationships of this quantity to molecular dynamics and beta relaxation are likewise unknown. Accordingly, results are presented of coarse-grained molecular dynamics simulations representative of the two biological systems mentioned above– a bulk glass-former and a multicomponent lipid membrane – in order to elucidate the connections between fast dynamics, the beta relaxation, dynamics heterogeneites, and biomolecular preservation by biological glasses.

Funding for this research has been provided by NIH Grant # R01 EB006398-01.

See more of this Session: Meet the Faculty Candidate Poster Session
See more of this Group/Topical: Education

(4bf) Development of Hydrogel-Based Kinase Assay for Monitoring Cancer and Developing Patient-Specific Treatments

Gargi Ghosh, Department of Chemical and Biological Engineering, University of Wisconsin-Madison, Madison, WI

Protein array/microarray systems have received considerable attention because of their potential high throughput and cost-effective characterization of protein concentration and activity. However, the major drawbacks associated with these systems are low surface concentration of substrates and non specific binding of sample components resulting in low signal to noise ratios. Moreover, dehydration and denaturation of protein substrates immobilized onto solid surfaces reduce protein activity. We have developed hydrogel based array where the substrate is embedded within the polyacrylamide hydrogel by copolymerization. Immobilization of peptides or fusion protein constructs within the hydrogel network helps maintain the native form of the protein by providing a hydrated environment for the immobilized substrate. Moreover, the characteristic three dimensional network of the hydrogel increases the substrate capacity to a great extent. These characteristics provide a probe with high binding capacity and sensitivity, ultimately improving signal-to-noise ratio.

Epidermal growth factor receptor (EGFR) signaling plays important role in cancer progression and therapeutics directed against the tyrosine kinase domain of EGFR have been developed. However, therapeutics targeted against EGFR have been only partly successful to reproduce promising preclinical model results in clinical settings due to incomplete assessment of EGFR status in cancer, inappropriate drug dosage and /or development of drug resistance during treatment. Hence, a diagnostic tool capable of quantifying the activity and inhibition of EGFR activity will be beneficial in identifying the optimal therapy for individual cancer patients and monitoring for treatment progression. We have developed a hydrogel based assay for quantitative and reproducible determination of the activity of EGFR directly from cellular extracts as well as for determining the sensitivity of different inhibitors to EGFR activity. The protein array has been able to identify EGFR upregulation in a mixture containing 7% EGFR-overexpressing cell lysate diluted in lysate from a cell line expressing low levels of EGFR. Moreover, this system is also capable of evaluating the efficacy of different inhibitors and screening for the most promising therapeutics to overcome the acquired drug resistance.

In future, this assay can be adapted to not only quantify activities of other kinases dysregulated in different cancers but can also be extended into a multiplexed assay with multiple substrates for molecular monitoring of cancers as well as other diseases affected by dysregulated kinases. Moreover, this hydrogel based array can also be used to determine the effect of inhibitors of different kinases on stem cell biology. This is particularly important as cancer stem cells as been implicated in cancer relapse and eradicating cancer stem cells is of profound importance in long term cancer eradication.

(4bg) Nano-Biomaterials: Sensors, Fuel Cells and Robust Hybrid Composites Using Room-Temperature CVD

Gautam Gupta, Center for Integrated Nanotechnolgy, Los Alamos, Los Alamos, NM

There is an essential need for a simple, reliable, and efficient technology to immobilize bio-molecules for sensor, fuel cell, energy harvesting, and bio remediation applications. Here I will go over successful different techniques to integrate soft materials (essentially biological components or carbon nanotubes) into/on hard materials (silica composites, carbon based systems, and Porous Metal films). Initially, I will show a unique tool to encapsulate almost any sensitive molecule using chemical vapor deposition of silica. The process is widely suitable to encapsulate a host of molecules including proteins, lipid bilayers, carbon nano tubes, organelles, and even whole cells. Unique interfacial interactions allow the entrapped species in a solid transparent silica matrix to perform for extended periods of time and even in harsh conditions. I will also illustrate enzyme and DNA immobilization on carbon and gold-based electrodes for sensor and bio-fuel cell application. Eventually, I will go over formation of fluid and resistive lipid bilayers on nano-porous noble metals and graphene.

See more of this Session: Meet the Faculty Candidate Poster Session
See more of this Group/Topical: Education

(4bh) Engineering Materials for Biomedical and Automotive Applications

Holly J. Martin, Center for Advanced Vehicular Systems (CAVS), Mississippi State University, Mississippi State, MS

Graduate/Post-Doctoral Advisor: Dr. Kirk H. Schulz – President, Kansas State University

Post-Doctoral Advisor: Dr. Mark F. Horstemeyer – Center for Advanced Vehicular Systems (CAVS), Mississippi State University

Corrosion, whether it is caused by the human body or the atmosphere, leads to the destruction of metals and ultimately, to metal failure. Engineering materials that can withstand these corrosive effects is the ultimate goal of my research. As a graduate student and as a post-doctoral researcher, I have had the opportunity to perform research in both biomedical implants and automotive metals. In addition to these two wide ranging research areas, which have produced five published journal articles and four manuscripts currently in review, I have also had the opportunity to teach classes and write proposals based on my research.

As a graduate student, I performed research in biomedical implant modification, in an effort to improve the adhesion of a biopolymer coating on titanium. By employing two silane molecules, aminopropyltriethoxysilane (APTES) and triethoxsilylbutyraldehyde (TESBA), dissolved in toluene, we were able to significantly increase the adhesion of chitosan to titanium without changing the nanomechanical properties of the biopolymer. Cell adhesion and growth studies showed no differences in cell adhesion between the control and the two silane bound chitosan coatings, although cell growth was decreased due to residual toluene.

As a postdoctoral researcher, I am performing research into the corrosion of magnesium. Understanding how pit characteristics, such as pit number density, area, and nearest neighbor distance, change after prolonged exposure to saltwater environments is the first step in developing a more corrosion resistance magnesium. Modeling the corrosive effects of two saltwater environments, immersion and cyclical salt spray, on multiple magnesium alloys, including as-cast AE44, AZ91, and AM60 as well as extruded AZ61 and AZ31, can lead to the development of a magnesium alloy that can be effectively used in the automotive industry.

As a graduate student, I had the opportunity to assist with several classes and teach two classes by myself. The first class I taught was an entirely new laboratory for an established class, focusing on process design simulation. As part of my responsibilities, I created all of the materials needed, including the syllabus, Powerpoint presentations covering individual unit operations, homework, and a final project combining the individual unit operations into a working flowsheet. I also was responsible for grading both the homeworks and the final projects. I taught this class for the Fall and Spring semesters and was evaluated by the students, receiving 4.7/5.0 for both semesters. I also taught Unit Operations Laboratory I and II in the Spring semester. I was responsible for teaching the class, including assigning the groups to the experiments each week, for grading the laboratory reports, and for assigning final grades. As with the class I developed, I was evaluated by the students, receiving a 4.5/5.0 average.

As an assistant professor, I plan to continue collaborating with professors in both my research area, biomedical implants, biomaterials, and magnesium, and in other research areas and departments. I truly enjoy teaching and look forward to teaching both established undergraduate and graduate classes and classes I create based on my research interests.

See more of this Session: Meet the Faculty Candidate Poster Session
See more of this Group/Topical: Education

(4bi) Molecular Modeling of Biological Interfaces

Mark J. Uline, Department of Biomedical Engineering, Northwestern University, Evanston, IL

This research project is concerned with investigating lipid bilayers and their interactions with various biologically relevant environments. The thermodynamics and kinetics of ordered domain formation in lipid bilayers is studied using a highly detailed three-dimensional molecular theory for the lipids that uses all of the physical conformations along with attractive and electrostatic interactions. The theory captures the shape of experimental phase diagrams of model lipid bilayers of a high melting point saturated lipid, a low melting point unsaturated lipid and cholesterol. Using the theoretically determined phase diagram, the partition coefficients of protein chain anchors into liquid-ordered and liquid-disordered phases is calculated as a function of temperature and degree of saturation of the chain anchor.

Extending this work, we can now take into account the exact molecular architecture of protein chain anchors and calculate the orientations and positions of proteins as they arrange themselves at the order-disorder phase boundary. In addition to providing these insights into important biological processes in the cell membrane, the theory is used to study the effects of tension on phase transitions in lipid bilayers, interleaflet monolayer coupling free energies, using lipids as a platform for biological sensors, and calculating the potential of mean force between transmembrane proteins. Finally, this theory is used in conjunction with current molecular models of micelle and liposome drug delivery devices that take advantage of the physical properties of tumor cells (in particular, over-expressed ligand receptors and excess negative charge from exposed glycolipids on the outer layer of the cell membrane). By putting these two models together, we calculate how the membrane and micelle/drug systems interact with each other in biological environments. Understanding these interactions on a fundamental level allows us to use the combined theory as a design platform for drug delivery applications. We are currently working in collaboration with David Thompson's group at Purdue University to use these design principles in drug delivery devices.

See more of this Session: Meet the Faculty Candidate Poster Session
See more of this Group/Topical: Education

(4bj) Silicon-Carbon Anode with Improved Structural Integrity for Lithium-Ion Batteries and Sulfur Cathode Encapsulated in Hollow Carbon Nanostructures for Lithium-Sulfur Batteries

Juchen Guo and Chunsheng Wang, Department of Chemical and Biomolecular Engineering, University of Maryland, College Park, MD

The recent development of silicon-carbon composite anode with high energy and high power densities will be presented. This porous Si-C anode is directly synthesized on the current collector by carbonization of a precursor consisting of Si nanoparticles, polyacrylonitrile (PAN) and polystyrene (PS). The structural integrity is improved by the carbon framework obtained by carbonization of PAN, and the porous structure is obtained by the decomposition of PS. To ensure the uniform distribution of Si nanoparticles in the carbon framework, Si particles are grafted with PAN through emulsion polymerization. Development of sulfur cathode materials for lithium-sulfur batteries will also be presented. Sulfur is encapsulated into hollow carbon structures to improve the electrical conductivity and more importantly, to improve the cyclability of the sulfur cathode. The carbon capsules include carbon tubes synthesized with anodic aluminum oxide (AAO) template method, hollow carbon nanofibers from electrospin technique, and hollow carbon spherical structures using emulsion polymerization with various templates.

See more of this Session: Meet the Faculty Candidate Poster Session
See more of this Group/Topical: Education

(4bk) Coarse-Grained Modeling of Polymers for Energy Applications

Lisa M. Hall, Sandia National Laboratories, Albuquerque, NM

Ionomers, or copolymers containing a small fraction of charged monomers, have potential application as safe, mechanically stable solid electrolytes in batteries. Since conduction of only the cations is desired, ionomers with covalently bound anions have an inherent advantage over polyelectrolyte systems in which both the anion and cation are mobile. However, due to strong electrostatic interactions in these materials, counterion diffusion can be slow. Simulations and theory can provide a molecular level understanding of ion transport as a function of controllable parameters such as polymer architecture to aid in the future design of improved materials.

The focus of this poster will be molecular dynamics simulations of ionomer melts with regularly spaced charged beads. The ionic aggregate structure and counterion diffusion has been studied at various values of dielectric constant, spacing of the charged beads on the polymer, and the sizes of the charged beads and counterions. Related advanced polymers such as polyampholytes and hydronium conducting ionomers for fuel cell membranes are amenable to study through similar methods. The application of density functional theory to such systems will also be discussed.

*Sandia National Laboratories is a multi-program laboratory operated by Sandia Corporation, a wholly owned subsidiary of Lockheed Martin Corporation, for the U.S. Department of Energy's National Nuclear Security Administration under contract DE-AC04-94AL85000.

See more of this Session: Meet the Faculty Candidate Poster Session
See more of this Group/Topical: Education

(4bl) Predictive Kinetics for Chemical Engineering

Richard H. West, Department of Chemical Engineering, Massachusetts Institute of Technology, Cambridge, MA

Although detailed chemical kinetics are commonly used in gas-phase systems (combustion, atmospheric chemistry) they have yet to be widely applied to predictively modeling liquid phase and heterogeneous systems. Such models could rationalize observations, provide insight, direct experimental research, and suggest new catalyst materials. Ultimately I aim to tackle problems such as "what temperature, pressure, flow rate, and catalyst alloy should I choose for my process to achieve product spectrum Y from feedstock X?"

See more of this Session: Meet the Faculty Candidate Poster Session
See more of this Group/Topical: Education

(4bm) Phase Behavior of Polymer/Nanoparticle Blends near a Substrate

Venkat Padmanabhan[1], Michael E. Mackay[1], Amalie L. Frischknecht[2] and Sanat Kumar[3], (1) Materials Science and Engineering, University of Delaware, Newark, DE, (2)Sandia National Laboratories, Albuquerque, NM, (3)Chemical Engineering, Columbia University, New York, NY

We use both fluids density functional theory (DFT) with mean-field attractions and molecular dynamics (MD) simulations to investigate the phase behavior of polymer/nanoparticle blends near a substrate. Earlier studies have shown that for an athermal system, with freely jointed chains, there is a first order transition, where the nanoparticles expel the polymer from the surface to form a monolayer at a certain nanoparticle concentration. This transition has been justified by nanoparticle segregation to the substrate observed in experiments. In the presence of attractions, the first order transition moves to higher nanoparticle concentrations as the strength of attraction is increased. Chains with limited flexibility show nematic ordering at higher densities in the bulk. Upon reducing the temperature, the system undergoes a sol-gel transition resulting in a nematic gel. We intend to study the phase behavior of blends with semiflexible polymer chains and nanoparticles in presence of a substrate. The primary objective of this work is to model the experimental system with PCBM in P3HT thin films used in solar cells. By modeling the experimental results, we can identify the stiffness of P3HT, which is currently unknown.

See more of this Session: Meet the Faculty Candidate Poster Session
See more of this Group/Topical: Education

(4bn) Sustainable Materials: From Energy Storage to Biomaterials and New Opportunities in Chemical Engineering Research

Santanu Kundu, Polymer Characterization and Measurement Group, Polymers Division, National Institute of Standards and Technology (NIST), Gaithersburg, MD

Energy, water, and healthcare are three major challenges humanity is facing today and it is widely accepted that the solutions to these problems need to be more sustainable, will use renewable resources, and at the same time will be affordable. Origin and development of the modern Chemical Engineering in the last century was mostly based on usage of petroleum based sources and sustainable, bio-based alternatives have not received the same attention in development. Social and political pressures are driving industrial interest in alternative materials, but transformative advances in measurements, methods and models are critical to success and Chemical Engineers will play an important role for this transformation. Specifically, for the development of alternative materials, it is important that we select suitable materials system, understand and optimize their properties as a function of structural hierarchy (from molecular to macro scale) using conventional and newly developed measurement techniques, develop and optimize processing strategies. In this spirit, as a new faculty, I would like to investigate the *synthesis/processing-structure-property* relationships for various sustainable materials, with the goal of applying these materials to different challenging applications, ranging from energy storage to new generation bio-materials.

Specifically, I would like to develop a research program in the following areas:

A.Mechanics of bio-based/bio-inspired soft materials

B.Carbon materials from natural resources with hierarchical pore structures for energy storage and water purification

The research program will be developed based on my expertise in chemical engineering and materials science and the program will emphasize a multidisciplinary approach that encompasses knowledge and skills covering the chemical, physical, and life sciences and engineering. The techniques/platforms will be used in my group are lab-on-a-chip, rheology, polymer processing (electro-spinning, extrusion), high temperature materials processing, etc.

Presently, I am a postdoctoral researcher working in the polymer characterization and measurement group at the National Institute of Standards and Technology (NIST), where I am developing microfluidics/lab-on-a-chip based platform to investigate the enzymatic polymerization of biodegradable polymers.[1] The goal is to provide systematic, rigorous and quantitative characterization for these poorly understood systems. We studied Candida Antartica Lipase B (CAL B) catalyzed ring opening polymerization of ε-caprolactone to polycaprolactone (a model system) in inexpensive metal microrecactors. Metal microreactors enabled us to perform polymerization reactions in organic media and at high temperatures. By implementing a suitable reactor design we have achieved (1) faster polymerization rate and (2) higher molecular weight of polycaprolactone in microreactors compared to that obtained in batch reactors at similar experimental conditions.

Before joining NIST, I was a post-doctoral research associate at the Polymer Science and Engineering Department of University of Massachusetts-Amherst. I was involved in investigating surface, near surface and mechanical characterization of gel/hydrogel/elastomeric materials.[2-4] The mechanical properties of gels present qualitatively contradictory behavior; they are commonly soft but also notoriously brittle. I investigated the elasticity and fracture behavior of swollen polymer networks using a simple experimental method to induce cavitation within polyacrylamide hydrogel, a common material used in many biological applications. A transition from reversible cavitation to irreversible fracture was observed with the increase of polymer volume fraction. Adapting scaling theories, it is shown quantitatively that the transition from reversible cavitation to irreversible fracture depends on the polymer volume fraction and an initial defect length scale.[2] Application of this simple technique (cavitation rheology) provides opportunities for studying mechanical properties in both synthetic polymer networks and biological tissues from molecular to macroscopic length scales at an arbitrary location.

In my Ph.D. research I have studied the flow and three dimensional microstructure of a liquid crystalline carbonaceous material (mesophase pitch) for different flow conditions, such as steady shear flow, dynamic flow, and processing flow using polarizing optical microscopy and X-ray diffraction.[5-9] The evolution of microstructure in different flow conditions was uniquely studied in three orthogonal planes. The systematic understanding of flow and its effect on microstructure helped us to predict/model the complex flow behavior and microstructural evolution of this material, which in turn will help to design carbon materials, such as carbon fibers, battery electrodes, carbon-carbon composites, and fuel cell separators in more efficient ways. My dissertation, 'Investigation of flow and microstructure in rheometric and processing flow conditions for liquid crystalline pitch', has been awarded the *Best Dissertation in Carbon Science, 2007* for *'outstanding scientific achievement'* by the Elsevier-Carbon Journal.

In addition to research, I am passionate about teaching and am inspired to educate and to mentor future scientists and engineers to meet the challenges of today's ever-changing technologies. My teaching philosophy has been developed through my experience as a lab and teaching assistant during my graduate school days, and as a supervisor during my industrial tenure. I can teach traditional undergraduate and graduate level chemical engineering courses. In addition, I intend to offer advanced graduate level courses in polymer adhesion, fabrication and characterization of nanostructured materials, viscoelasticity of polymers and biopolymers: theory and simulation.

References:

1.Kundu S, Bhangale A, William WE, Flynn KM, Gross RA, Beers KL. Immobilized enzyme catalyzed polymerization reactions in microreactors. Polymer Preprints 2010, 51(1), 745.

2.Kundu S, Crosby AJ. Cavitation and fracture behavior of polyacrylamide hydrogels, Soft Matter 5(20), 3963-3968 (2009).

3.Kundu S, Crosby AJ, Sharma R. Adhesion Behavior of Non-planar Wrinkled Surfaces, to be submitted.

4.Breid D, Kundu S, Denic V, Crosby AJ. Curvature effect on the wrinkle morphology, in preparation.

5.Kundu S, Grecov D, Rey AD, Ogale AA. Shear flow induced microstructure of a synthetic mesophase pitch, Journal of Rheology 53(1):85-113(2009).

6.Kundu S, Ogale AA. Rheostructural studies on a synthetic mesophase pitch during transient shear flow, Carbon 44(11): 2224-2235 (2006).

7.Kundu S, Ogale AA. Microstructural effects on the dynamic rheology of a discotic mesophase pitch, Rheologica Acta 46(9):1211-1222 (2007).

8.Kundu S, Naskar AK, Ogale AA, Anderson D, Arnold JR. Observations on a low-angle x-ray diffraction peak for AR-HP mesophase pitch, Carbon 46(8):1166-1169 (2008).

9.Kundu S, Ogale AA. Rheostructural studies of a discotic mesophase pitch at processing flow conditions, Rheologica Acta, In press, 2010.

See more of this Session: Meet the Faculty Candidate Poster Session
See more of this Group/Topical: Education

(4bo) Systematic Design, Implementation and Evaluation of Vitrification as a Preservation Method for An Encapsulated Cell System

Alison Lawson, Chemical & Biomolecular Engineering, Georgia Institute of Technology, Atlanta, GA and Athanassios Sambanis, School of Chemical & Biomolecular Engineering, Georgia Institute of Technology, Atlanta, GA

The primary goal of Tissue Engineering is to utilize current biological knowledge and engineering principles to "produce functional replacement tissue for clinical use" (MacArthur, Nature 2005). In order to achieve large-scale clinical use, it will be necessary to manufacture, test, store, transport and therefore preserve the product. Preservation will be critical in bringing most tissue-engineered constructs from the benchtop to the clinic. Currently, most of the methods being used in Cryobiology are developed through trial-and-error. In order to better understand and design methods for cryopreservation, it is important to systematically investigate all components of cryopreservation. This includes the investigation of the effects of cryoprotectants (CPAs), alone and as cocktails in solution, and the investigation of CPA addition/removal and cooling/warming protocols. A tissue-engineered pancreatic substitute consisting of encapsulated murine insulinomas was chosen as the test-bed for these studies. The systematic approach was as follows. (1) A mathematical model was used to design addition and removal protocols for cryopreservation. (2) The effects of cryopreservation on the encapsulated cell system—both cellular viability and function and biomaterial integrity and function—were determined in vitro. (3) The effects of cryopreservation on the in vivo functionality were determined in an induced-diabetic mouse model. The most common method of cryopreservation in use today is conventional freezing. Conventional freezing utilizes low levels of CPAs, slow cooling and rapid warming to preserve cells, tissues or constructs. The relatively low concentrations of CPAs are usually minimally cytotoxic to the cells. Unfortunately, these low levels allow ice formation in the extracellular matrix. This ice formation may be detrimental to the integrity, strength or even overall function of a construct. Vitrification was developed in response to this and utilizes high concentrations of CPAs with rapid cooling and warming to achieve a vitreous, or glassy, state. Successful vitrification eliminates ice formation and may maintain the biomaterial better than conventional freezing. However, the addition of high concentrations of CPAs also increases the overall cytotoxicity and requires multi-step addition and removal as osmotic excursions must be maintained within the cell's tolerable limits. These two types of cryopreservation address different concerns in the preservation of a construct. In order to gain insight into the preservation of an encapsulated system, both were used to cryopreserve the tissue-engineered pancreatic substitute. A previously developed mathematical model that describes the mass transfer and cell permeability was used to simulate the osmotic excursions that the cells within the construct undergo. This allows the design of protocols that achieve vitrifiable concentrations throughout the construct while minimizing the cell's osmotic excursions and exposure to the CPAs. This mathematical model was expanded to include heat transfer and CPA cytotoxicity. The expanded model allows for investigation of more complex conditions that may be difficult to achieve experimentally but that warrant further exploration. This mathematical model was used to design addition and removal protocols for vitrification. Critical to the success of a preservation method is the maintenance of cellular viability and function as well as biomaterial integrity and function. Encapsulated cells were cryopreserved using conventional freezing or vitrification and the results compared with encapsulated cells that did not undergo a cryopreservation procedure. The results of in vitro experiments and of in vivo studies involving implantation of encapsulated cells at sub-therapeutic amounts in normal animals indicate that the alginate matrix is preserved during vitrification but appears also not to be functionally damaged by the ice formed during conventional freezing. However, encapsulated cells that are vitrified may retain their in vitro function better than those that are conventionally frozen. Finally, the cryopreserved pancreatic substitute is assessed for in vivo performance by implanting therapeutic amounts in mice with chemically induced diabetes. The results of this systematic design of cryopreservation protocols, their in vitro assessment, and the in vivo performance and efficacy of the cryopreserved capsules will be discussed.

See more of this Session: Meet the Faculty Candidate Poster Session
See more of this Group/Topical: Education

(4bp) Nanocomposite Sorbents for High Efficiency CO2 Capture

Genggeng Qi, Material Science and Engineering, Cornell University, Ithaca, NY

Carbon dioxide has drawn significant attention for being one of the main anthropogenic contributors to climate change. Various approaches, such as solvents, membranes and solid sorbents, have been proposed to capture CO2 using. Among those, liquid amine scrubbing is the most widely studied technology for CO2 capture on industrial scale. Although effective at CO2 capture, amine scrubbing has a number of shortcomings including the corrosive nature, energy intensive regeneration process, and solvent degradation via oxidation and vaporization. In contrast to the liquid amine scrubbing technique, solid-supported amine sorbents offer significant advantages for CO2 capture, such as low energy requirement for sorbent regeneration, and potential elimination of corrosion problems. Meanwhile, solid-supported amine sorbents exhibit high selectivity and reversibility for CO2 capture due to the CO2-amine chemistry. Because of the numerous advantages, a great deal of effort has been directed towards the development of solid-supported amine sorbents for CO2 capture. However, the application of solid-supported amine sorbents is largely limited by their relatively low CO2 capture capacity. High capture efficiency sorbents with good recyclability are required in order to develop an economically feasible industrial-scale process. Here robust nanocomposite sorbents for high efficiency CO2 capture have been developed based on specially designed mesoporous silica impregnated with polyethylenimine (PEI). The newly synthesized composite sorbents exhibit extraordinary capture capacity up to 7.9 mmol of CO2 /g of sorbent under simulated flue gas conditions. The sorbents have fast CO2 capture kinetics and reach their 90% of the total capacities within few minutes. Other than the high capacity and fast sorption kinetics, the sorbents also show full reversibility and good stability during cyclic adsorption-regeneration tests. The new sorbents reported here offer many advantages over the amine scrubbing technique and show promise as a potential candidate for alternative CO2 capture technologies.

See more of this Session: Meet the Faculty Candidate Poster Session
See more of this Group/Topical: Education

(4bq) Soft and Biological Material Odes: Optics, Dynamics, Elasticity-Extensibility, and Self-Assembly

Vivek Sharma, Mechanical Engineering, Massachusetts Institute of Technology, Cambridge, MA

Polymers, colloids, surfactants, liquid crystals and gels are all termed as complex fluids or soft materials, for they respond to stress like semi-solids or squishy materials. A combination of spring-like elasticity and ability to flow makes their behavior viscoelastic. Their ability to flow is itself distinguished by behavior unlike that of common liquids or gases: the complexity comes from their time-dependent, or stress-dependent behavior. Many industrial and biological materials exist or are processed in the form of multicomponent, complex fluids. The ultimate function and use of these materials is dependent on both the physical properties of the constituents as well as on the microstructure formed by them. The relation of the optics, structure and dynamics of complex fluids to order and entropy, thermodynamics and kinetics of phase transitions, elasticity and extensibility and self-assembly is of both scientific and technological interest. In my graduate studies and post-doctoral research [1-4], I have focused primarily on the optics, dynamics, self-assembly and pattern formation in complex fluids, as described below:

1. Doctoral research [1-3]: "Colloidal gold nanorods, iridescent beetles and breath figure- templated assembly of ordered arrays of pores in polymer films". (Advised by Mohan Srinivasaro, Georgia Institute of Technology). (a) Breath figures formed on evaporating polymer solutions exposed to a blast of humid air involve growth and self-assembly of water drops that are non-coalescent, and pack into highly ordered arrays. These hexagonally ordered arrays of nearly monodisperse drops, eventually evaporate away, leaving a microstructured, porous polymer film. I elucidated the mechanism of this breath-figure-templated assembly, by accounting for various transport and thermodynamic processes that control the dynamics of drop nucleation, growth, noncoalescence and self-assembly. The theoretical framework developed in this study allows us to rationalize and predict the structure and size of pores. Further, we (together with M. Crne, J. Blair , J. O. Park , C. J. Summers and M. Srinivasarao) provide an exemplary study where we use coat these films with inorganic oxides and mimic the structural color of Papilio palinurus butterfly (b) Colloidal Gold Nanorods (with K. Park and M. Srinivasarao) By using hydrodynamic arguments, I determined the conditions required to separate nanorods from nanospheres. The monodisperse samples obtained using the separation strategy were then used to examine the optical response of the dispersions of gold nanorods, and their self-assembly over various substrates. (c) Structural origin of circularly polarized iridescence in jeweled beetles (with M. Crne, J. O. Park and M. Srinivasarao) We determined that the patterns on iridescent beetle Chrysina Gloriosa are structurally and optically analogous to cholesteric focal conics on a free surface.

2. Post-doctoral research: Extensional rheology of complex fluids and particle-filled fluids for improved spraying performance (with Arezoo Ardekani, James Serdy and Gareth McKinley, Massachusetts

Institute of Technology). I am developing experimental protocols for testing shear and extensional properties of associative polymer solutions and particle-filled systems over a wide range of shear and extensional rates, using microfluidic devices and by studying capillary break-up of liquid bridges and viscoelastic jets. Surface tension driven break-up of cylindrical fluid elements into droplets plays a crucial role in the use or processing of many multicomponent, microstructured complex fluids like paints, inks, insecticides and pesticides, cosmetics, food, etc. These industrial fluids are typically formulated using dilute polymer solutions, and are exposed to a wide range of shear (1 - 10^6 s^-1) and extension rates (in excess of 10^3 s^-1). Since the polymer solutions and the resulting dispersions have low viscosity (~10 mPas) and short relaxation times (<1 ms), their non-Newtonian behavior is not apparent in the conventional rheometric measurements. However, the presence of even a dilute amount of polymer alters the character of instability growth and capillary break-up during jetting. The interplay of capillary, inertial, elastic and viscous effects on small length and time scales typically leads to complex dynamics in a necking fluid thread and in some cases, the extensional stresses generated in the neck lead to formation of very thin and stable filaments between drops, or to 'beads-on-a-string' structures. I experimentally study the influence of both elasticity and extensibility on the growth of instability and capillary break-up of harmonically perturbed jets of the viscoelastic fluids. By combining experiments with simulations, we show how and when capillary thinning analysis can be applied to capillary break-up during jetting to measure rheological response of fluids. We also show that how the dynamics of the capillary break-up and satellite formation during jetting is influenced by the amplitude and frequency of the imposed disturbances.

In my poster, I will present snippets from my graduate research as well as from my current post-doctoral work. My goal is to establish a research program to develop multicomponent, multifunctional materials incorporating colloids, polymers and liquid crystals for use in applications ranging from fluids for spraying applications to understanding of multicomponent biological fluids. By understanding and using design principles observed in nature (like in iridescent beetles and in butterflies, or in silk fibers), I will develop materials that combine structural integrity and robustness with the required optical functionality (reflection, transmission, absorption) for harvesting solar energy. While the understanding of dynamics and principles of self-assembly is crucial for developing practically useful processing methods, the knowledge of interaction of light with matter is essential for both design of optical or photonic devices, and for observing changes in microstructure during flow.

References

1.V. Sharma, M. Crne, J. O. Park and M. Srinivasarao "Structural Origin of Circularly Polarized Iridescence in Jeweled Beetles," Science, 325, 449 (2009).

2.V. Sharma, K. Park and M. Srinivasarao "Shape Separation of gold nanorods using centrifugation," Proceedings of National Academy of Sciences, 106(13), 4981 (2009).

3.V. Sharma, K. Park and M. Srinivasarao "Colloidal dispersion of gold nanorods: Historical background, optical properties, synthesis, separation and self-assembly," Material Science and

Engineering Reports, 65, 1 (2009).

4.R.O. Grigoriev, M.F. Schatz and V. Sharma "Chaotic mixing in microdroplets," Lab on a chip, 6, 1369 (2006).

See more of this Session: Meet the Faculty Candidate Poster Session
See more of this Group/Topical: Education

(4br) Fundamentals of Gene Delivery From Tissue Engineering Scaffolds

Misael O. Aviles, Chemical and Biological Engineering, Northwestern University, Evanston, IL

My research focus has been the development of biomaterial scaffolds capable of localized gene delivery that can enhance the native ability of the body to regenerate or heal wounds. Porous scaffolds are widely used to mechanically support the injury, promote cell infiltration, and deliver tissue inductive factors. To locally deliver these inductive factors, an approach with the unique potential to target one or more processes is by delivering gene therapy vectors encoding for the factors. I have performed mechanistic studies of gene delivery from biomaterials, and applied these delivery systems to regenerate nerves following spinal cord injury. For the mechanistic studies, I investigated two approaches to improve the localization and duration of transgene expression: i) cationic polymers to modify the scaffolds to reversibly bind the vectors, and ii) hydrogels to fill the pores of the scaffold. For the first approach, I used scaffolds made from PLG microspheres modified with cationic polymers, such as PEI, PLL, and dopamine to reversibly associate with the DNA and delay release from the scaffold, which should increase the local concentration of the vector for enhanced delivery. The scaffolds modified with polydopamine enhanced plasmid incorporation and at the same time extended release. However, extended release did not impact the duration of transgene expression in vivo, suggesting that the initial response of the host to the implant may dictate the extent of transgene expression. In the second approach, collagen, fibrin, and alginate hydrogels were deposited within the pores of the scaffolds. These three hydrogels were selected as they have different rates of degradation and cell interactions, which were expected to influence the rate of cell infiltration and vector release. Both plasmid and lentiviral vectors were loaded into the hydrogels. The plasmid produced short-term expression, whereas expression following lentivirus delivery persisted for up to 4 weeks post implantation. These results indicate that expression can be localized and sustained for extended periods of time.

We subsequently investigated gene delivery from porous scaffolds as a means to improve regeneration following spinal cord injury. Lentivirus encoding for the neurotrophic factors NT3 and BDNF were complexed with nanoparticles, loaded into scaffolds, and implanted in a rat spinal cord hemisection spinal cord model. The spinal cords were retrieved at 1, 2 and 4 weeks post implantation and stained for neurofilament (NF200). Lentiviral delivered significantly increased neurofilament growth through the injury site. Gene delivery from scaffolds is a viable approach to enhance spinal cord regeneration, and does not require cell transplantation common to other approaches. Overall, these studies demonstrate that gene delivery from biomaterials can enhance regeneration and has potential for cell transplantation and gene therapy applications. I see opportunities to improve gene delivery applications with control release systems, targeting, and actuated release. The use of magnetic particles can be used to enhance delivery, as well as to study cell response to mechanical stimuli, which can alter regeneration. I have experience in magnetic targeting which will allow me to utilize magnetic systems applied to tissue engineering and gene delivery for future research.

(4bs) Systems Engineering for Sustainability

Arun Giridhar, Chemical Engineering, Purdue University, West Lafayette, IN

Systems engineering methodologies have a huge role to play in achieving sustainability. In this work, I summarize key findings from my doctoral and post-doctoral research in process synthesis, design, operations, and control, applied to reducing energy consumption, reducing cost, and increasing quality, in the chemical and pharmaceutical sectors. I identify and describe further lines of research in applying systems engineering methods to achieving sustainability in energy, water, and health systems.

See more of this Session: Meet the Faculty Candidate Poster Session
See more of this Group/Topical: Education

(4bt) Membrane and Catalyst Degradation in Polymer Electrolyte Fuel Cells

Panagiotis Trogadas, Chemical & Biomolecular Engineering, Georgia Institute of Technology, Atlanta, GA

Components of polymer electrolyte fuel cell (PEFC) membrane electrode assemblies (MEAs) deteriorate during long-term operation. Two of the most serious issues are the deterioration of perfluorinated ionomer membrane used as the solid electrolyte and electrocatalyst dissolution.

In the case of membrane degradation, it has been shown that reactive oxygen species (ROS) such as HO• and HO2• radicals (1-7) produced at the anode and cathode of the fuel cell as well as within the electrolyte membrane are responsible for the degradation reactions. The formation of hydrogen peroxide during fuel cell operation has been confirmed and degradation by-products have been detected in product water. Three approaches can be used to minimize the effect of ROS in a PEFC: (i) the use of free radical scavengers (8-10); (ii) the use of dispersed peroxide decomposition catalysts within the electrolyte (11) and (iii) the use of dispersed peroxide decomposition catalysts within the electrodes (12-14). The present study is aimed at validating these approaches by: 1) incorporating hydrogen peroxide decomposition catalysts (MnO2; WO3) in the electrodes; and 2) incorporating radical scavengers (CeO2; MnO2; metal nanoparticles) in the electrolyte. The efficacy of the hydrogen peroxide decomposition catalysts is evaluated in terms of: 1) their ability to mitigate the generation of H2O2; and 2) their ability to lower the rate of electrolyte membrane degradation. The former is estimated through rotating ring disk electrode (RRDE) experiments by estimating the percentage of H2O2 detected at the ring electrode during the oxygen reduction reaction. The latter is estimated through MEA studies, wherein the amount of fluoride ions released by the membrane is estimated after a specific operating period in an accelerated test. The efficacy of the radical scavengers incorporated within the electrolyte was evaluated through accelerated durability testing of MEAs.

In the case of Pt dissolution, it was revealed that the major cause of active Pt surface area loss is Pt coarsening (Pt dissolution and redeposition as in Ostwald ripening; 15-17) and Pt diffusion away from the cathode (18). Cathode catalyst degradation models were developed examining the effect of relative humidity (RH), hydrogen crossover and reactant gas partial pressure on Pt dissolution (19-22). RH has a drastic effect on Pt degradation; an approximate three- and two-fold increase of Pt mass and active surface area loss respectively at 100% RH (compared to 50% RH) was observed (20). Hydrogen crossover increases Pt mass loss from the particle surface resulting to an enhanced Pt surface area loss (22), while reactant gas partial pressure does not affect catalyst degradation rate (20). Despite these modeling attempts, there is partial information about the role of Pt particle size distribution (23), Pt+2 ions diffusivity and platinum oxide chemical dissolution reaction on catalyst degradation rate. This work aims to address the role of particle size distribution on Pt stability. A theoretical model is built based on modified Butler-Volmer equations (19, 21) to describe the electrochemical reactions of Pt dissolution and oxidation considering the size dependency of particle stability (Kelvin equation) and platinum oxide surface coverage. The introduction of volume-average particle radius equations in the model allows the use of different Pt particle size distributions in the model, examining their distribution trend and effect on Pt stability.

[1] P. Gode, J. Ihonen, A. Strandroth, H. Ericson, G. Lindbergh, M. Paronen, F. Sundholm, and N. Walsby, Fuel Cells, 3, 21 (2003). [2] A. Pozio, R. F. Silva, M. deFranchesco, and L. Giorgi, Electrochim. Acta, 48, 1543 (2003). [3] E. Endoh, S. Terazono, H. Widjaja, and Y. Takimoto, Electrochem. Solid-State Lett., 7, A209 (2004). [4] A. Bosnjakovic, and S. Schlick, J. Phys. Chem.. B, 108, 4332 (2004). [5] A. Panchenko, H. Dilger, J. Kerres, M. Hein, A. Ullrich, T. Kaz, and E. Roduner, Phys. Chem. Chem. Phys., 6, 2891 (2004). [6] A. Panchenko, H. Dilger, E. Moller, T. Sixt, and E. Roduner, J. Power Sources, 127, 325 (2004). [7] T. Kinumoto, M. Inaba, Y. Nakayama, K. Ogata, R. Umebayashi, A. Tasaka, Y. Iriyama, T. Abe, and Z. Ogumi, J. Power Sources, 158, 1222 (2006). [8] F. D. Comms, H. Liu, and J. E. Owejan, ECS Trans., 16, 1735 (2008). [9] P. Trogadas, J. Parrondo, and V. Ramani, Electrochem. Solid-State Lett., 11, B113 (2008). [10] P. Trogadas, J. Parrondo, and V. Ramani, Functional Polymer Nanocomposites for Energy Storage and Conversion, ACS Books, 187 (2010). [11] G. M. Haugen, F. Meng, F.; N. V. Aieta, J. L. Horan, M. Kuo, M. H. Frey, S. J. Hamrock, and A. Herring, Electrochem. Solid-State Lett., 10, B51 (2007). [12] P. Trogadas, and V. Ramani, J. Power Sources, 174, 159 (2007). [13] P. Trogadas, and V. Ramani, J. Electrochem. Soc., 155, B696 (2008). [14] N. R. de Tacconi, C. R. Chenthamarakshan, K. Rajeshwar, W. Y. Lin, T. F. Carlson, W. A. Wampler, S. Sambandam, and V. Ramani, J. Electrochem. Soc., 155, B1102 (2008). [15] P.J. Ferreira, G. J. la O', Y. Shao-Horn, D. Morgan, R. Makharia, S. Kocha, and H.A. Gasteiger, J. Electrochem. Soc., 152, A2256 (2005). [16] P. Bindra, S.J. Clouser, and E. Yeager, J. Electrochem. Soc., 126, 1631 (1979). [17] Honji, T. Mori, K. Tamura, and Y. Hishinuma, J. Electrochem. Soc., 135, 355 (1988). [18] M. Schulze, A. Schneider, and E. Gülzow, J. Power Sources, 127, 213 (2004). [19] R.M. Darling, and J.P. Meyers, J. Electrochem. Soc., 150, A1523 (2003). [20] W. Bi, Q. Suna, Y. Deng, and T.F. Fuller, Electrochim. Acta, 54, 1826 (2009). [21] W. Bi, and T.F. Fuller, J. Power Sources, 178, 188 (2008). [22] E.F. Holby, W. Sheng, Y. Shao-Horn, and D. Morgan, Energy & Environmental Science, 2, 865 (2009). [23] K. Kinoshita, J. T. Lundquist, and P. Stonehart, Electroanal. Chem. Interfac. Electrochem., 48, 157 (1973).

See more of this Session: Meet the Faculty Candidate Poster Session
See more of this Group/Topical: Education

(4bu) Rational Design of Advanced Organic Materials

Richard A. Lawson and Clifford L. Henderson, School of Chemical & Biomolecular Engineering, Georgia Institute of Technology, Atlanta, GA

In the late 1800's, Thomas Edison tried thousands of different materials to find a suitable filament material for his incandescent light bulb. Over a hundred years later, this "Edisonian" approach is still one of the dominant methods used in the development of new materials. While this method may eventually yield improved materials, it is far too costly and slow for many advanced applications such as fuel cell membranes, organic electronics, organic photovoltaics, photoresists, biomaterials, and many other important purposes. Modern materials require modern, interdisciplinary approaches to their design and development. It is no longer sufficient to make thousands of different materials to identify a few superior materials. Instead, it would be much preferred to utilize predictive models to guide material design and selection and utilize a more limited experimental data set to calibrate and refine these model predictions. This is a daunting task even at the level of simply predicting the bulk properties of a material, but unfortunately in many emerging applications (e.g. organic electronics) the manner in which the material is processed and used can strongly affect its physiochemical properties. Thus, coupling in this information about processing effects will also be important in many advanced material applications. Applying these fundamentals to the synthesis of new materials combined with intelligent synthetic schemes quickly yield new and better materials.

Our work has piloted the use of such a rational materials design scheme in the area patterning materials for microelectronics processing. More specifically, we have used a combination of atomistic and mesoscale modeling to guide the design of novel molecular glass photoresist materials that can be used for high resolution patterning in semiconductor fabrication. Conventional state-of-the-art photoresist materials are composed multi-component blends of polymers with small molecule photoacid generators, dissolution inhibitors, and base quenchers. While this design approach has worked for the patterning of multiple generations of smaller and smaller transistors, it can be used for future generations because this design is unable to print features small enough with sufficient smoothness (line edge roughness) and sensitivity. This is especially true since the patterned feature sizes have truly approached the scale of individual polymers. Custom written kinetic Monte Carlo mesoscale simulations of the lithographic process have allowed us to investigate multiple different materials design strategies and processing conditions that lead to greatly improved material performance and helped guide the design of better materials. These simulations not only accurately model current performance; they have successfully predicted the improved performance of new materials. We have also developed thermodynamic models of molecular glass materials that successfully predict the glass transition temperature and accurately predict the full range of dissolution properties of these materials. These models allow the reduction of thousands of candidate materials and conditions to tens of materials. Even if excellent materials are designed, they are only useful if they can be made and made economically. To that end, we have created synthetic strategies that have allowed us to synthesize several generations of these materials, each with improved performance. After modeling and synthesizing new materials, we have improved their performance even further thorough the use of fundamental reaction engineering. By modifying the kinetics of cross-linking of one negative tone resist, the resolution was improved down to 25 nm. This rational design approach has yielded truly superior performing materials.

See more of this Session: Meet the Faculty Candidate Poster Session
See more of this Group/Topical: Education

(4bv) Tailoring Structure, Thermodynamics, and Rheology in Surfactant-Colloid Mixtures for Soft Material Design

Matthew E. Helgeson[1], Eric W. Kaler[2], Norman J. Wagner[3] and Patrick S. Doyle[1], (1)Chemical Engineering, Massachusetts Institute of Technology, Cambridge, MA, (2)Stony Brook University, Stony Brook, NY, (3)Chemical Engineering, University of Delaware, Newark, DE

Surfactants and other amphiphilic molecules can self-assemble into a wide array of micellar structures in solution that can be used as templates and building blocks in more complex materials for nanotechnology, biology, and energy. The integration and interaction of these self-assembled structures with other colloidal entities such as nanoparticles, emulsions, macromolecules and proteins is particularly critical to the structure and functionality of materials in these emerging applications. However, gaps in understanding of mesoscale surfactant-colloid interactions and how they are manifest in macroscopic thermodynamic, rheological, and transport behavior is a significant roadblock in *de novo* engineering of new materials from these building blocks.

My research aims to advance the fundamental understanding of complex surfactant-colloid mixtures to enable new technologies and processes for the engineering of soft materials such as polymers, gels, and functional colloids. A platform that combines both experiment and theory, including microscopy, scattering, thermodynamics and rheology allows us to probe structure and dynamics at scales ranging from the molecular to the continuum level. Applying this approach to the toolkit of amphiphilic self-assembly results in new, flexible methodologies for material formulation and synthesis. This presentation summarizes two separate projects in this area from my doctoral research at the University of Delaware and postdoctoral work at MIT.

Controlling the structure and dynamics of wormlike micelles using colloidal particles (and vice versa)

Wormlike micelles (WLMs) are long, thread-like surfactant aggregates that exhibit dynamics and rheology similar to polymer solutions. The interactions and resulting structuring between colloids and WLMs is of great importance to applications ranging from consumer products to bioseparations to enhanced oil recovery. Here, we show that controlling interactions at the surfactant-colloid interface in mixtures of model WLMs and nanoparticles leads to unique structuring of the resulting fluid. Specifically, the fusion of WLMs with an equilibrium adsorbed surfactant layer at the nanoparticle surface gives rise to WLM-particle junctions. These junctions act as physical cross-links between micelles, resulting in significant gel-like viscoelasticity in an otherwise Newtonian fluid. This rheological enhancement is understood by the formation of a so-called "double network", where the viscoelasticity can be tuned through two energetic scales, the micellar end cap energy and micelle-nanoparticle junction energy, which are readily manipulated by adjusting solution conditions. This achieves sensitive control over the rheology of the WLM fluid, including the ability to suppress instabilities shear-induced phase separation and shear banding.

Conversely, the formation of micelle-nanoparticle junctions gives rise to new colloidal interactions between nanoparticles which can be used to tune the stability of the colloidal dispersion. A statistical mechanical model has been developed based on micellar bridging to describe these interactions in terms of experimentally measurable properties, allowing *a priori* predictions of the interaction potential between particles mediated by WLMs. Predictions of both the suspension microstructure and colloidal phase behavior are in excellent agreement with experimental measurements, and provide a firm basis for the model. These results show that the colloidal interactions can be easily tuned through small changes in temperature and composition via these WLM-mediated interactions, and provides an entirely new route to manipulating dispersion and stability of nanoparticles, emulsions, and proteins.

Formulation and properties of polymerizable nanoemulsions for the synthesis of polymer hydrogels with controlled porosity

Particulate hydrogels are finding a number of applications of biological importance, including tissue engineering, drug encapsulation and delivery, complex bioassays, and molecular separations. A commonality of all of these applications is the ability to control diffusion of molecular and colloidal species into and out of the hydrogel interior. As such, the ability to control the porosity and pore size distribution of the cross-linked polymer network is critical. Ideally, a method to control hydrogel porosity would enable simultaneous control over the volume fraction and size distribution of pores for average pore sizes at biologically relevant length scales. However, current methods for tuning the porosity of hydrogels represent tradeoffs between control over these various aspects of hydrogel porosity.

We have developed a synthesis method for particulate hydrogels with controlled porosity based on microfluidic preparation and polymerization of nanoscale polymerizable high internal phase emulsions (nano-polyHIPEs). In this method, nano-polyHIPEs are prepared by high-pressure homogenization, which yields nearly monodisperse droplets with controlled average size ranging from 50-500 nm at volume fractions exceeding 50%, properties that are unattainable using other methods. Subsequent photopolymerization of the continuous phase results in polymer microstructures with porous features templated by the nano-polyHIPE. Combining this polymerization method with the recently developed stop flow lithography (SFL) technique enables production of particulate hydrogels with exquisitely controlled size, shape, and porosity. This method is flexible toward a number of different polymer chemistries and microstructural motifs, including closed-cell and open-cell pores, as well as polymer composite structures. The resulting multi-functional hydrogel particles are capable of encapsulating both hydrophilic and hydrophobic species, and show promise as novel materials in pharmaceutical development and formulation.

See more of this Session: Meet the Faculty Candidate Poster Session
See more of this Group/Topical: Education

(4bw) Solid Polymer Electrolytes Derived From Ionic Liquids: From Synthesis to Applications

Yuesheng Ye, Department of Chemical and Biological Engineering, Drexel University, Philadelphia, PA

Solid polymer electrolytes (SPEs) are of profound interest to chemists and engineers, in part, due to the drawbacks of liquid electrolytes (e.g. leakage, flammability, toxicity, stability). The SPEs derived from ionic liquids (ILs) have received increased attention over the past years because of their excellent properties (e.g. high conductivity, electrochemical and thermal stability, and non-volatility), leading to a variety of applications for energy conversion and storage devices such as fuel cells, batteries, actuators and capacitors.

Our study focuses on the development of next-generation IL-based SPEs, targeting multifunctional properties that can be utilized for energy applications. Different from an IL, a PIL bears the moiety of an IL that is covalently bonded to a polymer chain by polymerization of an IL monomer through polymerizable functional groups, offering a new type of polymer electrolyte material. We started with synthesis of IL monomers and PILs, which allows to the most opportunity for the preparation of new IL-based electrolyte materials with tailored chemical structures and tunable properties. Our current work has shown that the anion type has a significant effect on the glass transition temperature of the polymer and subsequently the ionic conductivity. Ion transport in polymer electrolytes is also studied, which provides valuable insights into ion conduction in these ion-containing polymers and guides design and preparation of novel SPEs for energy applications.

Currently I am working with Professor Yossef (Joe) Elabd in the Department of Chemical and Biological Engineering at Drexel University (2008 - present). My current research focuses on the synthesis and characterization of a new type of polymer electrolytes (polymerized ionic liquids) for energy conversion and storage devices such as fuel cells, actuators, and batteries. My graduate study under Professor Kyu Yong Choi along with my side work with the department chair, Professor F. Joseph Schork, at University of Maryland at College Park (2002 – 2008) provided me with a strong and broad background in polymer science and engineering (polymer synthesis, characterization, kinetic and reactor modeling, and process scale-up). My Ph.D. work focused on the development of high molecular weight bisphenol A polycarbonate as well as kinetic and reactor modeling. New methods have been developed to prepare micron-sized polycarbonate particles, to prepare high molecular weight polycarbonates, and to reduce bisphenol A (BPA) content in polycarbonate resins. In the side work, new model framework based on the moment method has been developed to calculate and control sequence length for radical copolymerization processes.

See more of this Session: Meet the Faculty Candidate Poster Session

(4bx) First Principles Study of the Photodynamics of Oxidized Silicon Clusters

Benjamin G. Levine, Institute for Computational Molecular Science and Department of Chemistry, Temple University, Philadelphia, PA

The effect of quantum confinement on the properties of semiconductor nanoparticles is well known, but also of interest are the properties which arise from the high surface-to-volume ratio of these systems. It is widely believed that imperfect surfaces result in the trapping and localization of charge carriers at the surfaces, and subsequently in recombination or emission. By modifying the surface chemistry of these nanoparticles one can therefore tune there electronic and optical properties, but the dynamics of such trapping at the atomic level is poorly understood. We have applied first principles quantum chemical and molecular dynamics methods to small silicon clusters in order to understand the effect of various surface treatments. Specific silicon-oxygen bonding configurations at the surface are found to introduce conical intersections within the band gap. These intersections are dynamically accessible and facilitate fast nonradiative decay to the ground state.

See more of this Session: Meet the Faculty Candidate Poster Session
See more of this Group/Topical: Education

(4by) Convective Assembly of Nanoparticles Into Thin Structured Films

J. Alex Lee and Michael Tsapatsis, Department of Chemical Engineering and Materials Science, University of Minnesota, Minneapolis, MN

Convective nanoparticle film assembly is a simple process whereby particles from dilute liquid suspension assemble onto a substrate. Assembly occurs at the suspension-substrate-air contact line, where particles are carried toward it by convective currents set up in the meniscus region due to liquid evaporation. This process is explored as a potential method for the fabrication of highly ordered (i.e. uniformly oriented) nanocrystal films for application in zeolite membrane technology. Typical results indicate plate-like crystal nanoparticles do tend to orient preferentially in convectively assembled films, but that such films are often patterned (precluding their usefulness as membrane precursors). Specifically, the resulting films tend to have discrete banded morphologies; the bands are almost universally reproducible, and can be quite regular. Banding and film morphologies are studied using spherical nanoparticles, especially in the context of the competing rates of particle assembly and substrate withdrawal (contact line recession). Some attempts at modeling the band spacings are presented, as such descriptions may be helpful in modeling the process itself.

See more of this Session: Meet the Faculty Candidate Poster Session
See more of this Group/Topical: Education

(4bz) Biorheology with Applications - Adhesive Wall Climbing, Bacterial Motility, and Predatory Defense

Randy H. Ewoldt, Department of Chemical Engineering and Materials Science & Institute for Mathematics and its Applications, University of Minnesota, Minneapolis, MN

Many biological systems involve viscoelastic material behavior for their function, including (i) the wall-climbing locomotion of snails, (ii) the motility of ulcer-causing bacteria through a biopolymer gel, and (iii) the predatory defense mechanism used by hagfish. Rheology plays a critical role in achieving the desired function in each system. This poster summarizes efforts to fundamentally understand the complex fluid rheology in these three inspiring biological systems, including experimental measurement of large deformation rheology, modeling of structure-property relationships, and modeling of systems which include non-Newtonian fluids as a component. Bioinspired engineering efforts are also described which deliberately use and extend the relevant viscoelastic properties for novel functionality; wall-climbing robots and tunable magnetic fluid adhesion are examples. Future developments will be enabled by a better understanding and description of these ubiquitous but technologically under-utilized materials.

See more of this Session: Meet the Faculty Candidate Poster Session
See more of this Group/Topical: Education

(4c) Cancer Metastasis: Deconstructing Cell Motility On Micropatterned Islands and Tracks

Goher Mahmud, Chemical and Biological Engineering, Northwestern University, Evanston, IL

Cell motility is a process deriving from the synchronized dynamics of the cytoskeleton. In several important physiological processes—notably, cancer metastasis—the randomly moving cells can acquire a directional motility phenotype and bias their motions in response to environmental cues. Despite intense research, however, the current understanding of directional cell migration is incomplete and there is a growing need to develop systems that would enable the study and control of this process.

In particular, substrates prepared by the reaction–diffusion ASoMic (Anisotropic Solid Microetching) method localize cells onto transparent micro-islands or tracks surrounded by an opaque, adhesion-resistant background. ASoMic is compatible with several important imaging modalities (e.g. wide-field, fluorescent, TIRF and confocal microscopies), and can be used to study and quantify various intracellular and cellular processes related to cell motility. For cells constrained on the islands, the imposed geometry controls spatial organization of the cytoskeleton, while the transparency of the islands allows for real-time analysis of cytoskeletal dynamics.

See more of this Session: Meet the Faculty Candidate Poster Session
See more of this Group/Topical: Education

(4ca) Studying Cancer as a Metabolic Disease

Christian M. Metallo, Chemical Engineering, Massachusetts Institute of Technology, Cambridge, MA

Metabolism is at the core of cellular function, driving fundamental processes at the level of cells and entire organisms. A comprehensive understanding of the dynamic interplay between a cell's metabolic state, its microenvironment, and ultimate fate choices is necessary to successfully target metabolism in the treatment of disease or to engineer a microbe for production of a given product. How do metabolic fluxes regulate cellular behavior, signal transduction? In turn, how do these phenomena direct the metabolic phenotype of a cell or organism? The use of advanced computational and experimental techniques, such as metabolic flux analysis and chromatography coupled to mass spectrometry, enable us to address these fundamental questions in a variety of biological systems. Importantly, these analyses must be undertaken at the systems level to fully understand these global interactions. Through the use of stable isotopic tracers and metabolic flux analysis (MFA) we have characterized the metabolic phenotype of tumor cells from various tissues of origin. By quantifying the effects on tumor metabolism of specific genetic and pharmacological perturbations, we have expanded our understanding of how cells adapt their metabolic state to proliferate and respond to chemotherapeutic drugs. The integration of systems level analyses with cell biology and tissue engineering techniques will greatly improve our understanding of cellular physiology. These fundamental studies may then translate to the clinic in the form novel drug targets for cancer and other diseases.

See more of this Session: Meet the Faculty Candidate Poster Session
See more of this Group/Topical: Education

(4cb) High-Throughput Microrheology of Therapeutic Hydrogelators

Kelly M. Schultz and Eric M. Furst, Department of Chemical Engineering, University of Delaware, Newark, DE

The design of therapeutic hydrogelators enables the engineering of this material for applications such as wound healing and tissue regeneration. Development of these scarce materials is generally synthetically laborious with small yields. In order to address the need for rapid rheological characterization of therapeutic hydrogelators while conserving material, we use multiple particle tracking microrheology, which requires small sample volumes (4-40 μ L) while enabling rheological measurements of both the hydrogelation reaction and the final material properties. During the hydrogelation reaction, the material is monitored with time and the sol-gel transition is measured, yielding the critical gelation time and critical relaxation exponents. High-throughput microrheology, microrheological measurements in a microfluidic device, are then used to screen the equilibrated materials, vastly reducing the time of sample preparation and increasing the amount of samples evaluated. From the data, gelation state diagrams are created, which identify compositions where hydrogels form. Overall, our methods can be used to thoroughly screen the material rheological properties, yielding a high information density, but requiring only small volumetric amounts.

See more of this Session: Meet the Faculty Candidate Poster Session
See more of this Group/Topical: Education

(4cc) Novel Material Design through Multiscale Numerical and Analytic Approaches

Erin Lennon, Engineering Sciences and Applied Mathematics, Northwestern University, Evanston, IL

I discuss multiple approaches to multiscale modeling of novel materials that may be used in wide-ranging applications from biomaterial scaffolding to catalysis formation. The first uses a coarse-grained molecular approach to model the self-assembly of polymeric systems that have domains on the micron scale with interfaces of only a few nanometers. I have developed an extension to the typical mean-field theory approach to these models. Instead of relying on average polymer configurations, I allow for fluctuations around a minimal energy state. While previously intractable for large systems, I have developed a methodology for numerically resolving these systems. The progress has come primarily from a novel technique for finding the free energy of fluctuating complex systems as well as increasingly efficient computational algorithms. The second uses a primarily analytic approach to resolve the transport and reaction processes in stationary reacting waves. In such systems, the characteristic length scale of a typical reaction zone is asymptotically small relative to the characteristic length scale of conduction through the reacting media. Nonetheless, both scales must be considered to determine the behavior of coupled propagating reactions, including propagation velocity, extinction limits and conversion effectiveness. I have developed and applied approximation strategies to analyze the disparate length scales and develop relationships between reactant and material properties and the propagation of coupled reactions.

See more of this Session: Meet the Faculty Candidate Poster Session
See more of this Group/Topical: Education

(4cd) Biointerfacial Science and Engineering: From the Fundamental to the Applied

Roger L. York, Koch Institute for Integrative Cancer Research, Massachusetts Institute of Technology, Cambridge, MA

This poster will discuss my research into biointerfacial science and engineering. Sum Frequency Generation (SFG) Vibrational Spectroscopy will be highlighted in the context of the adsorption of peptide to model surfaces. The adhesion of bacteria to solids will be discussed in the context of biofilm formation. Lastly, the design of real biomaterials with well defined function will be mentioned.

See more of this Session: Meet the Faculty Candidate Poster Session
See more of this Group/Topical: Education

(4ce) Computational Molecular Science: Designing Improved Materials for Applications in Energy, Pharmaceutics and Desalination

Amish J. Patel, Chemical and Biological Engineering, Rensselaer Polytechnic Institute, Troy, NY

Computational molecular science has been invaluable in helping us understand the underpinnings of diverse phenomena, ranging from nanotechnology to biology, spanning a wide spectrum of time and length scales. With the tremendous growth in computing power now available, we are on the verge of being able to use computation as a tool for designing novel materials. Novel polymer electrolytes provide the promise of delivering a lithium battery with high energy-density that could revolutionize the way we consume energy. However, judiciously chosen candidates need to be examined, using state-of-the-art modeling techniques in order to guide the search for the optimal electrolyte material. Similarly, in order to design drugs that target a specific site, we need to develop a fast and robust technique that can scan a library of candidates and suggest one that will bind optimally. This requires a detailed understanding of hydration of ligands and proteins and their association in solution. Finally, the rapid growth in world population has put an enormous burden on one of its most precious resources: potable water. Sea water is available in abundance, but in order to make it suitable for human consumption, it has to be desalinated. Carbon nanotubes have been shown to be promising candidates to tackle this task. Increasing interest in this field has led to a rapid growth in the available chemistries with which these nanotubes can be decorated. We need to identify the underlying principles of ion solvation and exclusion from nanotubes in the presence of the decorated moieties, in order to design nanotubes with maximal separation at minimal cost. While these are challenging problems, it is also extremely important that we expend every resource in order to overcome these challenges. I propose to employ a combination of statistical mechanics and molecular computation in order to realize this goal.

-->

See more of this Session: Meet the Faculty Candidate Poster Session
See more of this Group/Topical: Education

(4cf) Use DNA to Probe the Molecular Deformation at Micro/Nanoscale and Design Novel Nanochip Devices for Drug/Gene Delivery and Biosensing

Pouyan E. Boukany, Nanoscale Science and Engineering Center for Affordable Nanoengineering of Polymeric Biomedical Devices, The Ohio State University, Columbus,, OH

Polymer behavior under different flow conditions has long been research interest due to its fundamental and practical applications of these materials in our daily life. The main aim in polymer rheology is to link the macroscopic properties with microscopic changes. DNA became one of the best candidates as a model system to examine flow features of entangled polymer including stress overshoot in startup shear and shear thinning phenomenon. Recently, stretching and aligning DNA molecules also provide new opportunities to develop/design nanotemplates for next generation biomedical devices and key steps in developments of single-DNA sequencing and gene delivery techniques. I have worked previously in two areas of interest related to understanding the flow behavior of DNA as polymer model and development of nanochip based devices by DNA combing and imprinting (DCI). First, my past research at the University of Akron, advised by Prof. Shi-Qing Wang in polymer science, focuses on investigating fundamental questions in polymer/DNA dynamics such as how entangled polymeric liquids response fast external deformation. By integrating confocal fluorescent microscope with rotational rheometer, the moleualr picture behind flow instabilities such wall slip and shear banding was captured, with simultaneous velocimteric and rheometric measurement. This unique set-up provides new insights into the phenomenology of entangled fluids in presence of flow [1-9]. Second, my postdoctoral research at the Ohio state University advised by Prof. James Lee in Nanoscale Science and Engineering Center (NSEC) focuses on the developing novel and unique approach to produce nanochip based devices for many challenging biomedical applications. In one of my unique nanostructure, arrays of DNA nanostrands were generated on micro-patterned PDMS stamp and they were transferred to confined nanoscopic elements by imprinting method, to fabricate sealed arrays of nanochannels/nanowires [10]. This fabricated nanochip was used for localized gene, drug and nanoparticle delivery to individual cells with precise control of dose, composition, and location. The delivery is achieved by the focused electric field through a nano-channel. By adjusting the voltage level, pulse duration and pulse number, dose and location control of the delivery can be precisely controlled. This approach is a benign process that can achieve perfect cell viability and transfection with much less gene/drug/nanoparticles than conventional delivery methods [11].

References: 1. Wang, S. Q.; Rvindranath, S.; Boukany, P.; Olechnowics, M.; Quirk, R. P.; Halasa, A.; Mays, J. Phys. Rev. Lett. 2006, 97, 187801. 2.Wang, Y.; Boukany, P.; Wang, S. Q.; Wang, X. Phys. Rev. Lett. 2007, 99, 237801. 3. Boukany, P. E.; Wang, S. Q. Macromolecules, 2008, 41(4), 1455. 4.

Boukany, P. E.; Hu, Y. T.; Wang, S. Q. Macromolecules, 2008, 41(7), 2644. 5. Boukany, P. E.; Wang, S. Q. Soft Matter, 2009, 5 (4), 780. 6. Boukany, P. E.; Wang, S. Q. J. Rheol. 2009, 53 (1), 73. 7. Boukany, P. E.; Wang, S. Q.; Wang, X. Macromolecules 2009, 42(16), 6261. 8. Boukany, P. E.; Wang, S. Q. Macromolecules 2009, 42 (6), 2222. 9. Boukany, P. E.; Hemminger, O.; Wang, S. Q.; Lee, J. L. Phys. Rev. Lett. 2010 (submitted). 10. Guan, J.; Boukany, P. E.; Chiou, N.; Hemminger, O.; Lee, J. L. Adv. Mater. 2010, (in press) 11. Boukany, P. E.; Mross, A.; Zhang, X.; Hu, X.;; Bo, Y.; Lafyatis, G.; Lee, J. L. 2010 (under preparation).

See more of this Session: Meet the Faculty Candidate Poster Session
See more of this Group/Topical: Education

(4cg) Combining Molecular Level Insights with Advanced Synthesis Strategies to Design (Photo) Catalytic Materials: Efficient Chemical Processing Utilizing Thermal and Solar Stimuli

Phillip Christopher, Chemical Engineering, University of Michigan, Ann Arbor, MI and Suljo Linic, Chemical Engineering, The University of Michigan, Ann Arbor, MI

Developing energy efficient, environmentally friendly methods of producing industrial chemicals is an extremely important area of research for the 21st century. Herein a cross-disciplinary approach for catalyst design combining theoretical chemistry, to develop molecular level insights, with advanced materials synthesis techniques, for the synthesis of uniform materials with tailored physical properties is presented. The approach is applied to the development of more efficient catalysts for industrial thermo-catalytic processes, and creates a foundation for incorporating solar energy as an energy input for activating important chemical reactions.

Recent advances in computational chemistry and nanomaterials synthesis were utilized to design new catalysts for the industrially important ethylene epoxidation reaction. Density functional theory calculations were used to identify promising surface facets of Ag catalysts and advanced synthesis techniques, developed in the nano-science community, were used to synthesize particles with the targeted surface facets. These studies show that catalytic particles of controlled size and shape represent promising heterogeneous catalysts for selective production of chemicals and act as critical platforms to study heterogeneous catalytic process and identify crucial factors that impact process selectivity.

The metallic nanostructures utilized in our initial studies efficiently concentrate light at their surfaces. This effect, known as Localized Surface Plasmon Resonance, results in either: the absorption of photons thereby heating the nanostructure, intense re-emission of the photons or a transfer of energetic electrons to surrounding medium. The intensity, resonance energy and dominant decay mechanism can be tuned by changing the size and shape of the metallic nanoparticles and the external environment. We have exploited the tunable parameters of these optical properties to develop versatile photocatalytic chemical processing platforms that have opened countless possibilities for the development of chemical processes based on the use of multiple energy stimuli for efficient solar/thermal chemical processing.

This research highlights the utility of molecular insights in concert with well-defined metallic nanostructures for the development of new catalytic and photocatalytic materials for energy efficient, environmentally friendly chemical processing. In light of these results future work utilizing a similar over-arching approach towards environmentally friendly chemical processing will be discussed.

(4ch) From Materials to Peptide Design: a Short Trip through Computational Techniques

Diego A. Pantano, Chemical and Biomolecular Engineering, University of Pennsylvania, Philadelphia, PA

Sustained advances in algorithms together with increased computer power and decreasing costs make computational techniques attractive for all types of molecular research. Today's simulations of millions of particles can span many microseconds, and although experimental investigations remain a primary source of information, computational studies are increasingly providing explanation of meso-scale properties and processes.

My research interests have always been focused on computational methods but in a range of topics from materials to peptide design. I will present my latest work going from a rough (coarse grained) to full atomic description of the systems. Coarse-grained molecular dynamics are used to describe a bilayer composed of polymer mixtures that exhibit lateral segregation and spatial colocalization of domains across the membrane (registration): simulations depict the mechanism by which the registration is achieved and suggest possibles causes. The concomitant use of different techniques such as normal mode analysis and molecular dynamics (MD) allow us to test hypothesized mechanisms for one membrane protein among the neurotransmitter sodium symporter (NSS) family. These proteins are of major biological importance as they are associated with various diseases and addictions. The results confirm the mechanism and show that the response of a protein to a mechanical perturbation is directly linked to its topology. Lastly I will show our efforts to design a peptide that functions as "self"-marker to macrophages. This peptide would make particles invisible to macrophages. The use of MD techniques together with docking algorithms lead us to a prediction of sequence that indeed shows, via affinity essays, to have the desired activity.

See more of this Session: Meet the Faculty Candidate Poster Session
See more of this Group/Topical: Education

(4ci) Quantitative Engineering Approach to Particle and Particulate System Design

Defne Kayrak-Talay, Chemical Engineering, Purdue University, West Lafayette, IN and James D. Litster, School of Chemical Engineering; Industrial and Physical Pharmacy, Purdue University, West Lafayette, IN

Particle and particulate systems are very important to a wide range of industries including pharmaceutical, food, detergents and consumer goods, specialty chemicals manufacturing, minerals and metals processing. Particle/particulate system formation processes need to be well understood at different levels in order to control these processes. My recent research is focused on granulation at both particle and process levels. In conventional granulators, the rate processes (nucleation, consolidation and growth, and breakage) take place simultaneously, which makes it difficult to control the process. At process level, our aim is to develop a new process (regime separated granulation) where each rate process takes place in separate equipment so that process control is easier, and tight control on product properties is possible. In the proposed regime separated granulation process, nucleation is the major process that determines the shape and size of the granules. Therefore, the interaction between the liquid drops and powder upon the first impact is also investigated (particle level). The study at the particle level aims developing a model to predict the granule shape and size based on particle and liquid properties, and process conditions.

See more of this Session: Meet the Faculty Candidate Poster Session
See more of this Group/Topical: Education

(4cj) Bio-Inspired, Smart and Functional Systems through Reaction-Diffusion-Convection Processes

Siowling Soh and Bartosz Grzybowski, Chemical and Biological Engineering, Northwestern University, Evanston, IL

Smart and functional systems detect cues from its surrounding, respond appropriately to these cues and perform their intended purposes. At present, development of these systems have mainly relied on electronic-based machineries, devices and robots for solving problems in all aspects of our lives. While this approach has largely been successful, it is still far less intelligent than naturally-occurring examples ranging from a single cell to any larger biological organisms. Despite their complexity, these systems from nature operate almost entirely out of chemical reactions coupled with diffusion and/or convection processes. Learning from these systems and engineering the appropriate reaction and transport processes provide the solution to non-electronic-based intelligent systems with limitless potential. Current research is also gravitating toward development of intelligent systems using these physical principles. Some demonstrations include (1) camphor "boats" which self-assemble dynamically on an air-water interface, swarm autonomously and display "chemotactic" behaviors toward gradients of concentration, temperature and depth of water. Directional and rotational control is also possible for this primitive system. Other examples are (2) oil droplet on an air-liquid interface solving for the shortest path in a maze, (3) core-shell catalytic hydrogels for performing sequential reactions, (4) at a micron-nano scale, ultra-sensitive chemical sensors through cross-linking the surface of a crystal composed of nanoparticles and so on. This research would involve interdisciplinary knowledge from physics, chemistry to biology. Besides manipulating with physical processes such as diffusion and convection, this effort will be coupled with knowledge of chemical reactions fueled by examining databases of organic molecules. Exploring these databases or the networks of organic chemistry will enable the identification of appropriate sets of reactions necessary for mimicking complexity found only in biology.

See more of this Session: Meet the Faculty Candidate Poster Session
See more of this Group/Topical: Education

(4ck) Binary Combinations of Lipid-Like Materials Act Synergistically to Improve siRNA Delivery in Vitro and In Vivo

Kathryn A. Whitehead[1], George Z. Li[1], Kevin Love[1], Robert Langer[1] and Daniel G. Anderson[2], (1) Chemical Engineering, Massachusetts Institute of Technology, Cambridge, MA, (2)Koch Institute for Integrated Cancer Research, Massachusetts Institute of Technology, Cambridge, MA

RNA interference therapeutics represent a new class of gene-suppressing drugs which have marked potential to cure a wide variety of human diseases, including viral infections, heart disease, and cancer. The primary obstacle to the clinical application of siRNA is the development of safe and effective delivery systems. Previously, lipid-like materials, termed 'lipidoids', have been synthesized in our lab, and several leading candidates have been employed for siRNA delivery applications. The vast majority of the lipidoids studied, however, were not capable of inducing gene silencing in vitro. In order to overcome the inherent limitations of single lipidoid delivery materials and to increase the material space available for therapeutic development, binary combinations of lipidoids have been assessed for improved transfection performance both in vitro and in vivo.

Thirty-six lipidoids were synthesized through the Michael addition of amines with acrylates or acrylamides, which were then combined to form 630 binary pairs. These combinations were each tested at six weight fractions each, resulting in 3,780 test formulations. The binary lipidoid formulations were screened for their ability to safely deliver siRNA into a modified HeLa cell line, and this high-throughput analysis resulted in the identification of a number of synergistic lipidoid combinations which demonstrated non-toxic, specific knockdown of firefly luciferase. Although the individual lipidoid components had no transfection ability on their own, some materials were able to achieve 90% knockdown in combination. In vivo testing was performed on the leading hits from the in vitro screen. Anti-Factor VII lipidoid-siRNA nanoparticles were injected in C57BL/6 mice, and Factor VII levels were measured in the blood serum 48 hours post-injection to quantify knockdown. Again, although the individual lipidoids were incapable of silencing the target gene on their own, several lipidoid combinations (dosed at 2 mg/kg of siRNA) produced dramatic synergistic effect in the form of near-complete Factor VII knockdown. Toxicity was preliminarily assessed by monitoring changes in mouse bodyweight, and no toxicity was detected for synergistic lipidoid combinations compared to the negative control.

Although the reason for such synergistic behavior is unclear, it is speculated that the phenomenon may involve the two-step siRNA delivery process. While certain lipidoids may be able to facilitate transport of siRNA across the cellular membrane, they may lack the ability to escape from the endosome once inside the cell. Similarly, other lipidoids may be skilled at endosomal escape, but are not able to gain cellular entry. It is speculated that the in vitro and in vivo screening analysis has enabled the identification of binary lipidoid combinations that are adept at different steps of the transfection process.

See more of this Session: Meet the Faculty Candidate Poster Session
See more of this Group/Topical: Education

(4cl) Exploring Interfacial Phenomena in Organic Photovoltaic Devices Using Block Copolymers

Bryan W. Boudouris and Rachel A. Segalman, Department of Chemical Engineering, University of California, Berkeley, Berkeley, CA

Polymer-based organic photovoltaic (OPV) cells have attracted much attention in recent years due to the promise of power sources that are flexible, lightweight, and inexpensive to fabricate. To date, the most efficient devices are bulk heterojunction cells composed of an active layer blend of an electron-donating polymer and an electron-accepting soluble fullerene derivative. By tuning the solution deposition conditions and post-processing treatments, power conversion efficiencies in these devices have reached values greater than 7%. Significantly, the performance of these cells is inherently linked to the structure at the electron donor-acceptor interface and at the electrode-active layer interfaces. However, these interfaces remain largely uncharacterized, which leads to a limited understanding of the structure-property relationships in these devices. Here we present a series of strategies for rationally controlling the: 1) intermolecular structure, 2) active layer phase separation, and 3) electrode surfaces in order to understand the correlation between these crucial materials junctions and device performance. Each of these approaches relies on the design of block copolymers with specific functionality. By tailoring the chemical structure of these macromolecules, we are able to systematically and rigorously characterize the self-assembled microstructures that are formed. And, finally, these materials are incorporated as functional components in photovoltaic devices. Importantly, because the resultant morphologies have inherently distinct domains, we are able to correlate interfacial structure with electronic performance in operational devices. These relationships, in turn, will provide a structured pathway for future plastic solar cell designs.

See more of this Session: Meet the Faculty Candidate Poster Session
See more of this Group/Topical: Education

(4cm) Rational Design of Photo-(Electro-) Catalysts: A Combined Theoretical/Experimental Route to Solar Fuels

David B. Ingram and Suljo Linic, Department of Chemical Engineering, University of Michigan, Ann Arbor, MI

As highlighted by the U.S. Department of Energy Solar Energy Technologies Program, "solar technologies diversify the energy supply, reduce the country's dependence on imported fuels, improve air quality, and offset greenhouse gas emissions". The efficient conversion of solar energy into chemical energy (solar fuels), through processes such as photo-catalytic water splitting and CO_2 reduction, is critically important for the development of sustainable, long-term solutions to society's energy and environmental problems. These and other photo-electro-chemical processes require the development of photo-electro-catalysts that efficiently make use of the solar spectrum and activate critical chemical transformations with low over-potential losses. Photo-electro-catalysts have a combination of thermal, electrical and solar stimuli, each of which must be investigated to provide insights into the performance and ultimately the design of novel photo-electro-catalysts.

Using a multi-scale approach, which combines a number of experimental and theoretical tools, we have begun to investigate and understand these complex systems. Several case studies are presented, including theoretical investigations of metal/oxide electro-catalysis (focusing on high temperature solid oxide fuel cells) and experimental studies of photo-catalytic reactions (photochemical water splitting and decomposition of organic molecules).

See more of this Session: Meet the Faculty Candidate Poster Session
See more of this Group/Topical: Education

(4cn) Assembly, Manipulation, and Controlled Release of Complex Nanocolloids by Dielectrophoresis

Victoria Froude, Chemical and Biomolecular Engineering, University of Notre Dame, Notre Dame, IN

Alternating current (AC) electric fields have been widely applied to particles and cells of micron size as an efficient and effective assembly strategy using typical electrokinetic techniques, such as dielectrophoresis. However, many technologies, such as targeted drug deliver and lab-on-a-chip diagnostics, will also require the manipulation of nanocolloids of sizes not previously studied. Here, I provide robust experimental evidence of the simple assembly and manipulation of nanoparticles, which has not been previously reported. In addition, the polarization of complex biocolloids, namely micelles and liposomes, has been studied with intriguing results, and I present the strong dependence of colloid concentration on AC-field frequency. Finally, I find a distinct frequency induced encapsulate release from micelle interiors with a dependence on the surfactant-encapsulate interaction. The fundamental understanding of polarization as well as the potential for this work in targeted, controlled drug delivery and medical diagnostics provides an exciting landscape for new materials in AC-electrokinetic studies.

See more of this Session: Meet the Faculty Candidate Poster Session
See more of this Group/Topical: Education

(4co) Interfacial Forces in Nanoparticle and Biological Systems at the Micro- and Nano-Scales

Kai Kristiansen, Department of Chemical Engineering, University of California at Santa Barbara, Santa Barbara, CA

The overall theme of my research is to understand the molecular and nanoparticle behavior at the interface of surfaces. As the size of particles reaches characteristic length scales, novel physical properties (compare to bulk properties) start to emerge [1]. The carbon nanotubes are one candidate and I will present a study of aqueous film of carbon nanotubes on mica surfaces and in particular its frictional properties.

Another interesting interfacial system is bio-membrane. We have focused on the myelin membrane. The myelin sheath is a multilamellar membrane surrounding the axon of neurons in the nervous system. This sheath acts as an electrical insulator, which allows for faster and more efficient conduction of nerve impulses than unmyelinated nerves. Myelin dysfunctions vary from deterioration of signal transduction to demyelinating diseases such as multiple sclerosis (MS). We want to understand the molecular synergistic interaction responsible for a healthy myelin membrane structure. Using a Surface Forces Apparatus we have studied the interaction forces and adhesion between supported myelin lipid bilayers mediated by myelin basic protein [2].

In the last part of this poster I want to present the development and implementation of new devices [3] for interfacial science and how to integrate them into a Surface Forces Apparatus (SFA) [4]. An electrochemical cell developed for the SFA for monitoring the thickness change of silica surfaces in a silica-mica system due to silica dissolution under influence of electrochemical potential will be shown [5].

[1] Min Y J, Akbulut M, Kristiansen K, Golan Y, and Israelachvili J; "The role of interparticle and external forces in nanoparticle assembly". *Nature Materials* **7** (2008) 527-538.

[2] Min Y, Kristiansen K, Boggs J M, Husted C, Zasadzinski J A, and Israelachvili J; "Interaction forces and adhesion of supported myelin lipid bilayers modulated by myelin basic protein". *Proceeding of the National Academy of Science* **106** (2009) 3156-3159.

[3] Kristiansen K, McGuiggan P M, Carver G, Meinhart C, and Israelachvili J; "3D Force and Displacement Sensor for AFM and SFA measurements". Langmuir **24** (2008) 1541-1549.

[4] Israelachvili J, Min Y, Akbulut M, Alig A, Carver G, Greene G W, Kristiansen K, Meyer E E, Pesika N, Rosenberg K, and Zeng H; "Recent advances in the surface forces apparatus (SFA)

technique". Reports on Progress in Physics **73** (2010) 036601.

[5] Kristiansen K, Greene G W, Valtiner M, Boles J, and Israelachvili; "Electrochemical interpretation of silica dissolution processes". Talk scheduled for AIChE annual meeting 2010 in session: 01C19 "Interfacial aspects in electrochemical systems".

See more of this Session: Meet the Faculty Candidate Poster Session
See more of this Group/Topical: Education

(4cp) Multicompartmental Microstructures Via Electrohydrodynamic Co-Jetting for Biomedical Applications

Srijanani Bhaskar, Macromolecular Science and Engineering, University of Michigan, Ann Arbor, MI and Joerg Lahann, University of Michigan, Ann Arbor, MI

We herein describe the synthesis of anisotropic colloidal microstructures via electrohydrodynamic-cojetting for a variety of applications such as targeted drug release, tissue engineering, diagnostics, sensing and display materials.1-8 In its simplest form, two polymer solutions are flown through a modified side-by-side capillary system. Application of an electric field results in the formation of an electrospray, and solvent evaporation results in particle or fiber formation, and the interface between two polymer solutions is sustained during jet fragmentation and size reduction. Because of its intrinsic simplicity and generality, the electrohydrodynamic co-jetting process can be applied to a wide range of specialty and non-specialty materials. Furthermore, simple variations of different solution and process parameters, such as concentration, flow rate, applied voltage, etc. provides access to a vast repertoire of shapes and sizes of particles and fibers. Such novel particle geometries enable independent control of key parameters, such as chemical composition, surface functionalization, biological loading, shape, and size for each compartment. We demonstrate the fabrication of multicompartmental particles and fibers from biodegradable and hydrogel polymers, namely poly(lactide-co-glycolide) (PLGA) and poly (ethyleneimine) (PEI) via electrohydrodynamic co-jetting. We then demonstrate the versatility of this process by fabricating a variety of non-equilibrium biphasic shapes, such as discs, rods and cylinders, in addition to spheres. Fabrication of organic-inorganic hybrid materials is demonstrated via incorporation of magnetite nanoparticles, carbon nanotubes, and optical pigments. Spatioselective control over particles' surface is demonstrated via anisotropic introduction of functional groups in particles and fibers, and their selective surface modification via "click chemistry". Sacffolds fabricated out of such selectively modified fibers are employed for spatial cell guidance. The selective surface modification is also employed in interparticle assembly. Anisotropic PLGA/PEI particles are also employed in drug and gene delivery.

See more of this Session: Meet the Faculty Candidate Poster Session
See more of this Group/Topical: Education

(4cq) Spatiotemporal Signaling During Cell Adhesion and Migration: Computational Models and Experimental Analysis

Erik S. Welf, Chemical and Biomolecular Engineering, North Carolina State University, Raleigh, NC

In order to further our understanding of how complex and dynamic interactions within a cell determine cell fate and function, we need a coordinated approach that combines experimental measurements and theoretical insights. In general, my research interests focus on the use of systems analysis, computational modeling, and quantitative experimentation to understand and manipulate complex biological systems by way of their cellular signaling systems. One specific aspect of cell signaling that interests me greatly is determining how cells coordinate and regulate the many signals required for cell migration. This process involves the precise regulation of creation, activation, degradation, and spatial localization of a multitude of proteins, small molecules, and lipids.

To date, my efforts have been divided between quantitative experimentation and computational modeling related to cell adhesion and migration. As a graduate student, I focused on the function of integrin proteins, which control cell adhesion and migration by binding extra-cellular matrix (ECM) proteins and transferring signals bi-directionally through the cell membrane. I studied the cellular response to the availability of ECM binding sites by confocal microscopy, and used probabilistic modeling to show how integrin cluster properties change as a function of ECM density. I also used computational modeling to evaluate the potential for a diffusing cascade of biochemical reactions to regulate integrin activation in space to create integrin clusters, and to postulate how integrin clustering may affect cell migration.

Recently I have been investigating the mechanisms that give rise to stochastic yet persistent behavior in migrating cells. Using stochastic computational modeling, I investigated how different feedback structures between adhesion dynamics and activation of the small GTPase Rac1 may be capable of tuning the frequencies of stochastic protrusion and adhesion events under various intracellular and extracellular conditions. Additionally, I used total internal reflection fluorescence (TIRF) microscopy to visualize proteins that shape cell adhesion and migration signals in migrating cells, and spatiotemporal analysis to understand how these signaling events may be related to cell morphology and migration behavior. This analysis approach provides specific insight into how cell signaling is influenced by cell morphology, and conversely how cell signaling may affect morphology and migration persistence by localizing cellular protrusion.

I am also especially interested in working with collaborators in the areas of cell biology, high-resolution microscopy, and biomaterials to develop cutting edge experimental and analytical approaches for understanding cell signaling and related physiological processes. Additionally, applications of engineering to biology and physiology provide exciting opportunities to infuse engineering education with topics that excite and engage students, and I am interested in developing ways to incorporate analysis of biological processes into classical chemical engineering coursework.

See more of this Session: Meet the Faculty Candidate Poster Session
See more of this Group/Topical: Education

(4cr) Computational High-Throughput Screening of Sorbents for Regenerable CO2 Capture at High Temperature

Ujjal Das and William H. Green, Department of Chemical Engineering, Massachusetts Institute of Technology, Cambridge, MA

Development of efficient computer screening methods has led to the discovery of many new materials. In this work, we present density functional theory-based high-throughput screening schemes to identify sorbents for carbon dioxide capture from syngas at high temperature. High temperature CO2 capture is particularly important because this process is more energy efficient under certain conditions. We have adopted a multi-step screening approach. At first we have computed reaction free energy of carbon dioxide in more than 300 reactions. While a large majority of these reactions are not suitable for reversible CO2 capture, we have also identified multiple reactions with favorable reaction thermodynamics. Further screening based on CO2 adsorption kinetics and sorbent capacity sorts out a few promising candidates that can be tested in experimental set up. The CO2 capture ability of one such sorbent is currently under study in our laboratory.

See more of this Session: Meet the Faculty Candidate Poster Session
See more of this Group/Topical: Education

(4cs) Engineering Proteins and Peptides for the Investigation and Treatment of Infectious Disease

Amy J. Karlsson, Chemical and Biomolecular Engineering, Cornell University, Ithaca, NY

My research interests are focused on engineering proteins and peptides that can be used to better understand fungal infections and study potential antifungal compounds. I plan to use two approaches to design and synthesize proteins and peptides for studying antifungal properties of the compounds and for studying the interactions of fungal pathogens with host cells. The first approach is to use organic synthesis to produce compounds that mimic and improve on the activity of naturally occurring antifungal peptides. Organic synthesis provides the ability to tightly control the peptide sequence and to easily incorporate non-natural amino acid building blocks into the peptides. During my PhD research, I used this approach to synthesize antifungal peptides composed of non-natural, β-amino acids that mimicked the structure of natural antimicrobial peptides. The designed peptides had both improved antifungal activity and improved specificity for fungal cells compared to natural antimicrobial peptides. The second approach to peptide and protein design and synthesis that I will apply to my research is directed evolution. The protein production machinery in bacteria can be harnessed to perform protein synthesis, and the proteins can be evolved to improve on their natural functions or to have novel functions. Directed evolution can be used without any knowledge of the mechanism of action of a protein or peptide and offers the ability to create large numbers of potentially improved sequences that can be screened for desired functions. My postdoctoral research utilizes directed evolution to engineer high-affinity antibodies that fold in the intracellular environment, providing the opportunity to study the effects of antibody binding to intracellular targets. My experience with both organic synthesis and directed evolution provides a complementary set of techniques that can be used to tailor a protein or peptide engineering approach based on what is known about a particular fungal pathogen. Thus, my future plans are to make use of this molecular toolkit to create proteins and peptides that can be used to address basic questions underlying fungal pathogenesis or that can inhibit key mechanisms in fungal disease.

See more of this Session: Meet the Faculty Candidate Poster Session
See more of this Group/Topical: Education

(4ct) Algal Derived Biofuels: A Systems Biology Approach to Increasing TAG Accumulation in *C. Reinhardtii*

Nanette R. Boyle[1], John A. Morgan[2] and Sabeeha S. Merchant[1], (1)Chemistry and Biochemistry, University of California, Los Angeles, Los Angeles, CA, (2)School of Chemical Engineering, Purdue University, West Lafayette, IN

Although algal derived biofuels have recently gained widespread attention for their potential as a source of renewable energy, the idea is not new. The aquatic species program (ASP), a federally funded program in the eighties, was charged with determining the feasibility of producing biodiesel from high-lipid accumulating algae (1). One unfortunate conclusion drawn from the program was that *Chlamydomonas reinhardtii*, a reference green algal species, was not capable of accumulating high levels of lipids and was therefore excluded from further study. While this conclusion may have been true for normal growth conditions, it has recently been shown that *C. reinhardtii* does accumulate lipids under certain nutrient limitations, such as nitrogen (2-4), sulfur (5), and phosphorous (2). The response to these nutrient limitations in *C. reinhardtii* is an increase in storage product accumulation, both starch and triacylglycerols (TAGs), but the exact distribution of carbon fluxes and the regulatory control behind the switch from balanced growth to accumulation of storage products is not yet known. Understanding how and why *C. reinhardtii* directs its carbon away from biomass production toward carbon storage may provide useful information on how to engineer the cell to accumulate higher amounts of lipids during normal growth and provide insights into the metabolism of other green algae.

Two approaches were taken to gain further insight into the control of carbon partitioning in *C. reinhardtii*. First, a large scale metabolic reconstruction was performed in order to formulate a flux balance analysis model (6), which is capable of predicting intracellular fluxes using linear programming. This model was then used to predict the effect of nitrogen deprivation on the intracellular fluxes of *C. reinhardtii* and how these predictions match reported lipid contents (3, 4). Temporal transcriptome analysis was also performed on nitrogen starved cells using RNAseq to eluciadate both transcription factors and potential gene targets to increase TAG yields and the results will be presented.

1. J. Sheehan, T. G. Dunahay, J. R. Benemann, P. G. Roessler, J. C. Weissman, U. S. D. o. Energy, Ed. (National Renewable Energy Lab 1998) pp. 328.

2. P. M. M. Weers, R. D. Gulati, Limnology and Oceanography 42, 1584 (1997).

3. Z. T. Wang, N. Ullrich, S. Joo, S. Waffenschmidt, U. Goodenough, Eukaryotic Cell, EC.00272 (October 30, 2009, 2009).

4. E. R. Moellering, C. Benning, Eukaryotic Cell 9, 97 (January 1, 2010, 2010).

5. M. Timmins et al., Journal of Biological Chemistry 284, 23415 (August 28, 2009, 2009).

6. N. R. Boyle, J. A. Morgan, BMC Systems Biology 3, 4 (2009).

See more of this Session: Meet the Faculty Candidate Poster Session
See more of this Group/Topical: Education

(4cu) Molecular Understanding and Design of Zwitterionic Biomaterials for Lubrication, Cartilage Tissue Repair and Drug Delivery Applications

Yi He and Shaoyi Jiang, Chemical Engineering, University of Washington, Seattle, WA

Research in biomaterials has shown significant growth in recent years. Poly-ethylene glycol (PEG) is one of most widely used biomaterials. However, PEG is known to be not suitable for long-term applications since it is subject to oxidation. My proposed research will take a different approach by focusing on developing zwitterionic polymer based biomaterials through a combination of molecular modeling tools and experimental methods. The major goal of my research program is to explore the unique feature of zwitterionic biomaterials and to apply the underlying design principles to various applications, including lubrication, cartilage tissue repair, and drug delivery.

See more of this Session: Meet the Faculty Candidate Poster Session
See more of this Group/Topical: Education

(4cv) Synthetic Extracellular Matrix (ECM) Hydrogels and Localized Gene Delivery for Stem Cells and Tissue Regeneration

Yuguo Lei, Chemical and Biomolecular Engineering, University of California, Los Angeles, Los Angeles, CA

My first project focused on developing synthetic ECM hydrogels for stem cell engineering. In the body, stem cells reside in dynamic and complex 3D niches with spatially and temporally well-regulated biochemical & biophysical cues that control cell maintenance, self-renew and differentiation. We are developing synthetic hydrogels to mimic these niches for 3D stem cell culture, delivery and tissue regeneration. Polyethylene glycol (PEG) and hyaluronic acid (HA) are cross-linked with protease degradable peptides in the presence of cells to form the hydrogels. Peptides derived from ECM proteins are introduced into the scaffold as the adhesion points for cells. Proteins factors can be incorporated into the hydrogel through physical or chemical immobilization to guide cell behavior and fates. PEG and HA scaffolds are ideal for tissue regeneration because they are safe and have low immunogenicity. In addition, they have a low biological background, thus allowing us to engineer the hydrogel bioactivity through introducing protein factors, DNAs or siRNAs. With these hydrogels, we built a 3D model for studying the calcification of vascular stem cells (VSCs), which may contribute to the calcific atherosclerosis in the body. Then we cultured mesenchymal stem cells (MSCs) in these gels and found the 3D environments including the mechanics, adhesion peptide identity, concentration and presentation significantly affected their proliferation, spreading and migration. Currently, we are using these hydrogels to culture and deliver neural progenitor stem cells (NPCs) to treat stroke with our collaborators, with the objective that these synthetic ECM can improve the cell survival rate and integration into the surrounding tissue.

My second project is to develop novel technologies that can deliver genes (encoding for growth, transcription factors) and siRNAs to the diseased sites locally in a method that is cell-controlled to promote wound healing, tissue regeneration and to reprogram cells. Our approach is through using hydrogels to carry and deliver DNAs, siRNAs. The challenge along with this approach is to incorporate non-viral gene delivery nanoparticles (polyplexes) into the hydrogels without losing their activity, which results from their aggregation or disintegration in the hydrogel precursor solution before and during the gelation. We successfully developed a universal process to load concentrated, un-aggregated and active polyplexes into various hydrogels (up to 5mg pDNA/mL hydrogel tested). Both in vitro and in vivo tests show the encapsulated polyplexes remain active. Currently, we are studying the kinetics of transgene expression after implanting these DNA loaded hydrogels in vivo with a mouse skin wound model and a mouse subcutaneous implantation model. Through this kinetics study, we hope to find out methods to control the gene transfer from hydrogels in vivo through tailoring the hydrogels properties. We are also investigating if hydrogels loaded with VEGF, PDGF genes can promote angiogenesis and wound healing in vivo.

My future research will focus on manipulating cell fates (stem cells to mature cells; mature cells to stem cells and mature cell type one to another cell type) in vivo through synthetic cell niche and gene delivery technology for regenerative medicine. The ultimate goal is to generate the degenerated cells in situ instead of delivering cells to the diseased sites.

See more of this Session: Meet the Faculty Candidate Poster Session
See more of this Group/Topical: Education

(4cw) Molecular Simulation Investigation Into Nucleation and Growth of Complex Structures

Sapna Sarupria, Department of Chemical Engineering, Princeton University, Princeton, NJ

Self-assembly of molecules in water results in the formation of materials involving various complex structures. These materials have found applications in various fields like medicine, energy and environment. The pillars upon which these complex structures can be designed and engineered include an understanding of both the thermodynamics and kinetics of their formation. In general, because the thermodynamic properties are related to the equilibrium state they are studied more in experiments, theory and simulations. The challenge faced in experimental studies of the kinetics is the time resolution of present experimental techniques. Although many ingenious methods have been developed to address this, nanoscopic timescales are still inaccessible in most experimental techniques. Molecular simulations are particularly helpful in this scenario in which the molecular motions of atoms are followed at nanometer length- and nanosecond time- scales. However, several computational challenges are encountered in studying the kinetics of complex structures formation. Specifically, most of these complex structures have large lengthscales and therefore the system sizes are computationally prohibitive. Considerable effort has been directed to develop different reduced models leading to coarse-grained force fields that remove the "less important" degrees of freedom. These approaches have focused mostly on the structure and stability of the system of interest. In addition, most of them are designed with treating water as a continuum. With kinetics perspective, several interesting questions can be raised: (1) Can we use the coarse-grained models developed using structure and stability to model kinetics? (2) What governs the nucleation and growth of these structures? (3) What is the role of water in the nucleation and growth of these complex structures? How can water be incorporated explicitly and efficiently into these simulations? I am interested in exploring these questions in my future research work. Viruses provide a perfect system for this study. They are complex assemblies composed of a multi-protein shell that encloses the genetic material of the virus. They vary in shape, composition and size which can range from tens to hundreds of nanometers. Due to their large sizes, atomically detailed simulations of viruses in explicit water are computationally prohibitive and therefore, several coarse-grained models have been developed to make computational simulations of viruses feasible. Most of these models are however based on the structure and stability of the viral capsids. Therefore they provide an excellent case study for the questions raised above. I will discuss strategies to address these questions for the virus systems. These strategies can be easily extended to other complex structures such as amyloids fibrils and self-assembled colloidal or nano-particles.

See more of this Session: Meet the Faculty Candidate Poster Session
See more of this Group/Topical: Education

(4cx) • Probing Equilibrium Phase Behavior of Asymmetric Block Copolymer Thin Films

Vindhya Mishra, Chemical Engineering, University of California Santa Barbara, Santa Barbara, CA, Glenn H. Fredrickson, Materials Research Laboratory, University of California, Santa Barbara, Santa Barbara, CA and Edward J. Kramer, Department of Chemical Engineering and Department of Materials Engineering, Materials Research Lab, Santa Barbara, CA

Confining block copolymers to thin films adds geometric frustration to the system and incorporates surfaces effects, which are absent in bulk. However the effect of these added energetic costs are not completely understood or quantified. We have probed how film thickness affects equilibrium behavior for highly asymmetric block copolymer systems with layering parallel to the substrate. In particular we have looked at the arrangement of microdomains in spherical morphology block copolymers and melting behavior of cylindrical morphology block copolymers. In multilayer films of spherical morphology diblock copolymers, competition between hexagonal packing of the microdomains preferred at the surfaces and the BCC (110) packing preferred by the internal 'bulk-like' layers leads to a transition in the packing symmetry from HEX to FCO to BCC as the film thickness is increased. We have found that this transition can be suppressed by blending a small amount of majority block homopolymer, which migrates to the highly stretched interstices of the respective lattices and relieves the packing frustration. Self-consistent-field theory simulations show that the homopolymer reduces the stretching of the PS block and the free energy penalty of HCP relative to BCC inner layers. The other phenomenon that we have probed is the smectic-nematic-isotropic transition in graphoepitaxially confined monolayers films of cylindrical morphology diblock copolymer. We find that in-plane defects can be suppressed and mediated by building multi-layer self assembled structures. We have combined quantitative AFM studies with grazing incidence small angle X-ray diffraction, which is a powerful tool to obtain sample information characteristic of large areas and is not limited to the surface, like most microscopy techniques. Diffraction lineshape analysis allowed us to quantify the decay of translational and orientational order with increasing temperature. The results have been interpreted in the context of the Toner–Nelson theory of melting for layered systems.

See more of this Session: Meet the Faculty Candidate Poster Session
See more of this Group/Topical: Education

(4cy) Controlled Release Films and Functional Surfaces for Applications in Medicine

Anita Shukla and Paula Hammond, Chemical Engineering, Massachusetts Institute of Technology, Cambridge, MA

The field of drug delivery is plagued by the lack of effective local delivery methods for multiple therapeutics over medically relevant timescales. Such capabilities are needed to treat cellular dysfunction while avoiding drug toxicity. A clear illustration of this problem is the systemic overuse of broad-spectrum antibiotics to treat bacterial infections, which has lead to a rise in antibiotic resistant bacteria worldwide. As an alternative to conventional systemic administration of therapeutics, we have designed layer-by-layer assembled degradable polyelectrolyte multilayer films as coatings to address infection. These films were engineered for the delivery of both existing and highly effective antibiotics as well as antimicrobial peptides, a novel class of therapeutics that do not cause antimicrobial resistant bacteria to arise. Film architecture, assembly technique, and film component secondary interactions were thoroughly explored and optimized to create active films which display a range of drug release profiles. These timescales and drug loadings are relevant to a wide range of medical needs, including those in which immediate eradication of an existing infection are required versus prophylactic administration. Recognizing the prevalence of infection and inflammation coexistence, degradable coatings for the concurrent release of antibiotics and anti-inflammatory therapeutics were also designed. All coatings were applied to medically relevant substrates, including bandages and sutures, and shown to be active *in vitro*. Additionally, to address the growing problem of biofilm formation on the surfaces of implants, non-degradable coatings were designed and shown to be highly effective in preventing bacterial attachment, the first critical step in biofilm formation. Together, these functionalized surfaces and drug releasing films have the potential to eliminate many of the current complications in treatment and prevention of infection.

See more of this Session: Meet the Faculty Candidate Poster Session
See more of this Group/Topical: Education

(4cz) Utilization and Modification of Municipal Sewage Treatment Plant Sludge and Wastewater Streams for the Production of Lipid-Based Biofuels

Andro Mondala, Chemical Engineering, Mississippi State University, Mississippi State, MS

Municipal wastewater treatment plant sludge and influent wastewater streams present an abundant source of lipids that could be extracted, refined, and converted to biofuels such as biodiesel and green diesel. Initial studies on primary and secondary or activated sludge have shown that fatty acid methyl esters (FAMEs) can be produced via *in situ* transesterification to maximum yields of 14.5 % and 2.5 % by weight of primary and secondary sludge, respectively. The biodiesel produced from primary sludge were thought to originate from free oils, grease, and fatty acid whereas those obtained from activated sludge were converted from phospholipids in the cell membranes of bacteria comprising the microbiota in activated sludge. In order to improve lipid yields and quality in terms of fatty acid composition, two options were considered. The first option was to inactivate the indigenous microflora in the wastewater using ozone followed by inoculation of the disinfected wastewater with oleaginous microorganisms, which are known to accumulate up to 80 % of its biomass as lipids. In this study, ozone was shown to be very effective in inactivating a majority of the naturally-occurring microflora in the wastewater by up to six log units. Ozone was also able to oxidize recalcitrant materials in the wastewater, improve its biodegradability, and increase dissolve oxygen content. This will be beneficial for the cultivation of oleaginous microorganisms in the wastewater in order to produce an activated sludge containing known oil-producing microbial strains such as *Rhodotorula glutinis* and *Rhodococcus opacus*. The second study, which is the focus of this paper is the attempt to trigger lipid accumulation in the activated sludge microbiota through cultivation in a high carbon-to-nitrogen (C:N) ratio medium. Activated sludge bioreactors were dosed with lignocellulose sugars using batch and semi-batch fermentation modes at varying levels of C:N ratio and substrate loading. Biomass production, lipid yields, fatty acid composition, and residual sugars were determined using conventional gravimetric and instrumental methods. Kinetic modeling of the fermentation data was conducted for process design and scale-up. In order to understand the dynamics within the activated sludge microbiota, variations in the microbial composition of the activated sludge were determined using sequence analysis of 16S rRNA gene clone libraries. These data will be useful in identifying and isolating potential lipid producers within the sludge microbiota. These technologies, which are being tested for possible integration in conventional wastewater treatment plant infrastructures could then generate biofuels that can be used to power energy intensive treatment processes such as disinfection and tertiary treatment.

See more of this Session: Meet the Faculty Candidate Poster Session
See more of this Group/Topical: Education

Model Membrane Platforms

Cellular membranes are the site of many important biological interactions and biochemical reactions. Depending on the type of cell, its membrane properties can greatly vary, leading it to recruit different proteins and other biomacromolecules. From an engineer's perspective, understanding how membrane properties control activities at its interface may lead to advances in medical diagnostics, drug delivery and pharmaceutical therapies. We are interested in developing model membrane platforms that recapitulate key features of cellular membranes to investigate questions related to lipid-protein and protein-protein interactions.

Bottom-up design strategies rely on the self-assembly of lipid vesicles upon interaction with solid supports. Depending on the support's properties, prior studies have demonstrated that the self-assembly pathway is surface-specific and results in the formation of a planar lipid bilayer or monolayer, or a layer of adsorbed, interact vesicles. To overcome the challenges of this surface-specificity—a notorious problem for the fabrication of high electrical resistance, lipid bilayer-based biosensors on industrially attractive titanium oxide and gold where intact lipid vesicles prefer to assemble—we have engineered solutions to promote the formation of planar lipid bilayers by self-assembly on these materials. Inspired by biology, a virus-derived peptide's interactions with a layer of intact vesicles on either titanium oxide or gold can disrupt the membrane structure and catalyze the formation of a planar bilayer. Recently, we have also been devising a simpler, pH-based scheme to control the lipid vesicle self-assembly pathway on oxide films that does not rely on the introduction of foreign materials.

While bottom-up design is an excellent approach for fabricating lipid assemblies such as a planar bilayer or intact vesicles on a solid support, it is difficult to recreate functional cell membranes to study protein-protein interactions. Therefore, we are also investigating the top-down fabrication of biomembrane-on-a-chip platforms composed of cellular derived membranes. Thus far, this platform has been used together with the model membrane platform to distinguish between lipid-protein and protein-protein interactions. However, there is also great potential to study more complex biological interactions such as vesicle fusion and cell-cell communication.

We are currently designing more complex model membranes that include biologically important membrane constituents such as cholesterol and sphingomyelin for lipid raft modeling, and phosphoinositides for cell signaling. Developing important, new biosensing applications of these engineered model membrane and biomembrane platforms is a primarily goal of our research group. This research will create and utilize novel platform technology that can be of tremendous benefit to the design of biologic membrane-based sensing devices and the development of new drugs for emerging infectious diseases and biodefense.

Integrated Biosensors

Biosensors encompass a wide range of sensitive analytical devices that can detect biological interactions via a physical or chemical response. Depending on the specific interaction of interest, certain property changes in one or both of the interacting bodies may elicit a measurable response. Our group focuses on the application of biosensors to understand the interactions of biomacromolecules such as viral peptides, interfacial enzymes and extracellular matrix proteins with model membrane platforms. In this regard, biosensors are needed for both characterization of the model membrane platform and detection of the subsequent interaction.

Platform characterization is important since a macromolecule's interaction with a model membrane depends on the latter's properties. Therefore, a host of surface-sensitive techniques are readily used in our lab, including atomic force microscopy (AFM), fluorescence recovery after photobleaching (FRAP), surface plasmon resonance (SPR), ellipsometry, and quartz crystal microbalance with dissipation monitoring (QCM-D), to probe membrane features such as morphology, lateral lipid mobility, mass, and film viscoelasticity. While individual sensor measurements are suitable for fully characterizing a fabricated model membrane platform, we are also interested in integrating multiple sensor techniques to simultaneously monitoring dynamic processes that occur in biological systems.

By simultaneously monitoring interfacial processes in situ with multiple sensors, we can obtain two or more measurements of a single process that rely on fundamentally different physical principles. Through consideration of the nature of each measurement, we can enhance our interpretation of complex biological processes. We have recently demonstrated the power of this approach with combined QCM-D and reflectometry measurements to study the mechanism of a viral peptide's interaction with intact vesicles, and are now extending this work for combined QCM-D and electrochemical monitoring of model membrane platforms on gold. In the case of a planar lipid bilayer on gold, its electrical insulating properties make this cell membrane-mimicking platform suitable for robust ion channel sensing applications.

Hepatitis C Virus Therapies

The Hepatitis C virus (HCV) is an important worldwide cause of liver disease. About 170 million people are infected with HCV and it is the leading cause of liver cancer and liver transplantation in the United States. Its 9.6kb genome encodes a single 3010 amino acid polyprotein that is proteolytically processed by cellular and viral proteases into structural (components of the mature virus) and non-structural (elements proposed to help replicate new virions) proteins. Like other positive-strand RNA viruses, HCV is believed to replicate in association with cytoplasmic membranes although how the RNA replication complex is assembled and maintained remains to be elucidated.

As the HCV genome replicates, there are mutations in the transcription process that give rise to mutant strains of HCV. This pattern has led to great variability in the HCV genome with at least six major genotypes (along with respective subtypes) having emerged over the years. Current interferon-based therapies are inadequate for most patients, and there is a need to develop and evaluate new potential medicines. It is believed that a patient's HCV genotype (or genotypes) may cause this variable response to interferon-based treatments. However, while the recent breakthrough of establishing a complete cell-culture system has permitted researchers to perform classical virologic studies on the complete life cycle of HCV, the restricted permissive Huh 7.5 model cells only allow for the study of a few specialized HCV clones. Therefore, there is an urgent need to understand how genotype-specific therapy is related to the molecular virology of HCV.

We are using engineering-based approaches to identify novel targets for specific antiviral treatments that would be of great benefit for improving these drug cocktails. Our study of isolated peptides derived from key regions of HCV non-structural proteins with model membrane platforms has led to new insight into the biophysical interactions that compose the HCV life cycle as well as new strategies, including small molecule inhibition, to target these interactions. Importantly, this approach is not restricted to a subset of HCV genotypes, allowing us to study the genotypes most prevalent in the United States as well as those found predominately in the developing world where current therapies are either too expensive for widespread use or largely ineffective due to genotype-specific efficacy. Moreover, we have successfully assembled a multi-protein HCV replicase complex with RNA polymerase activity for the first time on a model membrane platform.

Another active area of research in our group is the development of peptide drugs against HCV. In contrast to traditional drug discovery efforts, these peptide therapeutics are themselves derived from structural features of HCV proteins. Model membrane platforms and integrated biosensors have proven well-suited to study their modes of action. Continued use of these sensing platforms to characterize the biological activities of antiviral as well as antimicrobial peptides will provide us with an improved understanding of their modes of action, and lead to multi-functional, anti-infective drug design. Together, the aims of our project represent exciting new approaches to the development of anti-HCV therapies that may lead to more effective drug cocktails against HCV and potentially change the HCV treatment paradigm.

Engineering Solutions to Global Health

As major advances in the development of engineering and biomedical technologies help to create new therapies for pathogens, it is important to use these capabilities to address global health inequalities. The availability of a medicine does not guarantee its widespread use since its cost—in large part influenced by its production—can act as a barrier that creates health disparities. One possible solution is to adopt new approaches that can significantly lower the costs of drug discovery and development. With high palliative treatment costs, inadequate therapies and a global reach, HCV represents an attractive target for such approaches and has served as our prototype pathogen.

Our HCV drug discovery efforts are focused on understanding and targeting the fundamental macromolecular interactions of HCV non-structural proteins. So far, we have identified a novel class of small molecule inhibitors that target the HCV NS4B protein and also a heretofore unrecognized structural motif (the "BAAPP domain") within the HCV NS5A protein that mediates its binding of a specific phosphoinositide (PIP-2), an interaction which we also validated as necessary for the HCV life cycle. We are now searching for inhibitors of the PIP-2:BAAPP domain interaction. However, the common linguistics of infectious diseases make the engineering tools and strategies developed for our HCV studies capable of establishing a new technology platform to study a range of infectious, viral diseases.

Many pathogens that are found in the developing world have similar structural features and replication mechanisms to the prototype HCV. Indeed, BAAPP domains have also been identified in key proteins of Malaria, Dengue virus (DV) and Japanese encephalitis virus (JEV), suggesting that PIP-2 binding plays an essential role in a diverse group of important pathogens. Inhibition of the PIP-2:BAAPP domain interaction therefore might represent a novel antiviral strategy for a range of pathogens that represent critical health challenges for the developing world. While this strategy may represent a major breakthrough towards stopping viral genome replication, it is just one of many possibilities. We are interested in characterizing the biophysical properties of a wider range of infectious, viral diseases, with the goal of translating insight gained into the identification of new antiviral targets and ultimately more effective therapies. As such, there is great potential to study these pathogens with model membrane platforms.

(4da) Carbon Nanotubes as Optical Sensors and Polymeric Biomaterials

Daniel A. Heller, Department of Chemical Engineering, Massachusetts Institute of Technology, Cambridge, MA

Real-time, spatially resolved detection and identification of analytes in biological media or at the single-molecule level present worthy challenges for nanoscale sensors. Encapsulation of nanotubes in synthetic polymers and biopolymers creates a handle for the transduction of analyte binding. Encapsulation in short sequences of ssDNA allow detection of reactive oxygen species (ROS) via specific interactions with DNA, while nitroaromatic compounds are detected by the conformational change of a peptide. In both cases, analyte specificity is possible by variations in the nanotube response, resulting in distinctive optical fingerprints. Solvatochromic shifts of up to 50 meV and intensity modulation permit identification of analytes which are traditionally difficult to differentiate, such as certain ROS species. Single analyte molecules are detected via quantized quenching or solvatochromic shifting of the nanotube emission, permitting spatial mapping of analytes with unprecedented sensitivity.

See more of this Session: Meet the Faculty Candidate Poster Session
See more of this Group/Topical: Education

(4db) Engineering in the Microvasculature: The Mechanical Microenvironment's Control of Systemic Metabolism

Joseph M. Rutkowski, Touchstone Diabetes Center - Department of Internal Medicine, UT Southwestern Medical Center, Dallas, TX

Lying between the endothelial beds of the body's circulation is the interstial microenviroment wherein the complex biohphysical and biochemical interplay of cells, the extracellular matrix, and the molecules of interstital fluid control both immediate tissue homeostasis and systemic functions. The modulators of interstitial flow rates and shear, tissue hydration and composition, and cellular trafficking are the endothelial microvasculature of the blood and venous capillaries and the lymphatic system.

In my previous studies, we demonstrated that the primary modulator of interstitial drainage, the initial lymphatic vascualture, both controls the interstitim and cells found within and is itself modulated by the tissue mechnical environment. Generation of new functional lymphatic capiallaries, lymphangiogenesis, was found to be directional and fully dependent on interstitial flow. Blocked lymphatic drainage, clinically the pathology of lymphedema, resulted in hyperproliferation, yet poor organization and decreased function of lymphatic capillaries. In the pathology of lymphedema, where intersitial flow rates slow towards zero, inflammatory chemokines accumulated, interstitial immune cells failed to traffick, and adipose tissue depots expanded in adipogenesis. Indeed, we quantified lymphatic clearance rates in genetic models of lymphatic dysfunction and found a correlation with fluid transport and remodeling of the extracellular matrix and adipose expansion on tissue hydraulic conductivities and subsequent treatment and potential resolution of the disease. Blood capillary extravasation, as controlled by Starling's Law, was reduced in these instances due to the increased hydraulic and osmotic pressures of the interstitum. Concurrently, adipose expansion or hypercholesterolemia strongly reduced lymphatic drainage capability by inducing vessel degeneration. As lymphatic vessels are the primary route of lipid transport throughout the body, these studies combined to demonstrate the potential that the interplay of microvasculature cells with their surrounding tissues has to influence systemic metabolism.

My current research focuses on adipose tissue physiology and how the microvasculature may control its systemic functions via adipokine transport. The adipokine adiponectin has exhibited positive effects in regulation of systemic metabolism and insulin sensitivity through tissues such as liver, skeletal muscle, and centrally in the brain. We have recently demonstrated oligomer size-dependence on adiponectin clearance. The molecular transport sizes, particularly that of the high molecular weight form, present transport limitations across endothelial barriers with low intercellular permeabilities such as in muscle or in the brain. We have quantified size dependent exclusion of adiponectin oligomers in model endothelial cells *in vitro* and using genetic mouse models of receptor deletion *in vivo* to discern that oligomer function likely depends on microvasculature permeabilities. In chronic kidney disease, a common co-pathology of metabolic syndrome and obesity, we demonstrate peripheral effects of renal dysfunction in adipose tissue. Current studies are aimed at discerning the interplay between these two tissues and how renal and adipose endothelial functions affect their axis of communication. The biophysical environment of adipose tissue, such as matrix composition, interstitial flow, and lymphatic drainage, may influence adipocyte behaviors, and changes in cell infiltration and chemokine levels in the interstitial fluid have a direct effect on adipogenesis and metabolic lipid fluxes.

Miteva DO, **Rutkowski JM**, Dixon JB, Kilarski W, Shields JD, Swartz MA. Transmural flow modulates cell and fluid transport functions of lymphatic endothelium. Circ Res. 2010 Mar 19;106 (5):920-31.

Rutkowski JM, Markhus CE, Gyenge CC, Alitalo K, Wiig H, Swartz MA. Dermal matrix remodeling and fat accumulation control tissue swelling and hydraulic conductivity during murine primary lymphedema. Am J Pathol. 2010 Mar;176(3):1122-9.

Rutkowski JM, Davis KE, Scherer PE. Mechanisms of bbesity and related pathologies: the macro- and microcirculation of adipose tissue. FEBS J. 2009 Oct;276(20):5738-46.

Lim HY, **Rutkowski JM**, Helft J, Reddy ST, Swartz MA, Randolph GJ, Angeli VA. Hypercholesterolemic mice exhibit lymphatic vessel dysfunction and degeneration. Am J Pathol. 2009 Sep;175(3):1328-37.

Goldman JG*, **Rutkowski JM***, Shields JD*, Pasquier M, Cui Y, Schmökel HG, Pytowski B, Swartz MA. Co-operative and redundant roles of VEGFR-2 and VEGFR-3 signaling in adult lymphangiogenesis. FASEB J. 2007 Apr;21(4):1003-12. (* equal contribution)

Goldman J, Conley KA, Raehl A, Bondy DM, Pytowski B, Swartz MA, **Rutkowski JM**, Jaroch DB, Ongstad EL. Regulation of VEGF-C by interstitial flow. Am J Physiol Heart Circ Physiol. 2007 May;292 (5):H2176-83.

Rutkowski JM and Swartz MA. A driving force for change: Interstitial flow as a morphoregulator. Trends Cell Biol. 2007 Jan;17(1):44-50.

Rutkowski JM, Moya M, Johannes J, Goldman J, Swartz MA. Secondary lymphedema in the mouse tail: lymphatic hyperplasia, VEGF-C upregulation, and the protective role of MMP-9. Microvasc Res. 2006 Nov;72(3):161-71.

Rutkowski JM, Boardman KC, Swartz MA. Characterization of lymphangiogenesis in a model of adult skin regeneration. Am J Physiol Heart Circ Physiol. 2006 Sep;291(3):H1402-10.

Rutkowski JM, Santiago LY, Ben-Jebria, Ultman JS. Development of an assay for ozone-specific antioxidant capacity. Inhal Toxicol. 2003 Nov;15(13):1369-85.

See more of this Session: Meet the Faculty Candidate Poster Session
See more of this Group/Topical: Education

(4dc) Polymers for Advanced Energy Technologies

Daniel T. Hallinan Jr., Department of Chemical Engineering, University of California, Berkeley, Berkeley, CA

Polymers are poised to fulfill crucial needs in advanced energy technologies. My expertise in the applied fields of fuels cells and lithium batteries, as well as my fundamental knowledge of transport in polymers and block copolymer synthesis and characterization, place me in a position to be at the forefront of research in polymers for advanced energy applications.

I attended Lafayette College, where I obtained degrees in Chemical Engineering and Philosophy. I conducted doctoral work at Drexel University in Philadelphia, PA, under the tutelage of Prof. Yossef Elabd. I am currently a post-doctoral fellow at the University of California, Berkeley, with Prof. Nitash Balsara.

My doctoral work focused on the diffusion of water and methanol in polymer electrolyte membranes (PEMs) for fuel cells. I identified and modeled two different non-Fickian water diffusion regimes in PEMs that occur at critical hydrogen fuel cell conditions (extremely dry or fully saturated). I demonstrated that a better PEM for methanol fuel cells requires reduced methanol sorption as opposed to slower methanol diffusion within the PEM. My post-doctoral research pertains to lithium-air batteries, an advanced battery chemistry with potential to overcome the theoretical limits of lithium ion batteries. I use block copolymer electrolytes as a safe alternative to the flammable, reactive liquids used in existing battery electrolytes. Ionic conductivity and mechanical strength, which are important properties in polymer electrolytes, can be independently controlled with block copolymers. I have demonstrated the first instance, to my knowledge, of an all solid state rechargeable lithium-air battery.

See more of this Session: Meet the Faculty Candidate Poster Session
See more of this Group/Topical: Education

(4dd) Mercury Reaction Chemistry in Combustion Flue Gases From Experiments and Theory

Bihter Padak, Energy Resources Engineering, Stanford University, Stanford, CA

Coal-fired power plants are the greatest anthropogenic source of mercury emissions in the United States. Mercury exists in coal combustion flue gas in a variety of forms depending on the coal type and combustion conditions, i.e., elemental, oxidized, and particulate. Understanding mercury speciation during combustion and how the transformations occur between different forms of mercury is essential to developing an effective control mechanism for removing mercury from flue gas. Homogeneous oxidation of mercury in the flue gas of coal combustion utility boilers has been studied for many years to understand the speciation of mercury. In spite of a vast amount of experimental studies, supported by modeling efforts, there are still many questions to be answered and the speciation of mercury is not fully understood yet.

During my Master's in the department of Chemical Engineering at Worcester Polytechnic Institute and my PhD in the department of Energy Resources Engineering at Stanford University under the supervision of Prof. Jennifer Wilcox, I have worked on two projects focusing on mercury emissions from coal-fired power plants and have gained experience from both a theoretical and experimental perspective. The first project is investigating gas phase homogeneous mercury oxidation via bromine and chlorine in coal combustion flue gas to understand mercury speciation throughout the utility system. I have designed and built an experimental system consisting of a plug-flow reactor and a burner to generate laminar premixed methane flame to simulate both combustion and the flue gas environment. In this system mercury and chlorine are introduced into a laminar premixed methane-air flame to simulate the flue gas. Cooled flue gas is sampled and sent to the mass spectrometer to analyze the flue gas, with special focus to mercury species. One needs to be able to make precise mercury measurements to understand its speciation and accurately predict the extents of mercury oxidation. With this goal I have designed and custom-built an electron ionization quadrupole mass spectrometer (EI-QMS) to be able to directly measure mercury species on the order of parts per billion (ppb) in the flue gas. To accurately measure the low concentrations of different mercury species present in coal combustion flue gases, the EI-QMS must be sensitive to concentrations in the ppb range, which can pose a challenge. To increase the sensitivity, the system has been upgraded to have a supersonic beam and a skimmer is placed after the first orifice. In such a setup, scattering of the molecular beam is avoided and the amount of gas which reaches the ionization region, and subsequently the ion detector, is maximized, thereby improving the sensitivity of the instrument. Also, a tuning fork chopper has been implemented in the system along with a lock-in amplifier to enhance the signal-to-noise ratio. In addition, a custom flange has been designed with electrical feedthroughs to be able to heat the orifice in the vacuum chamber to prevent the formation of mercury clusters. After calibration of the mass spectrometer for mercury species, combustion experiments are conducted to speciate mercury in the flue gas environment. With this custom-built instrument, mercury species will be directly measured for the first time for high temperature combustion applications. By directly measuring mercury species accurately, one can determine the actual extent of mercury oxidation in the flue gas, which will aid in developing mercury control technologies.

Besides the experiments, my second project involves molecular modeling from first principle-based electronic structure calculations. From this study I have published two papers in the journals of Environmental Progress and Carbon in 2006 [1] and 2009 [2], respectively. I have employed ab initio studies to elucidate the binding mechanism of mercury on an activated carbon surface. Activated carbon, when injected into the gas stream of coal-fired boilers, is effective in capturing both elemental and oxidized mercury through adsorption processes. However, the mechanism by which mercury adsorbs on activated carbon is not exactly known and its understanding is crucial to the design and fabrication of effective capture technologies for mercury. The objective of the current study is to apply theoretical-based cluster modeling to examine the possible binding mechanism of mercury on activated carbon. The effects of activated carbon's different surface functional groups and halogens on elemental mercury adsorption have been examined. Also, a thermodynamic approach is followed to examine the binding mechanism of mercury and its oxidized species such as HgCl and HgCl2 on a simulated carbon surface with and without Cl. Energies of different possible surface complexes and possible products are compared and dominant pathways are determined relatively. In addition to studying activated carbon, we can also investigate different materials to find a better sorbent that is more effective and cheaper than activated carbon. Overall, this project will aid in preventing mercury emissions from flue gas.

In addition to the research that I have conducted within graduate school, I also have carried out a summer internship at Niksa Energy Associates LLC (NEA) and finalized NEA's reaction mechanism for the simultaneous oxidation and capture of mercury by chlorine and bromine species. I was able to validate NEA's most recent comprehensive mechanism for Hg/Cl/Br chemistry by interpreting three datasets from full-scale field tests within the measurement uncertainties. In addition to this milestone, I have finalized all values for the rate parameters, identified the dominant reaction pathways, and assessed the relative roles of homogeneous and heterogeneous conversion channels. At the end of three months I drafted a manuscript that was published in the journal, Energy & Fuels in 2010 [3].

Moreover, I have gained experience in teaching throughout my gradute study, which I think forms a strong basis for me to teach in the future. With the knowledge of topics like Combustion, Chemical Kinetics, Quantum Mechanics, Thermodynamics, Spectroscopy, Surface Science, Nanostructure and Characterization, Transport Phenomena, Fluid Mechanics and Fundamentals of Energy Process, I see my background as a good fit for a faculty position in either chemical engineering or another engineering program focused on energy. I am planning to continue doing research on mercury speciation in coal combustion flue gas and also apply my knowledge in combustion to a different application, specifically oxycombustion. Oxycombustion is a new technology developed to reduce carbon dioxide emissions from fossil-based energy resources. In an oxy-fired coal power plant, the coal is burned in an oxygen-rich environment instead of air and the flue gas is recycled back to the boiler, which in the end produces a higly pure carbon dioxide exhaust that can be captured at relatively low-cost and subsequently compressed for transport and sequestration. There is no commercial oxy-combustion plant operating yet because of the high cost of producing oxygen. Currently research is being conducted to develop new techniques for oxy-combustion of coal integrated with the oxygen production process and for testing and scale-up of oxy-combustion systems. As a faculty member I would like to be part of a global research project, which involves both academia and industry to make the oxy-combustion process more efficient.

References [1] Padak, B.; Brunetti. M.; Lewis, A.; Wilcox, J. "Mercury Binding on Activated Carbon" Environmental Progress, 25(4), 319-326, 2006. [2] Padak, B.; Wilcox, J. "Understanding Mercury Binding on Activated Carbon" Carbon, 47(12), 2855-2864, 2009. [3] Niksa, S.; Padak, B.; Krishnakumar, B.; Naik, C. V. "Process Chemistry of Br Addition to Utility Flue Gas for Hg Emissions Control" Energy Fuels, 24(2), 1020–1029, 2010.

See more of this Session: Meet the Faculty Candidate Poster Session
See more of this Group/Topical: Education

(4de) Engineering Nanostructured Materials for Green Energy

Sunho Choi, School of Chemical & Biomolecular Engineering, Georgia Institute of Technology, Atlanta, GA

Fossil energy is currently the most economic energy option and will continue to supply a significant portion of global energy demand for the next several decades. With a continued reliance on fossil energy, atmospheric CO2 levels will continue to rise, potentially contributing to continual climate change. Thus, the two seemingly contradicting societal needs of a growing energy demand and a desire for minimal environmental impact present a striking scientific and engineering challenge for our time. Clearly, it is critical to develop technologies to make fossil energy clean, while pursuing efficient means of using clean alternative energy.

This poster gives an overview of my recent studies on the design and development of nanostructured materials for gas separations that are critical to energy production. It includes investigations of (i) nanocomposite membrane-based H2/CO2 separations targeting pre-combustion CO2 capture, (ii) adsorptive CO2 removal from flue gases under post-combustion conditions using hybrid solid adsorbents, and (iii) selective capture of atmospheric CO2 from ambient air. Nanostructured materials, such as nanoporous layered materials and hyperbranched aminosilicas, are rationally engineered to facilitate these objectives.

Building on this past experience in fossil energy-related topics, my future work will focus on development of nanostructured materials for renewable energy applications. Among several alternative energy sources, my future research aims at developing clean and efficient routes for H2 production and purification. As an example, novel hollow-fiber photoelectrochemical cells are proposed for development. This system will generate H2 using sunlight and water as the inputs, in which the tubular architecture of proton-conducting materials prevents mixing of the gaseous products but allows the transport of oxidation products (H+) across the fiber walls, facilitating H2 generation in the inside of fibers. Once generated, H2 can be purified further using metal-filled nanotube membranes that enable synergistic separation of chemo-selectivity and size-selectivity. Hence, this work will contribute towards making H2 an alternative energy source that is more practical, economically feasible, and environmentally-benign.

See more of this Session: Meet the Faculty Candidate Poster Session
See more of this Group/Topical: Education

(4df) Design of Delivery Systems for Nanomedicine through Self-Assembly

Hitesh G. Bagaria, Dept of Chemical and Biomolecular Engineering, Rice University, Houston, TX

Nanoparticles – organic, inorganic and organic/inorganic composites - have proven potential for applications in nanomedicine. However, significant challenges still remain in realizing these applications. In most cases the solutions are sophisticated and there is a need to develop simpler, more practical methods. One of the ways to achieve the desirable complex assemblies is by making use of non-covalent forces or self-assembly. The development of self-assembly based nanoparticles is the motivation of my research.

See more of this Session: Meet the Faculty Candidate Poster Session
See more of this Group/Topical: Education

(4dg) Ultrafast Dynamics of Complex Fluid Interfaces

Valeria Garbin, Chemical and Biomolecular Engineering, University of Pennsylvania, Philadelphia, PA

In my research program I aim to measure the rheological properties of complex fluid interfaces subject to mechanical deformation on ultrafast timescales. To do so, I will combine experimental techniques borrowed from optics, acoustics and surface science, including high-speed imaging and ultrasound. My ultimate goal is to design responsive systems with tunable properties imparted by their interfacial rheology.

See more of this Session: Meet the Faculty Candidate Poster Session
See more of this Group/Topical: Education

(4dh) Metabolic Engineering for the Production of Biofuels and Chemicals

Patrick Suthers, Chemical Engineering, The Pennsylvania State University, University Park, PA

One of the key descriptors of cellular physiology is the set of flux values that flow through the cell's metabolic pathways. With the rapid generation of genomic data, creating such metabolic models in a systematic framework becomes crucial. Once constructed, these models can be used to propose experimentally testable hypotheses and guide modifications of the organism, e.g., to improve product yield. In this poster, I present 1) methodologies for quantifying fluxes in large and genome-scale models using isotope labeling data and 2) an optimization framework that uses these flux values to suggest engineering strategies for the overproduction of the product at the desired target.

Extending the scope of isotope mapping models becomes increasingly important in order to analyze strains and drive improved product yields as more complex pathways are engineered and as secondary metabolites are used as starting points for new products. Here I present how the elementary metabolite unit (EMU) framework and flux coupling significantly decrease the computational burden of metabolic flux analysis (MFA) when applied to large-scale isotope mapping metabolic models. The combined use of EMU and flux coupling analysis leads to a ten-fold decrease in the number of variables in comparison to the original isotope distribution vector (IDV) version of the Escherichia coli model. The observed computational savings reveal the rapid progress in performing MFA with increasingly larger isotope models and enable the ultimate goal of handling genome-scale models of metabolism.

Once obtaining flux values and ranges, I present an integrated computational base to support pathway identification and strain optimization, with an emphasis on C4+ biofuels and other biochemicals. I describe the application of the OptFlux computational framework to pinpoint engineering modifications (knock-outs/up/down) that are required for the targeted biochemical overproduction. This is accomplished by classifying reactions (and combinations thereof) in the metabolic model depending upon whether their flux values must increase, decrease or become equal to zero to meet the pre-specified overproduction target. A "force set" can then be extracted that contains a sufficient and non-redundant set of reactions that need to be directly changed to meet the production requirements. I apply the integrated framework for the production of succinate and 1-butanol in E. coli using the most recent in silico E. coli model, iAF1260. The proposed computational workflow not only recapitulates existing pathways and engineering strategies but also reveals novel and non-intuitive ones that boost production by using and performing coordinated changes on sometimes distant pathways.

See more of this Session: Meet the Faculty Candidate Poster Session
See more of this Group/Topical: Education

(4di) Biologic Conversion of Waste Streams to Renewable Fuel Sources

Sage R. Hiibel, Biochemistry and Molecular Biology, University of Nevada Reno, Reno, NV

A significant challenge facing the world in our immediate future is the need for clean, renewable energy sources. My work focuses on that challenge through the development and optimization of biological processes that convert waste streams into affordable and environmentally-responsible renewable fuels.

My past research efforts have given me a wide experience in microbiological systems and the necessary techniques for studying and controlling those systems. During my dissertation research I examined the biological remediation of acid mine drainage and focused on characterizing the members and functions of the microbial community involved in the treatment process. The research increased the understanding of the microorganisms that play a key role in this process, allowing for improved design and operation of these biological remediation systems. A novel biomolecular technique known as Active Community Profiling, which identifies the metabolically active members of a community was also developed. Other past research evaluated the cell signaling pathway and the cellular appendages responsible for *E. coli* swarming motility, while another project identified seven previously uncharacterized stress-related proteins in *E. coli* based on their sensitivity to *cis*-dichlorethylene.

My current research has both applied and fundamental aspects, and includes the cultivation and processing of algal species for the production of biodiesel, as well as genetic-level studies to elucidate the molecular pathways associated with lipid production in algae. This current work fits well with my broader research interests of converting waste product streams to renewable fuels and other value-added products through biologically-driven processes. This poster will summarize my past and current research, and will explore future research areas that I plan to pursue.

See more of this Session: Meet the Faculty Candidate Poster Session
See more of this Group/Topical: Education

(4dk) Ventures in Biofabrication: From Biological Nanofactories to Miniaturized Tools for Medicine

Rohan Fernandes, Chemical and Biomolecular Engineering, Johns Hopkins University, Baltimore, MD

Biofabrication involves the use of biological material such as cells, proteins and other bioactive elements as building blocks in creating tools for medical as well as non-medical purposes. In this work, we describe two different biofabrication techniques viz. the creation of biological nanofactories and miniaturized tools for medicine. Biological nanofactories are bio-inspired nanoscale factories that are comprised of multiple modules, each performing a particular function including (1) specifically targeting cells, (2) sensing and transporting raw materials present in their vicinity, (3) converting the raw materials to useful molecules, (4) transporting them back to the cell surface and (5) self-destructing upon completion of this sequence. Our nanofactories are engineered to contain modules that can target, sense and synthesize molecules that trigger communication between bacterial populations. The nanofactories are capable of eliciting the native signaling response in targeted bacteria in the absence of the native signaling molecules and can also trigger communication between bacterial populations that are otherwise non-communicating. Prospects for using these nanofactories in the creation of the next generation of anti-microbials that target bacterial communication rather than their viability are envisioned. Advances in engineering at small size scales have facilitated the creation of miniaturized tools for medicine. We describe methods to create such tools that involve the integration of cells and proteins with biocompatible materials. The tools comprise of spatially patterned cells or proteins that can be actuated to produce motion. Prospects for utilizing these miniaturized tools in minimally and non-invasive procedures are presented.

See more of this Session: Meet the Faculty Candidate Poster Session
See more of this Group/Topical: Education

(4dl) Photodegradation and Recovery of Pollutants in Aqueous Systems Using Magnetic Nanoparticles Coated with Photocatalyst Materials

Adriana P. Herrera, Chemical Engineering, University of Puerto Rico, Mayaguez Campus, Mayaguez, PR

Magnetic nanoparticles are the subject of intense research focusing on their synthesis, characterization, and functionalization. These nanomaterials are attractive in various novel applications including: nano-/bio-sensors, protein immobilization, MRI contrast enhancement agents, catalyst recovery, agents for treatment of localized cancerous tumors, and separation of pollutants in aqueous systems, among other. These applications require suitable nanoparticle surface modification, which provides colloidal stability in aqueous or biological fluids and improves the nanoparticle's transport and retention in specific areas.

From my doctoral thesis, I worked with the synthesis of multifunctional magnetic nanoparticles coated with silane-functional molecules and biocompatible materials such as carboxy-methyl dextran (CMDx) and polyethylene glycol (PEG). These nanoparticles showed a high stability in biological buffers mainly composed of electrolytes such as sodium chloride, phosphates, and proteins. I have also functionalized magnetic nanoparticles with a silane-acrylate molecule to graft thermo-sensitive polymers such as poly (N-isopropylacrylamide) (pNIPAM), which can be used for controlled drug delivery. Additionally, from my Postdoctoral Research Associate position I am working with the synthesis of novel magnetic fluorescent nanoparticles for neuron targeting, and the preparation of the laboratory guideline: Determination of Size and Size Distribution of Silica Nanoparticles Synthesized by the Stöber Method. From which I have had the opportunity to participate in symposiums, workshops, and projects related to addressing the challenges of nano-scale science on undergraduate education.

My previous experience with surface modification of magnetic nanoparticles gives me an excellent starting point to synthesize and functionalize nanoparticles for environmental applications such as the photodegradation and recovery of pollutants in aqueous systems. For this application, I propose encapsulate magnetic nanoparticles in a poly-siloxane matrix and then graft a layer of $TiO2$ molecules, which have been used for photodegradation of dyes under UV-Vis irradiation. After nano-photocatalyst process, the pollutants can be removed from the aqueous media by applying a magnetic field. I have found a funding opportunity from the National Science Foundation (NSF) division of Chemistry, and chemical catalyst for this nano-photocatalyst application.

See more of this Session: Meet the Faculty Candidate Poster Session
See more of this Group/Topical: Education

(4dm) High Throughput Polymer-Array Technology for Stem Cell Engineering

Ying Mei, MIT, Cambridge, MA

Human pluripotent stem cells hold enormous potential for applications in regenerative medicine and human disease modeling. They have the unique ability to undergo self-renewal indefinitely in culture and have potential to differentiate into almost all cell types in human body. Such cells can either be derived from human embryos or be 'reprogrammed' from patient's blood, skin, and fat samples. However, current methods to clonally grow them are inefficient and poorly-defined for genetic manipulation and therapeutic purposes. Here I report combinatorial development of synthetic biomaterials that support robust self-renewal of fully dissociated human pluripotent stem cells, preserve a normal karyotype, and maintain full differentiation capacity after prolonged cell culture. To the best of our knowledge, these materials provide the first chemically defined, xeno-free, feeder-free system to support the efficient clonal growth of human pluripotent stem cells at unprecedented high levels. Materials properties including surface topography after protein adsorption, surface wettability, molecular surface chemistry and reduced indentation modulus of all polymeric substrates were quantified using high-throughput methods to develop structure/function relationships between materials properties and biological performance. The studies described here provide a global understanding of relationships between human pluripotent stem cell growth and synthetic materials, as well as a general framework for the development of synthetic surfaces for stem cell culture.

See more of this Session: Meet the Faculty Candidate Poster Session
See more of this Group/Topical: Education

(4dn) Developing the Next Generation of Tools for the Study of Nucleation and Metastability in Liquids

Claudiu A. Stan, Chemistry and Chemical Biology, Harvard University, Cambridge, MA

Nucleation - the initiation of many types of phase transitions – is an important phenomenon, but one difficult to study experimentally. At the molecular scale, nucleation is a rare event that can involve tens or hundreds of molecules and we do not have yet good experimental and computational tools to investigate it. While the outcome of nucleation (creation of a new phase) is easily observable macroscopically, macroscopic investigations of nucleation are difficult because nucleation is often a stochastic phenomenon. A nucleation experiment must be repeated many times to acquire good statistical data. Microfluidic technology has characteristics that make it an attractive choice as a platform for the study of nucleation in liquids: it handles small amounts of liquids that can be kept very clean (creating reproducible experimental conditions), and the typical duration of experiments is short (allowing a high repetition rate). Several groups have already used microfluidics to investigate how crystals nucleate from solutions. We had chosen to look at a different class of problems – the nucleation of solids from supercooled liquids – and we have developed an instrument for the study of ice nucleation in supercooled water. This apparatus has the highest data throughput and state-of-the-art accuracy among instruments designed for the study of ice nucleation. We have used it to measure with the rate of homogenous nucleation in pure water, to study the effect of external electric fields on homogenous ice nucleation, and to measure the temperature dependence of heterogeneous ice nucleation due to ice-nucleating particles immersed in water. The apparatus can investigate the freezing of hydrocarbon liquids as well; we have supercooled and frozen drops of alcohols and alkanes. In all these measurements we were able to achieve well-controlled experimental conditions such as temperatures and cooling rates, high measurement accuracy, and better statistics than those achieved previously.

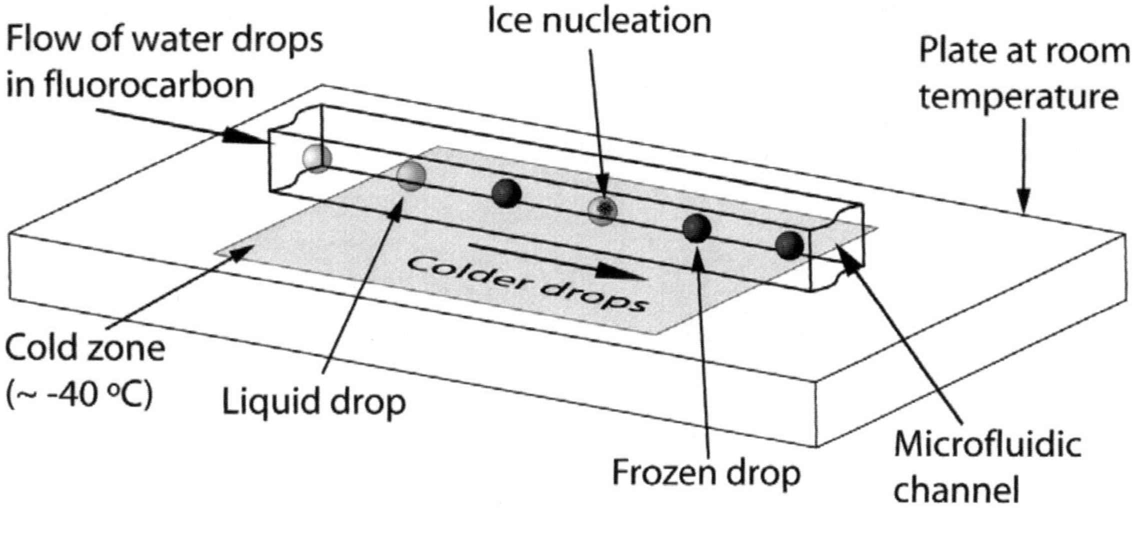

See more of this Session: Meet the Faculty Candidate Poster Session
See more of this Group/Topical: Education

(4do) Nano Materials for Energy Storage Applications

Surya Sekhar Moganty, Chemical and Biomolecular Engineering, Cornell University, Ithaca, NY

Secondary batteries containing Li metal as anode provide one of the highest known energy densities. They are therefore desirable candidates for electric and hybrid electric vehicles. However, current liquid electrolytes cannot be employed with lithium metal anode because they allow dangerous lithium dendrite growth during the charge discharge cycles. Solid polymer electrolytes, on the other hand, are known to be effective in reducing dendrite growth and hence, there has been significant effort devoted by research groups world wide to the development of polymer and composite polymer electrolytes for use in lithium batteries. The best known ionic polymer conductor, polyethylene oxide, is crystalline and has low conductivity at room temperature. While polyethylene oxide oligomers exhibit good ionic conductivity as well as chemical and thermal stability, amorphous low molecular weight electrolytes lack mechanical strength. The addition of free inorganic particles as well as inorganic networks to polymer electrolytes has been shown to improve both mechanical properties and conductivity. We are developing a new class of organic-inorganic hybrid materials as solid electrolytes for lithium metal batteries. These novel electrolytes are based on nanoscale organic hybrid materials (NOHMs) comprised of silica nanoparticle cores covalently bonded to polyethylene glycol oligomers and ionic liquids tethered to ziriconia (ZrO_2) nanoparticles. By varying the nanoparticle size and organic polymer molecular weight and grafting density, the volume fraction of the inorganic component can be systematically varied and mechanical properties of the hybrids manipulated over an unusually broad range. These novel organic-inorganic nanocompoite systens display rheology characteristic of yield-stress fluids termed soft-glasses. This presentation focuses on the conductivity and rheological properties of nanoscale hybrid electrolytes. The applicability of these hybrids as electrolytes in Li batteries will be discussed.

See more of this Session: Meet the Faculty Candidate Poster Session
See more of this Group/Topical: Education

(4dp) Understanding Transport Processes in Electrokinetics through Applied Mathematics: Challenges and Opportunities

Jennifer Pascal and Pedro Arce, Chemical Engineering, Tennessee Technological University, Cookeville, TN

Electrokinetics can be utilized in a wide variety of applications including environmental remediation, bioseparations for the development of novel pharmaceuticals, and drug delivery, just to name a few. In bioseparations, a technique that is commonly used to separate biomolecules, such as proteins, is electrical field flow fractionation (EFFF). In this approach, an electrical field is applied perpendicular to the direction of flow in a channel causing the biomolecules most susceptible to the field to move to the walls of the channel faster, thus creating a very efficient separation. Another area in bioseparations where electrokinetics is utilized involves nanocomposite gels and electrophoresis to separate proteins, antibiotics, and DNA, for example. It has been shown that the addition of charged nanoparticles into hydrogels has the potential to modify the internal structure (morphology) of the gel and can lead to enhanced separation efficiency.

In my doctoral research I have focused on the mathematical modeling of the aforementioned systems to study the effects of morphology, physicochemical properties (i.e. valence, mobility etc.), and applied electrical field on the determination of optimal times of separation. Predictions of optimal times of separation can aid in the improvement of the design of devices and materials, such as the ones mentioned above, and lead to better separation efficiencies. In order to accomplish the tasks mentioned above, I have studied a variety of systems with various types of geometries (rectangular, cylindrical, annular, axially varying) and types of flows (Couette, Poiseuille). In addition, I have examined the dynamics of the previously described systems in order to better understand the effects of these parameters on transient analyses.

In this presentation, I will summarize my doctoral research, the research program I plan to develop, and my philosophies about teaching. Understanding the behavior of the aforementioned biological systems is a complex task. Therefore, developing models to aid in this understanding can significantly advance the field. To accomplish this, a mathematical toolbox filled with a wide array of techniques from analytical (partial differential and integral equations, linear operator method approaches) to computational (finite element, finite difference, computer programming) is necessary. I hope to develop a research program that utilizes my knowledge in applied mathematics and computational strategies to develop models to predict the transport behavior in bioengineering systems, such as the ones mentioned previously. This broad area of research not only allows for versatile projects, but also lends itself to collaborations with colleagues from other disciplines, such as mechanical engineering, materials science, and mathematics, for example. In addition, collaborations with other engineers who perform experiments in the area to validate these models will be helpful. Potential sources of funding will also be identified and presented.

See more of this Session: Meet the Faculty Candidate Poster Session
See more of this Group/Topical: Education

(4dq) Closing the Carbon Cycle for Sustainable Energy and Environment:CO2 Capture Using Nanoparticle Ionic Materials (NIMs) and Methane Recovery in Clathrate Hydrates with Integrated Carbon Storage

Youngjune Park, Earth and Environmental Engineering, Columbia University, New York, NY and Ah-Hyung Alissa Park, Earth and Environmental Engineering & Chemical Engineering, Columbia University, New York, NY

Sustainability has become an urgent global issue with the exhaustion of fossil fuel and the climate debate. Thus, developments in alternative energy sources and capture/storage of carbon dioxide have been the focus of recent research in the energy and environment field. My research has specifically focused on two distinct engineered and natural materials: i) the synthesis of organic-inorganic nanoparticle ionic materials (NIMs) for CO2 capture and ii) the combined methane recovery and CO2 sequestration in complex structured naturally-occurring gas hydrate.

In order to capture carbon dioxide from flue gas streams, various absorbents have been suggested, and amine-based aqueous solution such as MEA solvent exhibits tremendous CO2 capture capacity. However, the undesired effects of amine-based solvents such as corrosion, vaporization/degradation loss, and large energy penalty during solvent regeneration have been obstacles for its large scale industrial application. As an alternative option to these amine-based solvents for CO2 capture, NIMs have been designed and synthesized with negligible vapor pressure. NIMs are organic-inorganic hybrid materials which comprise oligomer or polymer counterions tethered to a surface-modified nano-sized core by strong primary ionic bonds. Because it is the molecular analogs of ionic liquids, NIMs exhibit negligible vapor pressure with enhanced thermal stability. Since there is no solvent (i.e., water) associated with this CO2 capture medium, the energy requirement for NIMs regeneration is significantly lower than that of MEA. In addition, various physical and chemical properties such as rheological, optical, and electrical properties with the range from glassy solids to solvent-free nanoparticle suspension could be expected by tuning the core and canopy (ionically tethered corona and counterion species). It is proposed that ionic bonds located in corona chains would contribute electrostatic repulsion among corona chains and as the corona chains are frustrated by electrostatic forces the free volume within the system would increase. Therefore, CO2 capture by NIMs can be maximized by controlling both enthalpic and entropic effects.

Once CO2 is captured, it is important to store this carbon with long-term stability. Clathrate hydrates, so called gas hydrates, are non-stoichiometric crystalline compounds formed when guest molecules are incorporated in host water frameworks made up of hydrogen-bonded water molecules under the appropriated pressure and temperature conditions. Because of its various and unique physical and chemical properties, clathrate hydrates have been studied for energy gas storages such as hydrogen and for other material engineering applications. The total amount of natural gas hydrates on continental margins and in permafrost regions has been estimated to be about twice as much as the energy contained in fossil fuel reserves. Thus it could be serve as an alternative option as an energy source if efficient recovering way is provided. The option of swapping enclathrated methane and CO2 molecules is proposed for the combined methane production and CO2 sequestration. It is experimentally confirmed that ~ 90% of methane molecules in naturally-occurring gas hydrate can be recovered without structural decomposition while storing CO2 in place of methane. It is also found that due to the large and small cage structure of the gas hydrate for the maximum recovery of methane, both CO2 and smaller gaseous molecule such as N2 should be used simultaneously.

As shown in these studies, my research aims to close the carbon cycle by extracting maximum energy possible from the new natural resources while efficiently storing excess waste carbon in natural systems. The findings from these studies will provide valuable fundamental knowledge for the environmental field and energy sustainability.

See more of this Session: Meet the Faculty Candidate Poster Session
See more of this Group/Topical: Education

(4dr) Nanocatalysis for Aqueous Processing of Biomass Derived Feedstocks

Xiaoming Wang, Department of Chemical Engineering, Yale University, New Haven, CT

Catalytic conversions of biomass to different energy forms are promising technologies to bridge the gap between the large energy demand and the shortage of fossil fuels. Aqueous phase processing (APP) of biomass derived feedstocks has received much attention after the discovery of aqueous phase reforming (APR) in 2002. As most oxides gradually dissolve in the reaction conditions of APP, carbon is a perfect candidate for the catalyst support of APP, due to its insolubility in aqueous solutions.

Carbon nanotubes (CNTs), including both single-walled carbon nanotubes (SWNTs) and multi-walled carbon nanotubes (MWNTs), is a recently discovered novel carbon materials with well-ordered tubular structures. CNTs have been used as catalyst support shortly after its discovery, and have shown advantage over other forms of carbon due to its unique electronic, sorptive and structural properties.

My research has been largely focused on APR over CNT supported catalysts. We have demonstrated that our lab-made SWNTs is an efficient support for APR catalysts. Both monometallic Pt-SWNT and bimetallic Pt-Co-SWNT catalysts show higher catalyst mass time yield (up to ten times) than corresponding Pt and Pt-Co catalysts supported on oxide supports. We have developed two kinds of bimetallic catalysts, one with separate Pt and Co particles as bifunctional catalyst, and the other with Pt-Co alloy phase, while both shows much higher yield than alumina supported Pt-Co catalyst.

One of the advantage of carbon as a catalyst support is that its surface properties are tunable upon chemical functionalization. We studied the effect of CNT surface functional groups on the APR activity, and showed that the hydrophilicity of the carbon support is an important issue controlling the activity of aqueous processing. We have also studied other metal-support interactions in Pt-CNT system, such as the CNT radius of curvature effect, the relative location of the catalytic particles inside or outside CNTs, etc.

It should be noted that our lab-made SWNT shows much better performance as a catalyst support than all the commercially available CNTs. The reason is not clear but could be possibly that our SWNT has a surface area as high as $1800m^2/g$, the highest surface area ever reported for CNTs. We also studied the production of SWNT aiming to increase its yield and lower the production cost. We have demonstrated an SWNT yield (defined as the weight of SWNT produced per unit weight of catalyst) as high as 75%, which is ten times higher than the commercial CoMoCAT technology. Detailed spectroscopic study also showed that the formation of highly dispersed cobalt silicate species is the key step for successful preparation of catalyst for SWNT production.

(4ds) Sorption and Oxidation of Mercury in Flue Coal-Fired Power Plants

Erdem Sasmaz, Energy Resources Engineering, Stanford University, Stanford, CA

Activated carbon (AC) and flue gas desulfurization (FGD) processes are often incorporated to capture trace elements (TEs), such mercury (Hg), selenium (Se) and arsenic (As). Although these processes work fairly well for oxidized Hg capture for combustion applications, they fail for gasification fuel gases since gasification conditions require Hg capture at an elevated temperature, at which point activated carbon-based sorbents to break down. Hence, worldwide there exist no control technologies for TE species on current coal gasification plants used for energy generation. Also, elemental Hg proceeds through combustion flue gas uncaptured due to its low reactivity. Lastly, TE capture via FGD and activated carbon decrease coal's sustainability by compromising the recycling of fly ash and FGD waste byproducts. Recycling coal-combustion byproducts (CCBs) increases coal's sustainability through minimizing waste associated with the process. Typical CCBs are fly ash from loss on ignition and spent gypsum, which is obtained from calcium (Ca)-based sorbents in SO2 scrubbers. Traditionally, fly ash is used in concrete and asphalt manufacturing and spent gypsum is recycled through wallboard and stucco production. An unintended consequence of the recycling process is the existence of TEs in both byproducts. As the gypsum demand increases, it is essential to limit Hg capture in FGD processes to prevent the possible leaching of Hg within the high-temperature recycling processes of gypsum. AC that is used for Hg capture is collected in electrostatic precipitator units along with the fly ash for recycling processes. However, within the concrete manufacturing process the addition of spent activated carbon prevents concrete from meeting the freeze-thaw requirements, which compromises its functionality. Hence, the primary goal of this study is to design a multipollutant sorbent material that will be effective in capturing the TEs at elevated temperatures of both flue and fuel gases of coal-based energy production and to understand heterogeneous chemical pathway of Hg both considering AC and calcium oxide (CaO) sorbents. Results of this study will benefit the planet through the reduction of harmful emissions, and the effective recycling of fly ash and FGD by allowing developed and developing countries to prosper economically from sales associated with the manufacture of coal combustion byproducts, and environmentally through the prevention of waste materials filling landfills.

Although the adsorption mechanism of Hg is complicated, a fundamental understanding can be achieved by investigating both energetic and electronic interactions of Hg on CaO and noble metals surfaces using first principles quantum mechanical calculations. Preliminary theoretical investigations have been carried out and it has been found that mercury strongly binds to CaO surfaces as Hg2+ and forms a stable complex on the surface. SO2 also binds strongly to CaO surfaces with chemisorption being the likely adsorption mechanism. Both of these results indicate that both Hg and sulfur (S) bind to Ca-based sorbents in wet FGD systems. Palladium (Pd) alloys and overlays have also been investigated as a replacement for Ca and organic-based mercury sorbents, and the results show that the addition of a small amount of gold (Au) and silver (Ag) is able to enhance the sorbent reactivity. It is important to note that

Hg binding is sensitive to the specific local concentration of Au and Ag atoms and it would be difficult to optimize the structure for fabrication due to the invariance of the atomic configuration. Further preliminary studies revealed that the overlay substrate takes the lattice constant of the underlying metal which leads to an enhancement of the Hg binding compared to the Pd alloys of Pd deposited on another metal having a larger lattice spacing. The use of overlays removes the dependence of the random atomic arrangement, making the fabrication process easier, increasing the potential for these novel materials to be scaled up. Both of these CaO and noble metal, i.e. Pd, Au, Ag, studies have been completed and published in several manuscripts, i.e., Journal of Physical Chemistry C and Main Group Chemistry.1-3

During my PhD in the department of Energy Resources Engineering at Stanford University under the supervision of Prof. Jennifer Wilcox at Stanford University, I also designed a burner and fixed bed reactor to investigate Hg adsorption and possible oxidation on noble metals and AC. Quantum chemistry modeling tools guided the experimental research by helping to understand the physical mechanism and reducing the time and cost typically required for material fabrication. AC samples functionalized with halogens and sulfur are tested and the effects of functional groups and active sites on the binding of Hg are considered. Well-known AC samples assist in understanding the adsorption and oxidation mechanisms of Hg on the sorbent surface. In addition, Pd-Au sorbents have been fabricated by Johnson Matthey Technology Centre in Reading, UK according to the first principles quantum mechanical results. These sorbents are tested in the packed bed reactor (PBR) and the mechanism of Hg binding is investigated. To simulate the flue gas condition, methane is combusted in a laminar flow tubular burner and mixed with mercury and chlorine to simulate the flue gas environment. Mercury adsorbed sorbent materials are characterized using X-ray Photoelectron Spectroscopy (XPS) and X-ray absorption fine structure (XAFS) spectroscopy. Characterization experiments before and after the tests are necessary to identify the oxidation state of Hg on the carbon surface. The innovative nature of the project is evident as it is the first attempt to use molecular modeling techniques to propose a mechanism for the adsorption of the TEs onto Ca-based sorbents and Pd alloys and overlays. To further increase the technical merit of the proposed research, the adsorption strengths of these novel sorbents with TEs can be experimentally tested against qualitative model interpretations in order to further validate the model simulations.

As a PhD student and a researcher, I would like to take a part to improve clean energy technologies by combining first-principle computational tools with experiments. I have a strong background in Thermodynamics, Fluid Mechanics, Mass and Heat Transfer, Reaction Kinetics, Surface Science, Catalysis and Spectroscopy. I believe that the knowledge I gained from my education combined with the laboratory experience has adequately prepared me to train for a faculty position in an engineering program focused on energy.

References 1. Sasmaz, E.; Wilcox, J. Phys. Chem. C, 2008, 112, 42. 2. Aboud, S.; Sasmaz, E.; Wilcox, J. Main Group Chem. 2008, 7, 205. 3. Sasmaz, E.; Aboud, S.; Wilcox, J. Phys. Chem C 2009, 113, 7813.

See more of this Session: Meet the Faculty Candidate Poster Session
See more of this Group/Topical: Education

(4dt) Reprogramming Human Somatic Cells to Pluripotency for Disease Modeling and Therapy

Krishanu Saha, Whitehead Institute for Biomedical Research, Cambridge, MA

Currently, new cellular models of human disease are emerging through the use of reprogramming technology. Direct reprogramming of human somatic cells to embryonic-like induced pluripotent stem (iPS) cells uses readily accessible skin, fat, blood, and hair cells. A primary use of such technology is to reprogram somatic cells from patients who are classified into a disease group, thus creating iPS cell lines popularly labeled as "diseases in a dish." The disease-in-a-dish agenda thus frames disease as a phenomenon to be seen at the cellular level by laboratory scientists. This work will examine the assumptions built into "diseases in a dish" and further develop numerical models associated with this engineering endeavor. Applications in regenerative medicine, drug discovery, and disease modeling will be thoroughly discussed.

See more of this Session: Meet the Faculty Candidate Poster Session
See more of this Group/Topical: Education

(4du) Novel Polymeric Materials for Bioanalytical Separations and Microfluidic Systems for in Situ Detection of Organic Biomarkers of Extant or Extinct Life On Mars

Thomas N. Chiesl, Chemistry, University of California Berkeley, Berkeley, CA, Annelise E. Barron, Bioengineering, Stanford University, Stanford, CA and Richard A. Mathies, Department of Chemistry, University of California, Berkeley, Berkeley, CA

I will present work from my Ph.D. and post doctoral research which has focused on creating advanced polymeric materials and microfluidic devices to obtain sample in answer out capabilities for biomedical diagnostics and the chemical exploration of the solar system.

In order for micro total analytical systems to realize full maturity, the advancement of soft polymeric materials used for purification, concentration, and analysis must advance beyond current levels and be fundamentally well characterized. Here, I present the synthesis and characterization of families of poly(acrylamide-co-alkylacrylamide) that accomplish on-chip purification of DNA from protein and advance bioanalytical separations. Hydrophobic alkylacrylamide monomers were synthesized and subsequent families of poly(acrylamide-co-alkylacrylamide) were polymerized via "micellar" and free-radical synthesis. Monomers and copolymers were characterized by techniques such as: HPLC, gel-permeation chromatography, 1H NMR, dynamic light scattering, polymer solution rheology, and differential scanning calorimetry. Remarkably, the inclusion of $<$ 1% of the alkylacrylamide in the copolymer leads to a significant improvement in separation performance and provides enough adsorption sites to enable DNA purification. Along with the traditional polymer characterization techniques, electrophoresis of DNA was visualized via single-molecule video microscopy. DNA migration physics were studied in both standard polymer and the copolymers systems where a previously unkown mechanism was discovered. The transitions in separation modality were correlated to polymer overlap and entanglement concentrations.

In my post doctoral work, I developed microfabricated biochemical analysis systems for in situ detection of organic biomarkers of extant or extinct life on Mars and other solar system bodies as a part of the ExoMars mission on the European Space Agency. Ultra-sensitive (sub pptr) pseudo-2D analysis techniques were created for amino acids and validated on Martian analog samples (e.g. meteorite, Atacama Desert) and a simulant of an astrochemistry reaction producing complex organics including amino acids and dipeptides. Derivatization and separation chemistries were developed for analysis of aldehydes, ketones, carboxylic acids, and polycyclic aromatic hydrocarbons. Advances in analytical protocols and buffer chemistry improved the robustness of the CE system when analyzing highly saline and acidic samples as found in Rio Tinto sediments and Saline Valley brine samples. Validation of these methods is shown through the analysis of Martian analog samples (e.g. Murchison Meteorite, Atacama Desert), an environmental sample for PAH analysis, and a simulant of an astrochemistry reaction producing complex organics such as amino acids and dipeptides. Automation and integration of on-chip functionality was explored through a multichannel microchip electrophoresis system with a common sample bus and devices incorporating metering, mixing, serial dilutions and reaction networks.

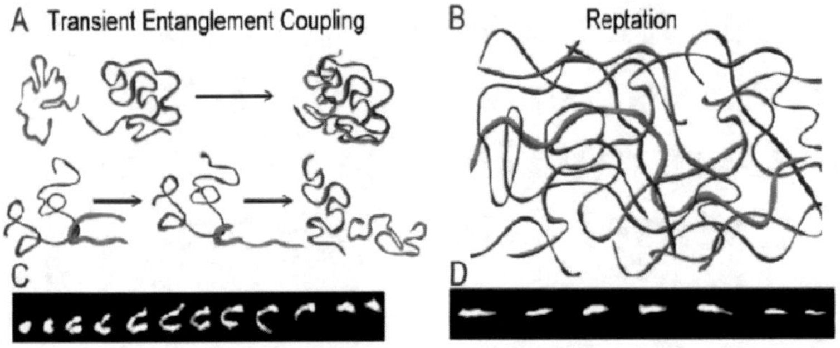

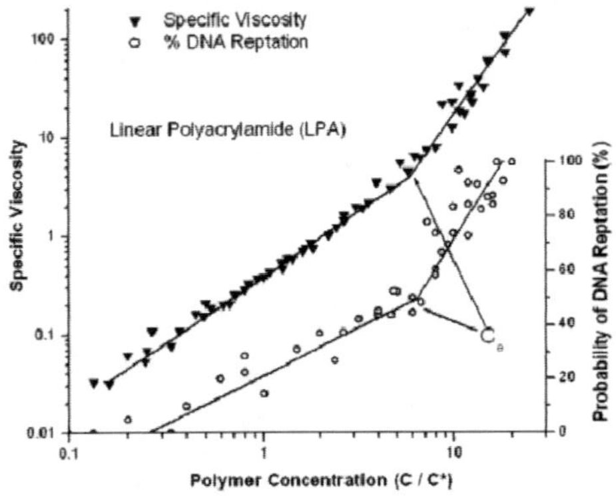

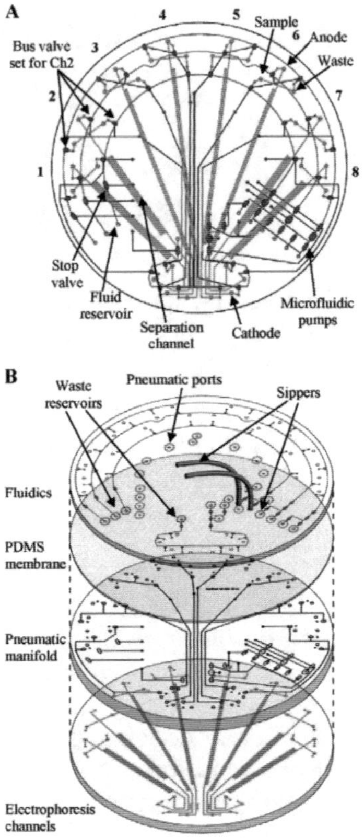

See more of this Session: Meet the Faculty Candidate Poster Session
See more of this Group/Topical: Education

(4dv) DNA Separation Using Nanoscale Organic Hybrid Materials

Henry W. Lau and Lynden A. Archer, Department of Chemical and Biomolecular Engineering, Cornell University, Ithaca, NY

Nanoscale Organic Hybrid Materials (NOHMs) are a recently-developed class of materials synthesized by covalently linking a corona of organic molecules to the surface of inorganic nanoparticle cores. By varying the dimensions and the chemical character of either the corona or the core, the physicochemical properties of such systems can be readily tuned. In particular, precise control over surface charge, hydrophobicity or reactivity can be achieved through the selection of appropriate corona constituents.

For this poster presentation, I will discuss our efforts to apply these materials to effect electrophoretic biomolecular separation. Specifically, we focused on systems consisting of silica nanoparticle cores functionalized with a corona of polyethylene glycol oligomers. Compared to those from a conventional polyacrylamide (PA) gel, results obtained using a silica NOHMs-PA hybrid gel showed an enhancement in the electrophoretic mobility of DNA molecules. In order to understand the physical mechanism for the observed behavior, we evaluated the influence of additive concentration, core size, and surface charge on the electrophoretic dynamics of DNA in these systems. Overall, I will discuss the experimental results in the context of my past work on the electrophoresis of DNA as well as potential future work on applying organic-inorganic hybrid molecules for biomolecular sensing.

See more of this Session: Meet the Faculty Candidate Poster Session
See more of this Group/Topical: Education

(4dw) Nanoengineered Materials for Sensing and Energy Conversion

Carlos Hangarter, Metallurgy Division, NIST, Gaithersburg, MD

Nanoscale structures, such as nanoparticles (NPs), nanowires (NWs), and nanotubes (NTs), are attractive materials for electronics because of their potential for device miniaturization and unique properties which arise from surface/interface dominated behavior and size dependent quantum confinement effects. These building blocks of can be further complexed with axial and radial interfaces, to create multilayered, core/shell, and nanopeapod structures. However, to fully exploit these higher level configurations, nanomaterials must be engineered for specific properties and interfacial characteristics that give rise to enhanced functionalities for communications, data storage, logic operations, sensing, and energy conversion. This arduous task requires an intimate understanding of processing condition on surface/interfacial quality and crystallographic features and their subsequent impact on physical properties of interest. In this work I will describe a variety of strategies developed to synthesize, probe, and generate understanding of conducting polymer, magnetic, Pt alloy and chalcogenide systems at the micro- and nano-level.

See more of this Session: Meet the Faculty Candidate Poster Session
See more of this Group/Topical: Education

(4dx) Design of Respiratory Nanoparticle Delivery Vehicle and Evaluation of Cellular Toxicity

Timothy Brenza, Pharmaceutical Science and Experimental Theraputics, The University of Iowa, Iowa City, IA

The extracellular fluid lining the respiratory tract presents a significant barrier to the effective delivery of therapeutics to and through the respiratory system. However, the ability to bypass first pass metabolism and locally delivery therapeutics to the site of action for respiratory ailments is significant. The ability of the innate defenses of the respiratory tract to remove exogenous materials reduces the time available for the action of therapeutic agents. The residence time of the delivery vehicle will be defined by its ability to move through the extracellular fluid and interact with the underlying epithelium. In order to move through the extracellular fluids the particle must be small enough to pass through the mesh of the mucus gel layer and then avoid adhesive interactions with this network. My current research has been focused on developing a particle surface which allows the particle to overcome these extracellular fluid interactions while maintaining the ability to interact with the underlying epithelium layer. This has led to the development of particles with a polyethylene glycol (PEG) shell and the use of lipooligosaccharide (LOS) glycoforms naturally expressed on bacterial cell surface to facilitate particle-epithelial interactions. To determine the efficacy of these particles an in vitro model of the respiratory system was developed using the Calu-3 cell line, a human bronchial cancer line with secretory properties. The presence of mucus on the surface of these cells which made this a desirable system to study also caused difficulties in processing the cells for analysis. Therefore, several mucolytics were analyzed before settling on one which did not alter the properties of the cell monolayer. This model has been used to assess particle attachment and internalization to cells and cell viability after exposure to particles.

See more of this Session: Meet the Faculty Candidate Poster Session
See more of this Group/Topical: Education

(4dy) Atom-by-Atom Metrology of Materials for Microelectronics, Energy and Biology

Domingo Ferrer, Microelectronics Research Center, University of Texas at Austin, Austin, TX and Sanjay K. Banerjee, Microelectronics Research Center, The University of Texas at Austin, Austin, TX

My research efforts focus on clarifying the fundamental mechanisms governing the synthesis of size-, shape- and composition-controlled nanoscaled systems for information storage nanoelectronics, energy sciences and biology. Metallic and semiconductor inorganic nanostructures hold an array of unique magnetic, electronic and photonic behavior that often differs from their bulk counterparts. The overarching goal of the research proposed herein is elucidating the synthesis-structure-performance relations of atomic-scaled materials by advanced microcharacterization techniques to engineer properties such as tunneling magnetoresistance (TMR), photovoltaic conversion, energy storage and catalytic activity. Nanofabrication techniques are pursued for nanoelectronics applications involving the demonstration of memory spin-based devices that are scalable to their ultimate physical limits. Diverse routes undertaken to fabricate spin-transfer-torque random access memories (STT RAM) are presented. Monolayer assemblies of FePt nanostructures have been fabricated with different geometries, including nanorods, oval-shaped particles and spherical particles for rational control of their magnetic properties. The employed wet-chemistry technique emerges as a low-cost alternative for fabricating magnetic nanostructures with much smaller dimensions than conceivable with lithographic methods. The coercive field of these FePt nanostructures can be tuned by engineering their shape. Micromagnetic calculations acquired using the code OOMF (Object Oriented MicroMagnetic Framework) explain the enhancement of the coercivity as consequence of geometry elongation in the colloids. Our research group endeavors will also serve to answer critical issues in the design and development of low cost, highly efficient nanomaterials that can ease widespread use of clean energy technologies such as solar cells, electrochemical capacitors and biofuels to address the world's energy and environmental challenges. The students of my research group will be skilled on the fabrication, characterization and application of inorganic nanomaterials to meet the research goals needed to ensure that viable fossil fuel alternatives are developed. The biodistribution of silver nanostructures injected into animals in vivo is currently unknown, remaining as a fundamental issue for potential therapeutic applications. Nanotoxicology efforts will aim at injecting metallic nanocrystals in live rats to elucidate their fate in several organs including liver, heart and brain. Very significant accumulations of nanoparticles were confirmed by inductively coupled plasma mass spectroscopy (ICPMS) and transmission electron microscopy (TEM) techniques on the liver and heart. In contrast, the brain tissue did not reveal evidence of particles content. Our results suggest that Ag^+ permeated across the blood–brain barrier (BBB), and followed swift clearance from the organ.

See more of this Session: Meet the Faculty Candidate Poster Session
See more of this Group/Topical: Education

(4dz) Networked and Distributed Predictive Control: Enabling Smart Manufacturing and Smart Renewable Energy Generation

Jinfeng Liu, Department of Chemical and Biomolecular Engineering, University of California, Los Angeles, Los Angeles, CA and Panagiotis Christofides, Department of Chemical and Biomolecular Engineering and Department of Electrical Engineering, University of California, Los Angeles, Los Angeles, CA

Traditionally, process control systems rely on centralized control architectures utilizing dedicated, wired links to measurement sensors and control actuators to regulate appropriate process variables at desired values. While this paradigm to process control has been successful, we are currently witnessing an augmentation of the existing, dedicated control systems, with additional networked (wired and/or wireless) actuator/sensor devices which have become cheap and easy-to-install. Such an augmentation in sensor information, actuation capability and networked-based availability of data has the potential to dramatically improve the ability of process control systems to optimize closed-loop performance and prevent or deal with abnormal situations more effectively. However, augmenting dedicated control systems with real-time sensor and actuator networks poses a number of new challenges in control system design that cannot be addressed with traditional process control methods, including: a) the handling of additional, potentially asynchronous and/or delayed measurements in the overall networked control system, and b) the substantial increase in the number of process state variables, manipulated inputs and measurements which may impede the ability of centralized control systems (particularly when nonlinear constrained optimization-based control systems like model predictive control are used), to carry out real-time calculations within the limits set by process dynamics and operating conditions.

To address these key challenges, we have focused on the development of rigorous, yet practical methods for the design of networked and distributed model predictive control systems for chemical processes described by nonlinear dynamic models. Specifically, we have developed a two-tier networked control architecture which naturally augments dedicated control systems with networked control systems. The key idea is the design of a networked controller, via Lyapunov and predictive control techniques, which takes advantage of additional, potentially asynchronous and/or delayed measurements, to maintain closed-loop stability and significantly improve closed-loop performance. In addition to networked predictive control, we have developed methods for the design of distributed model predictive control systems for nonlinear processes that utilize a fraction of the time required by the respective centralized control systems to compute optimal manipulated input trajectories and cooperate in an efficient fashion to achieve desired plant-wide performance, stability and fault-handling objectives. We will present applications of the developed networked and distributed predictive control systems to chemical processes and wind/solar energy generation systems using detailed nonlinear models.

(4e) Surface and Interface of Soft Materials

Wei Zhang, Chemical and Life Science Engineering Department, Virginia Commonwealth University, Richmond, VA

Surface and interface properties of soft materials (e.g. polymers) are governed by the first few nanometers on their surfaces or interfaces. These very small fractions of a material not only affect the bulk properties, but also dominate the surface energy, wetting behavior, adhesion, biocompatibility, and so on. A modification or treatment of the surface and interface can lead to dramatic changes in these properties. In this paper, I will demonstrate the importance of the surface and interface to a material with several research projects I have been involved. My future research interests revolving around these surface and interface phenomena will also be discussed. The examples will be presented in this paper are as follows: 1. Very strong dependence of the magnetic fluids and magnetic polymer particles containing magnetic particles on the dispersion, surfactant and the surface morphology of the particles was discovered. 2. Polymer blend or composite requires a compatiblizer or a surfactant to achieve the desired miscibility or dispersion for the target application. Block copolymers were designed to maintain a certain degree of immiscibility for control of morphology. 3. Particle formation using supercritical fluids utilizes the super low surface tension of supercritical fluids to prevent the aggregation of free particle products with very high surface area. 4. Other examples of the surface and interface phenomena will be discussed together with applications in polymer welding, adhesion, autohesion, lamination, and polymer coatings. 5. My current research involves polymer surface modification using a semi-fluorinated polymer or copolymer, which renders the surface a low surface free energy and tunable morphologies and topologies. Applications of such surfaces can be found in antimicrobial, antifouling, control of hydrophobicity and oleophobicity, biocompatible and biostable biomedical devices, etc.

See more of this Session: Meet the Faculty Candidate Poster Session
See more of this Group/Topical: Education

(4ea) Zeolite Films and Membranes for Computer Microprocessors and Biofuel Production

Christopher M. Lew, Chemical Engineering and Materials Science, University of Minnesota, Minneapolis, MN, Yushan Yan, Department of Chemical & Environmental Engineering, University of California, Riverside, Riverside, CA and Michael Tsapatsis, Department of Chemical Engineering and Materials Science, University of Minnesota, Minneapolis, MN

Zeolite thin films and membranes have been widely investigated over the last 15 years, mainly as separation membranes and membrane reactors. Their uniform microporosity and the ability to fine-tune their chemical composition and physical architecture lead to important macroscopic properties that can be exploited in a growing number of applications. The diversity in their use is demonstrated in this work through the exploration of zeolites as low-dielectric constant (low-k) films for computer microprocessors and catalytic membrane reactors for the production of biofuels.

Next-generation microprocessors require smaller feature sizes, but the slow development of new low-k films with better material properties is reducing progress toward improved computing performance. Pure-silica-zeolites (PSZ) are a strong candidate for replacement low-k materials because their intrinsic porosity and crystallinity gives them a low k value and strong mechanical properties. However, several other characteristics, including hydrophobicity, wet-etch chemical resistance, and pore size distribution, were not previously evaluated or optimized for low-k applications. Successful device integration of a viable new low-k alternative requires that all of these properties meet minimum specifications.

PSZ films were carefully engineered through different synthesis and modification techniques to obtain the desired aforementioned properties. PSZ MFI and MEL frameworks were functionalized with fluoroorganic groups through a direct synthesis method. The resulting hydrophobic films reduced moisture adsorption, which mitigated significant increases in the k value after exposure to ambient conditions. Wet-etch chemical resistance was imparted to PSZ films by annealing them in hydrophobic silanes. By silylating the films before cross-linking of the silica matrix occurred, a high carbon content was incorporated throughout the bulk of the films, thus making them water- and wet-etch chemical-resistant. The mesopore diameter and pore size distribution was controlled and narrowed through the addition of tetraethylorthosilicate to the spin-on zeolite nanoparticle suspension. After calcination and silylation, the induced changes in porosity and carbon content dramatically improved the mechanical properties and wet-etch chemical resistance. The results obtained during this process also indicate that zeolite can be crystallized "on-wafer" under ambient pressures. As a result of the methods developed in this work, PSZ low-k films are a leading low-k candidate and considerably outperform other alternative materials.

Studies are currently being performed to develop catalytic coatings of exfoliated and pillared zeolites for the production of hydrocarbons from biomass. These materials are derivatives of a certain class of layered materials that consist of microporous, zeolitic layers. They can, in principle, be pillared by inserting silica or other oxide pillars in the spaces between the layers (often called the gallery). They can also be exfoliated to yield single zeolitic layers of nanometer thickness. The pillared zeolites contain tunable mesoporosity in the gallery, while the single exfoliated layers allow for external surface catalysis at sites with catalytic activity similar to that of zeolite pores.

See more of this Session: Meet the Faculty Candidate Poster Session
See more of this Group/Topical: Education

(4eb) Remembering Kinetics: Studying the Dynamic Protein Activations and Protein Regulations That Form the Molecular Basis of Learning and Memory

Tamara L. Kinzer-Ursem, Biology, California Institute of Technology, Pasadena, CA

In the brain, changes in the strength of synaptic connections underlie our ability to form memories and to learn. These changes in synaptic strength are mediated at the molecular level by dynamic regulation of the spatial location of proteins and by the kinetics of protein-protein interactions. Two classes of transmembrane receptors, G protein coupled receptors (GPCRs) and NMDA receptors, mediate the activation of signal transduction cascades that work in concert to maintain the normal function of our synapses. Disruptions in these signaling pathways almost always lead to disease. All known drugs of abuse mediate their effects via GPCRs and change the dynamics of synaptic response to dopamine in the dopaminergic reward system of the brain. Many proteins involved both GPCR and NMDA signal transduction are disregulated in genetically encoded diseases of the brain, such as depression, schizophrenia, and Alzheimer's.

My research program will be focused on how these signal transduction cascades work together dynamically to control synaptic function and how disregulation of proteins in these pathways can lead to instance of disease. I will apply the tools of chemical engineering, mathematical modeling, molecular biology, and protein biochemistry that I have learned in my Ph.D. and postdoctoral training to perform multi-scale studies of synaptic regulation – from investigations into protein structure-function relationships, measuring protein binding kinetics and protein activation, modeling signal transduction cascades, to measuring large scale changes in protein regulation in synapses.

See more of this Session: Meet the Faculty Candidate Poster Session
See more of this Group/Topical: Education

(4ec) Biochemical Engineering of Cancer Immunotherapeutics

Susan N. Thomas, Institute of Bioengineering, École Polytechnique Fédérale de Lausanne, Lausanne, Switzerland

Despite the success of modern surgical techniques and chemotherapeutic strategies in treating primary tumors, the insidious process of cancer metastasis remains the most elaborate barrier to overcome in the cure of the disease. With accumulating evidence of inflammation underpinning metastatic progression, the integration of engineering fundamentals with biochemistry, biomaterials and molecular biology as it pertains to the inflammatory mediators of cell migration in immunity and metastasis is essential to cancer therapeutic development.

Cancer metastasis is a highly coordinated process in which cancerous cells "hijack" the homing mechanisms of immune cells in order to invade and colonize distant tissues. Selectins facilitate this process by mediating specific interactions between selectin-expressing immune and vascular cells and selectin ligands presented by tumor cells. My research led to the discovery of new surface adhesion molecules CD44v, carcinoembryonic antigen (CEA), and podocalyxin-like protein (PCLP), which are glycoproteins irregularly expressed by metastatic cells and function as selectin ligands in dynamic shear environment of the vasculature. Furthermore, we characterized and correlated the efficiency of selectin binding activity with their "metastatic-glycan fingerprint". Stable, specific knockdown of CD44v and CEA mediated by siRNA led to a significant reduction in interactions between colon carcinoma cells and immobilized selectins, implicating these molecules in metastases-targeting therapeutic applications.

Contemporary therapies to combat metastases increasingly focus on harnessing the immune system. However, methods to elicit potent adaptive immune responses that are cancer cell-specific yet safe remain elusive. Hence, I am currently applying engineering fundamentals to the development of cancer immunotherapeutics using a nanobiotechnology approach for optimal lymphoid tissue targeting and adjuvancy. First, using a mouse model of impaired fluid drainage and immune cell migration, we establish that delivery of antigen to the lymph node is crucial for efficient humoral but not T cell immunity in response to vaccination as well as to the process of toleragenesis. These data implicate that optimal priming of cancer specific cytotoxic T cells may be achieved via splenic targeting of cancer vaccines while strategies to reverse tumor tolerance should target the lymph nodes. Next, accumulating evidence suggests that complement is an important regulator of adaptive immune responses to vaccination as well as tumor immunity. Harnessing the inherent biochemistry of polymeric nanoparticle surfaces, activation and regulation of complement is achieved such that we might exploit its immunological functions for use in material-based vaccines. Indeed, immunization of mice with OVA-conjugated nanoparticle with surface chemistries that induce efficient complement opsonization enhanced anti-OVA immune responses. Cumulatively, these studies underscore how rational design of targeting strategies and biomaterial properties are fundamental to the development of effective immunotherapeutic strategies.

1. **Thomas, S.N.**, O'Neil, C., van der Vlies, A., Yu, S., Giorgio, T.D., Swartz, M.A., Hubbell, J.A., 2010. Engineering complement activation by polymeric nanoparticles. Submitted to **Biomaterials**.

2. **Thomas, S.N.**, Rutkowski, J., Pasquilier, M., Alitalo, K., E, Randolph, G., Swartz, M.A. 2010. Lymphatic drainage as a regulator of inflammatory and toleragenic immunity. Submitted to **Journal of Experimental Medicine**.

3. Dallas, M., **Thomas, S.N.**, Konstantopoulos, K. 2010. Carcinoembryonic antigen expression in hematogenous metastasis. In preparation.

4. Velutto, D., **Thomas, S.N.**, Simeoni, E., Swartz, M.A., Hubbell, J.A. 2010. Mixed micelles as 30 nm Unimolecular Nonviral DNA Vectors: Design and Demonstration in Tumor Immunotoxicity. In preparation.

5. Konstantopoulos, K., **Thomas, S.N.** 2009. Hematogenous Metastasis: Roles of CD44v and Alternative Sialofucosylated Selectin Ligands. **Advances in Experimental Medicine and Biology: Molecular Immunology of Complex Carbohydrates 3**. In Press.

6. Hubbell, J.A., **Thomas, S.N.**, Swartz, M.A., 2009. Materials engineering for immunomodulation. **Nature**. 462 (7272): 449-460. Epub Nov 26.

7. **Thomas, S.N.**, Tong, Z.Q., Stebe, K., Konstantopoulos, K. 2009. Identification, characterization and utilization of tumor cell selectin ligands in the design of colon cancer diagnostics. **Biorheology**. 46 (3): 207-225.

8. Konstantopoulos, K., **Thomas, S.N.** 2009. Cancer Cells in Transit: The Vascular Interactions of Tumor Cells. **Annual Review of Biomedical Engineering**. 11:177–202. Epub 2009 July 11.

9. **Thomas, S.N.**, Schnaar, R.L., Konstantopoulos, K. 2009. Podocalyxin-like protein is an E-/L-selectin ligand on colon carcinoma cells: Comparative biochemical properties of selectin ligands in host and tumor cells. **Am J Physiol Cell Physiol**. 296: C505-C513. Epub 2008 Dec 31.

10. **Thomas, S.N.**, Zhu, F., Schnaar, R.L., Alves, C.S., Konstantopoulos, K. 2008. Carcinoembryonic antigen and CD44 variant isoforms cooperate to mediate colon carcinoma cell adhesion to E- and L-selectin in shear flow. **Journal of Biological Chemistry**. 283(23): 15647-55. Epub 2008 Mar 27.

11. Alves, C.S., Burdick, M.M., **Thomas, S.N.**, Pawar, P., Konstantopoulos, K. 2008. The Dual Role of CD44 as a Functional P-selectin and Fibrin Counter-receptor. **Am J Physiol Cell Physiol**. 294(4): C907-16. Epub 2008 Jan 30.

12. **Napier, S.L.**, Healy, Z.R., Schnaar, R.L., Konstantopoulos, K. 2007. Selectin ligand expression regulates the initial vascular interactions of colon carcinoma cells: the roles of CD44v and alternate sialofucosylated selectin ligands. **Journal of Biological Chemistry**. 282(6); 3433-3441. Epub 2006 Nov 29. (Noted as recommended reading and a paper of special interest: Taylor, M. E., Drickamer, K. 2007 Current Opinion in Cell Biology 19: 572-577.)

13. Hanley, W., **Napier, S.**, Burdick, M.M., Schnaar, R., Sackstein, R., Kostantopoulos, K. 2006. Variant isoforms of CD44 are P- and L-selectin ligands on colon carcinoma cells. **FASEB J**. 20(2): 337-9. Epub 2005 Dec 13.

14. Ahn, K.C., Jun, A.J., Pawar, P., Jadhav, S., **Napier, S.**, McCarty, O.J., Konstantopoulos, K. 2005. Preferential Binding of Platelets to Monocytes over Neutrophils Under Flow. **Biochem Biophys Res Commun**. 1;329(1): 345-55.

15. Cantor, R.M., Yuan, J., **Napier, S.**, Kono, N., Grossman, J.M., Hahn, B.H., Tsao B.P. 2004. Systemic Lupus Erythematosus Genome Scan: Support for Linkage at 1q23, 2q33, 16q12–13, and 17q21–23 and Novel Evidence at 3p24, 10q23–24, 13q32, and 18q22–23. **Arthritis and Rheumatism**. 50: 3203-3210.

16. Cantor, R.M., de Bruin, T., Kono, N., **Napier, S.**, van Nas, A., Allayee, H., Lusis, A.J. 2004. Quantitative Trait Loci for Apolipoprotein B, Cholesterol, and Triglycerides in Familial Combined Hyperlipidemia Pedigrees. **Arterioscler Thromb Vasc Biol**. 24: 1935-1941.

See more of this Session: Meet the Faculty Candidate Poster Session
See more of this Group/Topical: Education

(4ed) Investigating Case II and Anomalous Penetrant Transport in Glassy Polymers

Adam K. Ekenseair and Nicholas A. Peppas, Department of Chemical Engineering, The University of Texas at Austin, Austin, TX

Glassy polymers are increasingly used in the construction and fabrication of high-tech devices. These devices are often used in applications, such as microfluidics, membrane separations, food packaging, artificial tissues, bones, and organs, and controlled drug delivery, where the polymers come into contact with fluids capable of penetrating the polymer network. To examine the behavior of such systems both on the macroscopic and molecular level, it is important to study the penetrant diffusion process. In such transport, the macromolecular chains rearrange toward new conformations, where the rate of relaxation depends on the penetrant concentration. The relative rates of penetrant diffusion and macromolecular chain relaxation determine the nature of the transport process and lead to a wide variety of penetrant transport phenomena, such as Fickian, anomalous (non-Fickian), and Case II sorption behavior. While previous models account for anomalous behavior, there is still a disconnect between theory and experiment, as data must be fit to the model with previously determined independent parameters. With trends leading to smaller device scales and increasingly complex polymer structures, there is a need for a quantitative understanding of the manner in which a polymer's network structure alters both the rate and the mode of penetrant transport. To this end, poly(methyl methacrylate) (PMMA) was synthesized in an iniferter-mediated radical polymerization process to produce a variety of well-characterized polymer network structures. The effects of basic network parameters, including the degree of crosslinking, polymer mesh size, and the crosslink size, on the transport process were studied. The effects of sub-Tg annealing/aging, temperature, the penetrant hydrodynamic radius, the penetrant-polymer thermodynamic compatibility, and the presence of un-reacted monomer were also investigated. Utilizing both gravimetric sorption data and in-situ ultra-high-resolution X-ray computed tomography studies, the transport dynamics of methanol and other solvents in PMMA discs were investigated. Controlling the relative timescale of the relaxational process by altering the polymer network structure is shown to directly influence the Case II front propagation velocity and the nature of the observed transport behavior.

See more of this Session: Meet the Faculty Candidate Poster Session
See more of this Group/Topical: Education

(4ee) Development of Novel Components for Next-Generation Microfluidic Systems

Minsoung Rhee, Anesthesiology, Brigham and Women's Hospital, Boston, MA

Although microfluidic systems are advantageous platforms for biological assays, their use in the life sciences is often limited by limited functionality of components, the required level of operational knowledge by non-experts, and the level of fabrication expertise required for construction. Here, I present novel microfluidic components with their simplest operational forms and a new concept system in microfluidics. First, a novel assembly approach for microdevice construction using prefabricated microfluidic components is presented (Fig. 1). Assembly blocks act as basic building units to form custom devices. Each Microfluidic Assembly Block (MAB) has a unique function such as inlet/ outlets, valves, straight/curved/bifurcated channels, and chambers. The MAB methodology allows for full flexibility in planar configuration. I have used MABs to prototype a variety of microfluidic devices including basic fluidic channels, a PDMS pneumatic valve to control fluid flow. The MAB system provides a simple way for non-fluidic researchers to rapidly construct custom, complex microfluidic devices. The proposed MAB methodology is thus suitable for constructing a variety of microfluidics devices. Second, I present here microfluidic digital pneumatic circuits that operate upon serially encoded pneumatic signals. Digital pneumatic components we constructed include flip-flops, logic gates, and shift registers and they are integrated to form advanced microprocessors capable of performing complex parallel operations with a single input source. Two distinct pressure levels are used for digital information: atmospheric pressure (value=0) and vacuum (value=1). The fundamental building unit is the asymmetric microfluidic inverter employing two-layered PDMS microvalves (normally closed but opens with vacuum input). With simple modifications of the symmetric inverter, NOR gates and NAND gates were constructed. Furthermore, with these cascadable universal gates along with the inverter, all sorts of logic gates were constructed including AND, OR, XOR, and XNOR. Such pneumatic logic gates comprise the building blocks for advanced circuits including bistable flip-flops, and micropneumatic shift registers (MSR). Serial-to-parallel converters are built by connecting multiple shift registers together. With n shift registers concatenated, n+1 bits of information are available at the parallel outputs. Fig. 2 shows that a 3-bit multiplexer controls individually accessible valves used for multiplexing eight different flow channels. The miniaturization and on-chip integration of control elements has the potential to significantly reduce the complexity of external controllers, interconnections and auxiliary equipment. Proposed digital pneumatic microprocessors can be a potent universal on-chip platform to autonomously manipulate microfluids in a high throughput manner. Third, I present a novel design for efficient 3D hydrodynamic focusing that has a simple and straightforward structure (Fig. 3). Hydrodynamic focusing in a microfluidic channel provides homogeneous reaction conditions that allow for various synthesis applications including polymeric nanoparticle synthesis by nanoprecipitation. Although microfluidic platforms have advantages of rapid mixing and controlled precipitation resulting in homogeneous particles, control of aggregation is a nontrivial issue since polymer particles tend to stick to channel walls and rapidly agglomerate inside, blocking the entire channel. Polymer aggregation can be avoided by 3D hydrodynamic focusing where the polymer stream is constrained both horizontally and vertically. Our system is constructed on a single PDMS layer consisting of three sequential inlets to a conventional 2D hydrodynamic focusing system. The sample flow is first vertically focused by two vertical sheath flows and then horizontally squeezed into a narrow stream flowing in the outlet channel. Computer simulations showed that the size and position of inlet holes play a critical role in determining the shape of hydrodynamic focusing in the channel. When the inlet hole size is comparable to the channel width, the sample stream is successfully constrained in a horizontal band. In case of such complete vertical lamination, two-dimensional analytical models become sufficiently accurate. One can also simply increase the channel height to achieve similar focusing results. Molecules in the vertically focused stream gradually diffuse into both sheath streams, eventually reaching the top and bottom walls. Simulated bottom wall concentrations perfectly matched with the predicted values from two-dimensional mathematical modeling. The performance of 3D focusing was also confirmed experimentally using Poly(lactic-co-glycolic acid) (PLGA) particles known to agglomerate rapidly. Our 3D hydrodynamic focusing can prevent the precipitation at the wall. In addition, confocal micrographs with fluorescent dyes show cross-sectional views with vertical focusing. We have synthesized various polymeric nanoparticles that could neither be synthesized by 2D hydrodynamic focusing nor by bulk mixing in a conventional way. The sample stream containing polymer precursors dissolved in acetonitrile (ACN) becomes vertically squeezed by two vertical ACN sheath streams and then horizontally squeezed by water sheath streams, producing nanoparticles by nanoprecipiation. Finally, a novel single-channel multiple 3D hydrodynamic focusing device for high-throughput systems is presented here. The device consists of two PDMS layers; the bottom layer contains a single channel for multiple focusing and the top layer governs fluidic supply and arrangement, respectively. Our prototypic device can produce 10 multiple focused streams in a parallel manner in a single channel. Each of focused streams can be adjusted to a different flow rate, resulting in a differently squeezed stream. Such multiple parallel focusing is ideal for high-throughput synthesis system where particles are synthesized by diffusive mixing that relies on hydrodynamic focusing. sson one layer and ed PDMS layer consisting of three sequential inlets to a conventional 2D hydrodynamic focusing system. In conclusion, I present here an innovative microfluidic fabrication method, a novel air-operable microfluidic circuit, and new designs for single and multiple 3D hydrodynamic flow focusing systems that have a variety of applications in biological sciences including flow cytometry, drug delivery, drug synthesis, and screening.

Figure 1

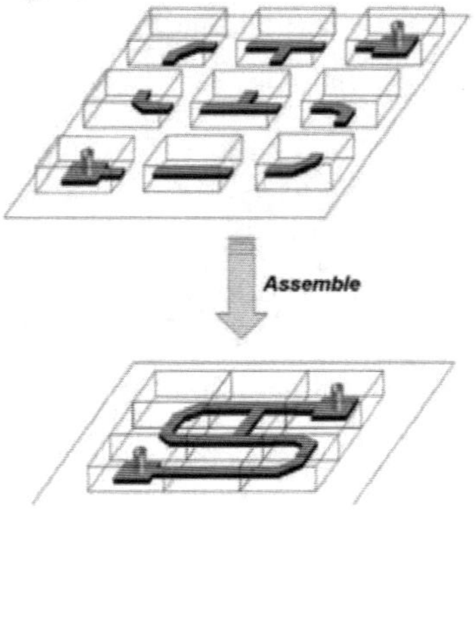

Figure 2

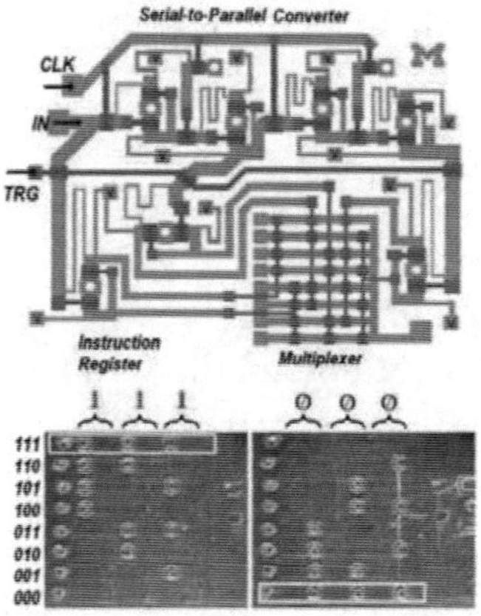

Figure 3

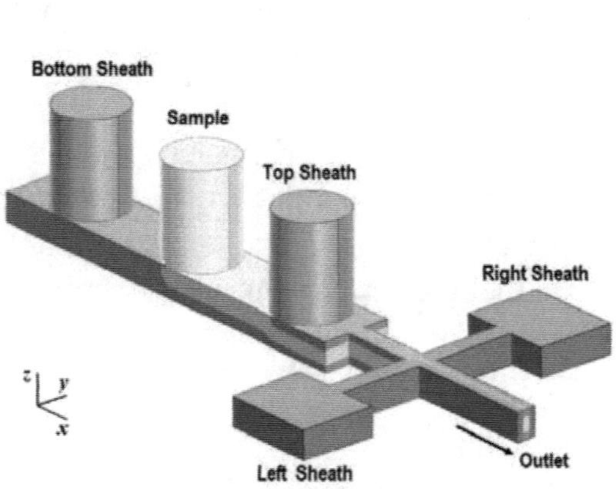

See more of this Session: Meet the Faculty Candidate Poster Session
See more of this Group/Topical: Education

(4ef) Chemical Descriptors and Quantitative Structure Activity Relationships for Catalyst and Materials Design

Thomas A. Manz, School of Chemical and Biomolecular Engineering, Georgia Institute of Technology, Atlanta, GA

My research involves the use of quantum chemistry methods like density functional theory (DFT) to study processes at the atomic level. Several quantitative descriptors like net atomic charges and bond orders are important for understanding all materials and chemical processes at this level. Other descriptors are important to broad classes of materials. For example, the amount of steric hindrance in transition metal complexes can be quantified by ligand cone angle and solid angle descriptors. The net magnetization associated with each atom is an important descriptor for magnetic systems.

In some cases, an adequate method for calculating one of these descriptors is lacking, and this slows the progress of science. For example, in molecular systems reasonably accurate bond orders are given by the Wiberg bond index array in the natural atomic orbital basis, but an analogous method does not yet exist for periodic systems. Current methods for periodic systems are based on projection of the local density of states (or similar techniques) which give bond populations not bond orders. Consequently, the mention of bond orders is ubiquitous in molecular chemistry but almost entirely absent in the chemistry of non-molecular crystals. An extension of the natural bond orbital (NBO) method to periodic systems would remedy this problem. I believe that such an extension should be possible and am interested in pursuing research along these lines. Recently, I have initiated collaboration with one of the developers of the NBO method to investigate this.

Judging from the more than twenty thousand citations to Shannon's seminal paper[1], ionic radius is another important concept to many researchers. However, ionic radii are currently available for only about half of the known oxidation states of the elements. Recently, I developed an accurate and universal method for computing the ionic radius of any element in a negative oxidation state, which is a step towards solving this problem. Unfortunately, a corresponding method for the positive oxidation states is still lacking.

With Professor David Sholl (my post-doctoral advisor) at the Georgia Institute of Technology, I recently developed the Density Derived Electrostatic and Chemical (DDEC) charge method.[2] DDEC is the first method applicable to both periodic and nonperiodic materials that simultaneously optimizes the net atomic charges to reproduce each atom's chemical state and the electrostatic potential outside the electron distribution in such a manner that constraints are not required to treat systems with buried atoms. DDEC should find widespread use as a general purpose method for calculating net atomic charges. With Professor Sholl, I also developed a dimensionless reaction coordinate for quantifying the relative lateness of transition states.[3] (A transition state is said to be late if it resembles the products more than the reactants.) A descriptor of this type is important for applying the Hammond-Leffler, structure sensitivity, and reactant sensitivity postulates that describe correlations between activation energies and system structure.[3]

My research group will use computational chemistry methods like DFT to study reaction mechanisms and to compute chemical descriptors that are combined with available experimental reaction rates to construct quantitative structure activity relationships (QSARs). These QSARs will be used to identify changes in the material's structure that are predicted to improve performance. These materials will then be synthesized and tested by experimental collaborators working in the areas of homogeneous and heterogeneous transition metal catalysis and advanced materials.

Under the direction of Professors James Caruthers and Kendall Thomson, this type of process was used in my doctoral research at Purdue University to design improved olefin polymerization catalysts.[4,5] These catalysts were homogeneous Ti and Zr complexes containing mixed cyclopentadienyl/aryloxide ligation. QSARs were developed for each of the main reaction steps. A new type of catalyst system was designed that exhibited superior performance due to opportunistic bonding of the ligand to the metal.[5] A more complete description of this work can be found in the abstract of my doctoral dissertation.[6]

I am particularly interested in using these methods to study problems in the area of green chemistry. For example, antifreeze is commonly made from ethylene glycol which has a sweet taste and is toxic to animals. Every year this leads to the death of many pets and other animals that ingest spilled antifreeze. Propylene glycol is a nontoxic alternative, but due to its more difficult synthesis is not as widely used. Computational modeling could be used to look for better propylene glycol synthesis routes. I have some specific ideas along these lines which could be pursued further.

References:

(1) Shannon, R. D. "Revised Effective Ionic-Radii and Systematic Studies of Interatomic Distances in Halides and Chalcogenides," Acta Crystallograph. 1976, A32, 751-767.

(2) Manz, T. A.; Sholl, D. S. "Chemically Meaningful Atomic Charges that Reproduce the Electrostatic Potential in Periodic and Nonperiodic Materials," J. Chem. Theory Comput., submitted.

(3) Manz, T. A.; Sholl, D. S. "A Dimensionless Reaction Coordinate for Quantifying the Lateness of Transition States," J. Comput. Chem. 2010, 31, 1528-1541.

(4) Manz, T. A.; Phomphrai, K.; Medvedev, G.; Krishnamurthy, B. B.; Sharma, S.; Haq, J.; Novstrup, K. A.; Thomson, K. T.; Delgass, W. N.; Caruthers, J. M.; Abu-Omar, M. M. "Structure-activity correlation in titanium single-site olefin polymerization catalysts containing mixed cyclopentadienyl/aryloxide ligation," J. Am. Chem. Soc. 2007, 129, 3776-3777.

(5) Manz, T. A.; Sharma, S.; Phomphrai, K.; Novstrup, K. A.; Fenwick, A. E.; Fanwick, P. E.; Medvedev, G. A.; Abu-Omar, M. M.; Delgass, W. N.; Thomson, K. T.; Caruthers, J. M. "Quantitative Effects of Ion Pairing and Sterics on Chain Propagation Kinetics for 1-Hexene Polymerization Catalyzed by Mixed Cp'/ArO Complexes," Organometallics 2008, 27, 5504-5520.

(6) Manz, T. A. "Quantitative Structure Activity Relationships for Olefin Polymerization Catalyzed by Ti and Zr Complexes with Mixed Cyclopentadienyl/Aryloxide Ligation," Ph.D. dissertation, Purdue University, 2009. See my homepage at http://www.prism.gatech.edu/~tmanz3/ for a link to the dissertation abstract.

See more of this Session: Meet the Faculty Candidate Poster Session
See more of this Group/Topical: Education

(4eg) Multiscale Modeling of Biomimetic Nanostructures

Dina T. Mirijanian, The Molecular Foundry / Materials Sciences Division, Lawrence Berkeley National Laboratory, Berkeley, CA

Peptoids are sequence-specific, oligo-*N*-substituted glycine polymers that can mimic the structural motifs and functionalities of proteins. Recently, novel sheet-like nanostructured materials have been self-assembled from peptoids under physiological conditions. These structures are biocompatible and may be selectively functionalized. We have constructed atomistic models of peptoids using high level *ab initio* calculations to guide the parameterization of a classical force field based on the CHARMM22 peptide force field. Atomistic molecular dynamics simulations show the accessible configurations of peptoids in water to be markedly different from those of peptides. We have also used coarse-grained linear polymer models, whose architectures are inspired by those of peptoids, to identify possible pathways of sheet self-assembly.

See more of this Session: Meet the Faculty Candidate Poster Session
See more of this Group/Topical: Education

(4eh) Interrelations Between Morphology and Rheological Properties of Microstructured Polymer Blends During Coarsening

Carlos R. López-Barrón, Chemical Engineering, University of Delaware, Newark, DE

Cocontinuous blends are used in a number of applications, including porous media for filtration, dessicant entrained polymers and substrates for drug delivery devices. A major drawback of these materials is that they are thermodynamically unstable, which implies that their morphology evolves into coarser structures when they are above their glass transition (or melting) temperatures. The mechanisms involved in the coarsening process are not fully understood yet. My research was focused on three aspects of the coarsening process: (1) thorough characterization of the microstructure during coarsening via the implementation of novel 3D imaging techniques, (2) modification of interfacial properties via addition of block copolymers in order to hinder the coarsening, and (3) determination of the connection between morphology and viscoelastic response during the coarsening process. We used laser scanning confocal microscopy (LSCM) to image fluorescently labeled polystyrene (FLPS)/styrene-ran-acrylonitrile copolymer (SAN) blends. A methodology to obtain 3D micrographs of the blends and analyzed them to extract information of the geometry (size and local curvatures), topology (connectivity) and anisotropy (normal vectors) was implemented. Symmetric diblock copolymers (BC) made of polystyrene (PS) and polymethyl methacrylate (PMMA) were used to compatibilize the blends. A dramatic decrease in the rate of coarsening was observed after adding just 1\% of BC. The stabilization was particularly good for BC with an intermediate molecular weight. This result was explained with a theory based on the equilibrium between micelle dissolution and formation and adsorption and desorption of BC from the interface. Small amplitude oscillation measurements were performed on both the compatibilized and the non-compatibilized blends. A noticeable extra contribution to the elastic modulus was observed for all the blends with bigger values for the compatibilized blends. A decrease in the elasticity was observed during the coarsening of the blends. This was explained by the decrease in interfacial area. This decrease was slower for the blends with BC, which evinces stabilization of the microstructure. In an attempt to find a quantitative relation between morphology and rheological response of these materials, a simplification for small deformation to the Doi-Ohta model [Doi, M.; Ohta, T. \textit{J. Chem. Phys.} \textbf{1991}, 95, 1242] was proposed. A key parameter in Doi-Ohta's theory is the degree of anisotropy. During my PhD research I developed a method to compute the anisotropy tensor from the analysis of the 3D images. Predictions from the simplification fit rheological data of blends with low interfacial tension in the late regime of coarsening.

See more of this Session: Meet the Faculty Candidate Poster Session
See more of this Group/Topical: Education

(4ei) Novel Design of Gene/Drug Delivery Systems by Virus-Polymer Hybrid

Kye Il Joo, Chemical Engineering and Materials Science, University of Southern California, Los Angeles, CA

My research focus has been the design and characterization of novel and efficient gene delivery systems by engineering viral vectors to target specific cell types. Currently, I am working on a rational design of virus-polymer hybrid to achieve more stable and enhanced gene/drug delivery system in vitro and in vivo.

(1) Targeted Gene Delivery with Viral Vectors

One of the most challenging problems in virus-mediated gene therapy is how the gene can be specifically delivered into the target tissues or cells. Our lab has developed a novel and efficient strategy that allows the retro/lentiviral vectors to enter specifically into the target cell. This innovative design is based on engineering retro/lentiviral vectors by altering the viral envelope glycoprotein (Env), the protein that is responsible for binding the virus to cell surface receptors and for mediating entry. Compared with conventional methods that manipulate the delicate coupling interactions of the binding and fusion domains of Env glycoprotein, which usually causes decreased viral infectivity, our engineering approach involved the incorporation of a targeting antibody and pH-dependent fusogenic protein as two distinct molecules on the viral surface. A major advantage of this scheme over others where the viral protein is engineered with a foreign binding component is that the fusion protein maintains its full biological activity so that viral titer is not killed for increased specificity. This targeting methodology is flexible and can be extended to other forms of cell type-specific recognition with the different targeting antibody (or other cell-binding protein) to mediate targeting. The flexibility and broadness of this method will facilitate the application of targeted gene delivery for therapy and research.

(2) Development of Quantum dot labeling of gene delivery vectors for visualization of viral transduction pathways

During viral entry, the virus interacts with various cellular structures and takes advantage of their environments to optimize the delivery of the viral genome to the nucleus and promote efficient viral replication. Improved understanding of their interactions can provide crucial insights for preventing virus-triggered diseases, as well as enhancing the efficacy of virus-mediated gene delivery. The ability to track individual viruses is a powerful tool for investigating viral infection routes and characterizing the dynamic interactions between viruses and target cells. As quantum dots have several advantages of remarkable photostability and brightness over conventional organic dyes, they are emerging as a fluorescent probe for biological imaging and medical diagnostics. We have established a general method of labeling enveloped viruses with semiconductor quantum dots (QDs) for use in single virus trafficking studies. Retroviruses, including human immunodeficiency virus (HIV), could be successfully tagged with QDs through the membrane incorporation of a short acceptor peptide (AP) that is susceptible to site-specific biotinylation and attachment of streptavidin-conjugated QDs. We have also demonstrated a strategy for linking adeno-associated virus (AAV) with quantum dots through a coupling reaction. These labeling methods can provide an attractive tool for elucidating intracellular processes of many kinds of viral and non-viral vectors in a detailed manner.

(3) Sustained and Enhanced gene/drug delivery systems by virus-polymer hybrid

Currently, in my postdoctoral research, I am exploring a novel design of virus-polymer hybrid system for enhanced stability and gene transduction efficiency in vivo. The design and evaluation of gene/drug delivery systems for sustained and localized release profiles to target sites of interest is desirable for the prolonged treatment and the precise introduction of therapeutic products, alleviating the concern of an off-targeting effect. Biomaterial scaffolds can provide an environment that facilitate tissue/cell adhesion (mechanical support) and allows drug/gene delivery vectors to be employed for the sustained release in a spatially controlled manner. In this study, we have investigated a strategy based on a porous scaffold of PLGA polymer associated with other immunostimulatory molecules (i.e. cytokines) and foreign substances (antigen) or viral vectors such as adenovirus and lentivirus to recruit host dendritic cells and further educate them in situ for developing immune defenses. In two weeks after implantation, the recruitment of dendritic cells (CD11c+) was observed in the implant. Antigen-specific immune responses with different combination of cytokines and antigens or viral vectors will be addressed in more detail.

References

Kye-Il Joo and Pin Wang. Visualization of targeted transduction by engineered lentiviral vector. Gene Ther. 2008, 15: 1384-96

Kye-Il Joo, Yuning Lei, Chi-Lin Lee, Jonathon Lo, Jiansong Xie, Sarah F. Hamm-Alvarez, and Pin Wang. Site-specific labeling of enveloped viruses with quantum dots for single virus tracking. ACS Nano, 2008, 2: 1553-1562

Kye-Il Joo, April Tai, Chi-Lin Lee, Clement Wong, and Pin Wang. Imaging multiple intermediates of single-virus membrane fusion mediated by distinct fusion proteins. Microsc. Res. Tech. 2010, Accepted.

See more of this Session: Meet the Faculty Candidate Poster Session
See more of this Group/Topical: Education

(4ej) Bio-Inspired Energy Systems and Programmable Materials

Ian Wheeldon, Department of Medicine, Harvard Medical School, Boston, MA

With design through evolution and molecule-by-molecule construction Nature creates materials with complex functions and that are embedded with detailed molecular information. Here, we present two engineering examples of materials that mimic biological functions to produce multi-functional hydrogels for electrode and surface modifications and hydrogels with programmed biological signaling. 1) Using a general method of enzymatic and multi-functional hydrogel design we create biocatalytic electrodes and oxidative enzymatic hydrogels for the potential use in alternative energy systems. The hydrogels are created with a 'bottom-up' approach through the self-assembly of chimeric fusion proteins. Self-assembly functionality is achieved by appending leucine zipper domains to the termini of the desired functional proteins, and compatible crosslinks allow for the co-assembly of two or more different building blocks into multi-functional hydrogels. The physical properties and enzymatic activities of the hydrogels are tunable, and we demonstrate control of the intra-gel spacing between functional building blocks. 2) With a biomaterials screen we reveal extracellular matrix (ECM) compositions that can be used to program hydrogels to promote epithelial-to-mesenchyme transition in progenitor cells. A microarrays of nano-liter sized hyaluronic acid (HA)-based hydrogels containing ECM components identified from the proteomic analysis of the developing atrioventricular (AV) cushion were screened for endothelial progenitor cell (EPC) adhesion and the expression of EMT markers. Confirmation of the screening results in the macroscale reveals functional EMT of transformed EPCs on HA-fibronectin hydrogels. These two bio-inspired engineering examples are strong answers to the question: how does Nature create complex functional materials and how can we learn from these examples to inspire new energy systems and programmable materials?

See more of this Session: Meet the Faculty Candidate Poster Session
See more of this Group/Topical: Education

(4f) Zwitterionic/Mixed Charge Polymers as Next-Generation Biomaterials

Zhiqiang Cao, Department of Chemical Engineering, University of Washington, Seattle, WA and Shaoyi Jiang, Chemical Engineering, University of Washington, Seattle, WA

We have demonstrated a class of "stealth" zwitterionic materials that are highly resistant to nonspecific protein adsorption, bacterial adhesion, and biofilm formation. Among zwitterionic materials, poly (carboxybetaine) (PCB) can be uniquely functionalized by biomolecules. PCB-coated nanoparticles (NPs) are ideal for targeted therapeutics and diagnostics with long circulation time. Cationic ester derivatives of PCB are developed for gene delivery with high efficacy and low toxicity. Those derivatives also enable the attachment of super-hydrophilic PCB to hydrophobic moieties, resulting in amphiphilic polymers with the sharp hydrophobic/hydrophilic contrast between two moieties. Bio-degradable NPs and liposomes assembled from those amphiphilic polymers exhibit unusual properties that are not found in any current NP systems including polyethylene glycol (PEG) modified NPs. Similar to zwitterionic materials, it has been demonstrated that mixed-charge materials such as natural peptides with alternating charges are also "stealthy".

See more of this Session: Meet the Faculty Candidate Poster Session
See more of this Group/Topical: Education

(4g) Novel Biomaterials for Immunotherapy Applications

Samantha A. Meenach, College of Pharmacy, Ohio State University, Columbus, OH

Acetalted dextran nanoparticles have been investigated for their use in various biomedical applications, specifically for the treatment of autoimmune diseases such as multiple sclerosis or celiac disease. The nanoparticulate materials would be involved in an immunotherapy application by inducing and/or enhancing an immune response in the body to help treat the disease. In addition to autoimmune diseases, these materials are being investigated for their ability to treat cancer via immune response mechanisms. The potential for using these materials in this applications will be outline in the poster.

See more of this Session: Meet the Faculty Candidate Poster Session
See more of this Group/Topical: Education

(4h) Designing Catalysts for Energy Applications Using Insights From Molecular Simulation

Rachel B. Getman, Department of Chemical and Biological Engineering, Northwestern University, Evanston, IL

Effective catalysts are needed to promote many essential processes, including fuel production, fuel utilization, and emissions control. Even seemingly simple reactions can exhibit complex catalytic behaviors, and chemical engineers must understand these complexities in order to design catalytic systems that optimize performance. Here I discuss prior theoretical research on NO_x remediation and hydrogen storage and propose new research on biomass to fuel conversion and fuel cell design. I discuss quantum chemical and classical simulation methods for analyzing catalytic phenomena under realistic reaction conditions. Finally, I introduce a "whole catalyst system design" method, whereby the results are used to synergistically design the catalyst system, including material, support, and structural features in order to optimize catalytic efficiencies. The ultimate goal of this research is to develop criteria for catalyst designs based on reactant and product properties and reaction conditions. The methods are highly versatile and applicable to a wide variety of other problems as well.

See more of this Session: Meet the Faculty Candidate Poster Session
See more of this Group/Topical: Education

(4i) Designing Bio-Nano Materials through Modeling and Simulation

Meenakshi Dutt, Chemical and Petroleum Engineering, University of Pittsburgh, Pittsburgh, PA

In recent years, the fast emerging field of bio-nano/bio-hybrid materials is seen to have great promise in diverse applications such as photovoltaics, bio-sensors and drug delivery vehicles. Creation of these materials from synthetic/ pre-assembled protein nanotube (or particles), with lipid, or polymer membranes, require approaches such as self-assembly and insertion of the synthetic moeity into the membrane. The transport selectivity is introduced by suitably functionalizing the nanotube. The nanotube is embedded in a low bending stiffness fluid environment such as a membrane. These approaches require a fundamental understanding of the interfacial properties resulting from the systems chemistry, and the mesoscopic properties of the membrane. Another class of bio-hybrid materials are granules or compacted blocks of bioactive powders which can be used as spacefillers and implant coatings. The properties of these materials depend upon the particle size, hardness and surface properties. In the poster, I will discuss the design and characterization of these materials.

See more of this Session: Meet the Faculty Candidate Poster Session
See more of this Group/Topical: Education

(4j) Directed Catalytic Materials Design through a Combined Theoretical and Experimental Approach

Siris Laursen, Department of Chemical Engineering, University of Michigan, Ann Arbor, MI

The bulk of chemical reactions that concern society today center about the production and consumption of energy or the synthesis of consumer product precursor chemicals. The production/operation of many of these chemical reactions would simply not be energetically possible or economically viable unless facilitated or accelerated by a catalytic material. The aim of my previous and future research projects is to couple newly available state-of-the-art theoretical calculations with tried-and-trued experimental techniques to further the fundamental understanding of catalytic materials, and develop new low temperature energy and atom efficient materials. Three examples of how this technique has been fruitful in producing novel catalytic materials will be touched upon:

i) The Unprecedented Room Temperature Catalytic Activity of the Most Noble Metal of the Periodic Table, Gold: Through well-defined theoretical and experimental studies we were able to isolate the elusive highly active catalytic Au sites on the Au/TiO2 catalyst surface, address the fundamental question whether cationic Au could be catalytically active, and rationally design catalysts with superior catalytic activity at room temperature and temporal stability.

ii) Developing Catalytic Materials that Exhibit Exceptional Catalytic Selectivity Through Theory-driven Experimental Oxide Nano-particle Design: The selectivity of a complex organic synthesis, catalyzed by oxide nano-particles, was rationally tuned using molecular insights obtained from quantum chemical calculation. Additionally, light was shed upon the role of oxide surface oxygen vacancies, a popular fundamental question that is still being researched.

iii) Mechanistic Understanding of Heterogeneously Catalyzed Carbon-Carbon Bond Formation in Fine Chemical Synthesis: Fundamental insight into the connection between homogenous and heterogeneous catalytic C-C bond formation enabled the rational design of a highly active and selective heterogeneous catalyst.

See more of this Session: Meet the Faculty Candidate Poster Session
See more of this Group/Topical: Education

(4k) Advanced Membrane Materials for Refinery Separations by Pervaporation

Claudio P. Ribeiro Jr. and Benny D. Freeman, Department of Chemical Engineering, University of Texas at Austin, Austin, TX

Refinery streams can contain a mixture of aromatic and aliphatic hydrocarbons, with volumes that can reach 1600-8000 $m^{3,}$/day. Many of these mixtures are difficult to separate because they contain close boiling compounds that generally have similar physical and chemical properties with respect to each other. The conventional technologies to perform these separations, such as extractive distillation, azeotropic distillation and liquid-liquid extraction, are energy intensive and expensive. Membrane processes, being more economical, energy-saving and ecologically friendly, constitute a promising alternative to conventional separation technologies. Pervaporation, in particular, has been intensively investigated for the separation of azeotropic, close-boiling and isomeric mixtures.

Aromatic polyimides are well known for their excellent thermal stability, mechanical strength and good chemical resistance. Thus, these polymers have attracted considerable interest as membrane materials for organic-organic pervaporation. Nevertheless, a suitable polymeric membrane material to enable commercial application of this technology in refineries is yet to emerge. The challenge is to develop a membrane with suitable transport properties that, at the same time, can withstand the aggressive feed streams and harsh operating conditions involved.

In this contribution, we report a structure/property study of a series of aromatic polyimides for the separation of toluene/n-heptane and benzene/n-heptane mixtures by pervaporation. Homo- and copolyimides were synthesized by the two-step polycondensation of three aromatic dianhydrides and four diamines (two containing a hydroxyl group relative to the amino moiety, one containing a fluorine moiety, and one containing methyl groups in the aromatic ring). Pervaporation experiments were performed at 80°C with feed streams containing 40 wt% aromatics. Pure-liquid sorption experiments at 25°C were also conducted to investigate the solubility selectivity of each polymer. The long-term chemical stability of the polymers was assessed in exposure tests to benzene in a Soxhlet extractor for at least a week.

Contrary to most studies on pervaporation conducted so far, the individual contributions of the membrane material and the driving force to the permeate flux were properly evaluated. It was observed that changes in the chemical structure of the polyimide led to an increase of up to 3 orders of magnitude in the permeability of the hydrocarbons.

See more of this Session: Meet the Faculty Candidate Poster Session
See more of this Group/Topical: Education

(4I) Sustainable Biomass Feedstock Production for Bioenergy: Is It Possible and How Will It Happen?

Yogendra Shastri, Department of Agricultural and Biological Engineering, University of Illinois at Urbana-Champaign, Energy Biosciences Institute, Urbana, IL

An efficient and sustainable biomass feedstock production system is critical for the success of the burgeoning biorefinery industry. There are two major challenges in our path towards achieving this goal. First, the techno-economic feasibility of biomass production and provision must be evaluated. The existing technological barriers, beneficial management strategies and optimal operating practices must be identified in the context of economic and environmental criteria. Second, implementation of these optimal practices in real world must be encouraged and ensured. This includes understanding the development of the biomass market in the presence of multiple stakeholders that must participate and yet compete for a fraction of the profit. This work will showcase research conducted towards achieving both these objectives. BioFeed is an optimization model that optimizes the design and management level decisions for feedstock production and provision. It has been used here to evaluate the techno-economic feasibility of Miscanthus and Energy cane as feedstock. Results for Miscanthus show that throughput capacities of various pre-processing options are the bottlenecks. Moreover, centralized processing facilities shared by farms are beneficial to ensure a uniform biomass form and quality at the refinery gate. In order to study the biomass market development, this work develops an agent-based simulation model. Each class of stakeholder within the system, such as farmers, refinery, transport companies, storage elevators, farm consultants and equipment renters, are modeled as a set of agents. The behavior and decision making of each class of agents is modeled with a set of rules, which are based not only on the economic rationale such as profit maximization, but also on unique mental models and cognitive processes such as risk aversion. The role of the government, assumed to be very important to stimulate growth in this field, is also modeled in the form of policies. The model will be used to carry out simulation studies over multiple years to predict different possible development paths for the biomass feedstock market.

See more of this Session: Meet the Faculty Candidate Poster Session
See more of this Group/Topical: Education

(4m) Advanced Dynamic Optimization and Control for Large-Scale Systems

Rui Huang, Chemical Engineering, Carnegie Mellon University, Pittsburgh, PA

Air Separation Unit (ASU) is an energy intensive process. It is economically attractive to apply advanced control techniques to the ASU. Current advances in dynamic optimization algorithms enable us to consider more complicated model and to improve controller's performance. This work addresses the following problems:

(1) Reducing computational delay: Implementing nonlinear model predictive control (NMPC) based on first principle dynamic model exhibit many advantages against traditional linear MPC based on empirical model. However, computational delay associated with solving large-scale optimization problem degrade the performance of the controller. Advanced step NMPC algorithm [1] is utilized to implement NMPC on the ASU problem and the computational delay is reduced by 200 times. (2) Offset-free in the presence of plant model mismatch: Offset-free behavior which requires the output of the process to be exactly at the set point is an important requirement in control practice. An NMPC framework based on various state estimators is proposed and implemented on the ASU to achieve offset free even in the presence of plant model mismatch. (3) Robust NMPC: It is important to guarantee robust stability of NMPC in the presence of large uncertainties. An multi-scenario formulation is proposed to guarantee the robust stability of NMPC algorithm. It is implemented on the ASU process. (4) Economic-oriented control: With the first-principle dynamic model, it is preferred to directly solved the NMPC problem to maximize the economic profit of the process. An infinite-horizon formulation with guarantee stability is proposed to solve the economic-objective control problem for the ASU.

It is worth pointing out that the proposed methods are not restricted to the ASU problem. On the contrary, they are expected to work for any large-scale processes.

Reference: [1] V. M. Zavala and L.T. Biegler. The advanced step NMC controller, optimality, stability and robustness. Automatica 2009, 45: 86.

See more of this Session: Meet the Faculty Candidate Poster Session
See more of this Group/Topical: Education

(4n) Lasers, Electrons and Nanotubes: A Roadmap to Affordable Solar Technology

Christiaan Richter, Department of Chemistry, Yale University, New Haven, CT

There is currently a world wide effort to make use of advances in nanotechnology to design next generation solar cells that could make economical solar energy a reality. One strategy pursued actively is using nanowires, nanotubes or nanorods in solar cells as efficient conduits for electron transport and extraction. However, these 'electron highways' have not always lived up to expectation. I use ultrafast far-infrared laser pulses to directly probe free carriers in nanomaterials on femtosecond time scales. Some of my most recent work revealed why TiO_2 nanotubes, used to improve electron transport in dye sensitized solar cells, are not yet significantly outperforming nanoparticle films and how they potentially could. Far-infrared laser spectroscopy could also supply unique information addressing core challenges facing several other promising renewable energy technologies. Some of these future directions will be discussed briefly.

See more of this Session: Meet the Faculty Candidate Poster Session
See more of this Group/Topical: Education

(4o) Chemical Modification of Stem Cell Membranes for Targeted Delivery of Cells and Therapeutics

Hao Cheng[1], Marta Byrska[2], Christian J. Kastrup[2], Robert Langer[1] and Daniel G. Anderson[2], (1) Department of Chemical Engineering and The David H. Koch Institute for Integrative Cancer Research, Massachusetts Institute of Technology, Cambridge, MA, (2)The David H. Koch Institute for Integrative Cancer Research, Massachusetts Institute of Technology, Cambridge, MA

Stem cell based therapies hold great promise for curing many diseases. Certain types of stem cells are known to have an innate ability to migrate to tumors. The cell membrane of stem cells is important and allows stem cells to recognize their microenvironment and receive signals. Here we describe a general technique to conjugate ligands on the cell membrane to enhance the homing ability of cells to areas of inflammation, which are associated with many diseases. Anchoring nanoparticles to these cells is a promising modality to target the delivery of therapeutics to tumors. We show that the nanoparticles can remain for up to 2 days on the cell membrane. Human bone marrow derived mesenchymal stem cells with anchored nanoparticles were able to retain their inherent tumoritropic properties, as shown using a tumor model in a 3D extracellular matrix. Our results provide a novel strategy to actively deliver cells and therapeutics to areas of inflammation via chemical modification of the cell membranes.

See more of this Session: Meet the Faculty Candidate Poster Session
See more of this Group/Topical: Education

(4p) Dynamic Relaxation Characteristics of Polymer Nanocomposites and Aromatic Polyimides

Anthony C. Comer, Chemical and Materials Engineering, University of Kentucky, Lexington, KY

Interest in polymer nanocomposite technology has grown because of the enhanced properties that can be achieved in the nanocomposite over those of the pure polymer. Particle-polymer interactions, as well as physical confinement, can lead to changes in polymer chain dynamics that are manifested in key performance properties, creating opportunities for new and unique material applications. In this work, the dynamic relaxation characteristics of both glassy (amorphous) and rubbery polymer nanocomposites have been examined; matrix polymers included poly (etherimide), poly (methyl methacrylate), and cross-linked poly (ethylene oxide) in combination with native and surface-modified metal oxide nanoparticles. Key material variables included nanoparticle size, surface chemistry, and loading. For nanocomposite formulations with favorable polymer-particle interactions, the inclusion of moderate to high levels of nanoparticles led to the emergence of a dual-T_g response that encompassed a bulk T_g event (matching T_g for the unfilled polymer), and a second elevated T_g corresponding to relaxations involving chain segments constrained in the vicinity of the polymer-particle interface. The combination of dynamic mechanical and dielectric methods provided a comprehensive view as to the influence of the particles both with respect to local sub-glass motions, as well as the larger-scale processes inherent to the glass transition.

A new class of thermally-modified aromatic polyimides (API's) has recently been identified as a promising material for membrane design. These membranes are based on soluble aromatic polyimides with *ortho*-positioned functional groups; exposure of the polyimides to thermal rearrangement at elevated temperature leads to fully-aromatic, insoluble polybenzoxazoles (PBO) with exceptional thermal and chemical resistance characteristics. The thermal conversion step produces fundamental changes in molecular connectivity and conformation that govern the distribution of free volume elements, and the ultimate transport properties of the membranes. A comprehensive study has been undertaken to examine the dynamic relaxation properties of the API's as a function of backbone structure and degree of thermal rearrangement, leading to insight as to the relative flexibility of the polymers, their local relaxation environment and corresponding free volume, and the influence of thermal rearrangement on segmental mobility. These results can be correlated with the static properties of the polymers and their separation characteristics for both gas separations and pervaporation applications.

See more of this Session: Meet the Faculty Candidate Poster Session
See more of this Group/Topical: Education

(4q) Nanoparticle-Based Thin Films for Tribology Control in MEMS

Kendall M. Hurst, Christopher B. Roberts and W. Robert Ashurst, Department of Chemical Engineering, Auburn University, Auburn, AL

Due to inherent interfacial forces that cause strong adhesion and friction in microelectromechanical systems (MEMS), only the most mechanically simple MEMS devices are currently used commercially. In the past, several approaches have been examined to overcome the inherent forces leading to stiction and friction. Chemical surface alterations have been extensively studied in order to reduce the surface energy, increase hydrophobicity, and reduce unwanted adhesion. Another approach has been to create surface roughness *in situ* during the microfabrication process and effectively reduce the real area of contact between contacting components on a MEMS device. Both methods have been proven to successfully reduce microstructure adhesion, however, some issues with film durability still remain.

This work utilized a novel CO_2-expanded liquid nanoparticle deposition process to conformally coat silicon surfaces and MEMS surfaces with rough nanoparticle coatings. Various liquid- and vapor-based film deposition techniques were also used to couple nanoparticle-based rough surfaces with low surface energy thin films, effectively reducing the surface energy even further. A unique water erosion technique was used to examine the durability of these nanoparticle-based composite thin films, while tribological devices and interferometric techniques were incorporated to examine the effect of the thin films on microstructure adhesion in MEMS. The results conclude that rough nanoparticle surfaces do indeed provide relief to the strong adhesion exhibited in MEMS devices.

See more of this Session: Meet the Faculty Candidate Poster Session
See more of this Group/Topical: Education

(4r) Multiscale Modeling of Biophysical and Biochemical Aspects of Viral Life Cycles

Eric R. May, Chemistry & Biophysics, University of Michigan, Ann Arbor, MI

Understanding viral life cycles is essential for the development of antiviral therapies, but also for the development of nanotechnologies based upon viral capsids and protein cages. My current and previous research have laid a foundation for understanding key components of viral systems (single viral proteins, virus capsids, membranes), but moving forward I want to integrate these components into a more holistic understanding of viral processes. I am interested in determining material properties of biological systems and understanding how these properties affect the phase and shape behavior of the system. As well as studying the coarse properties of these systems, I also want to understand the specific (local) interactions that play a crucial role in imparting bulk characteristics to the system. Identification of these "key" interactions can lead to testable hypotheses (i.e. mutagenesis studies) about the significance of the identified interactions. This information can lead to targeted therapies, design principles and control mechanisms (i.e. pH triggered uncoating) for nanotechnologies. In this presentation I shall cover results related to membrane and virus material properties, viral protein structural predictions, protein-protein interaction predictions and free energy calculations on transition pathways. Taken individually, these different studies each probe an important aspect of viral systems and many of the methods employed extended across multiple spatial and temporal scales. Taken as a whole, these different components and methodologies provide a multiscale approach for studying the complex processes associated with viral life cycles.

See more of this Session: Meet the Faculty Candidate Poster Session
See more of this Group/Topical: Education

(4s) Development and Characterization of Remendable Polymer Composites Using the Diels-Alder Reaction

Amy M. Peterson, Department of Chemical & Biological Engineering, Drexel University, Philadelphia, PA

Materials that can repair cracks and recover from mechanical failure are desirable. Because remendable materials both repair and prevent the propagation of cracks on the micro scale, they offer the potential for increased durability, safety, and cost efficiency for many applications. The focus of my work is to understand the kinetic and thermodynamic parameters that control thermoreversible Diels-Alder bond formation in different types of healing polymeric systems. Such studies are necessary to optimize the healing ability of Diels-Alder-based systems and also contribute to the understanding of networks containing equilibrium controlled bonds. I have developed three healing systems for epoxy-amine thermosets based on the thermoreversible Diels-Alder reaction of furan and maleimide. In one, crack healing of a traditional epoxy-amine thermoset is induced by thermally reversible cross-linking of a secondary phase. In another, a furan-functionalized epoxy-amine thermoset can be healed with a bismaleimide solution at room temperature and minimal pressure and significant load recovery is possible multiple times in a given location. The third system allows for interfacial healing of glass fiber-reinforced epoxy-amine composites via compatible functionalization of glass fibers and the polymer network. Although the Diels-Alder reaction was discovered almost 70 years ago, there has not been an exhaustive study of reaction kinetics and thermodynamic parameters for furan and maleimide. Kinetic parameters have been determined for a number of model compound systems. Preliminary work in characterizing the reaction kinetics of a reversibly cross-linking system and relating mechanical properties to the Diels-Alder reaction will be presented as well.

See more of this Session: Meet the Faculty Candidate Poster Session
See more of this Group/Topical: Education

(4t) Systems and Synthetic Biology in Bacterial and Human T-Cells

Wilson W. Wong[1], James C. Liao[2] and Wendell Lim[1], (1)Cellular and Molecular Pharmacology, University of California, San Francisco, San Francisco, CA, (2)Chemical & Biomolecular Engineering, University of California, Los Angeles, Los Angeles, CA

The main focus of my dissertation work is to uncover operating principles in Escherichia coli metabolism and regulation. Due to the highly complex nature of biological systems, fundamental principles are often masked by secondary effects (1-2). To uncover various design principles, different research approaches were taken: systems and synthetic.

Using the systems approach, I uncovered a relationship where the growth rate is less than or equal to the square-root of the product among the biomass yield, the protein synthesis rate of the carbon source transporter and its turnover number. Through theoretical analysis, this relationship appears to be an evolutionary optimal. I further demonstrated that adaptively evolved strain, knockout mutants, and ethanol-challenged strains also follow the square-root relationship. Moreover, this relationship provides an explanation as to why the maximal growth rate does not occur at the maximal yield, and a design principle for engineering product formation strains (3-4).

Through the synthetic approach, a gene-metabolic oscillator, termed "metabolator", was designed and constructed. Autonomous oscillations found in neuronal, cardiac, metabolic, and gene expression systems have attracted significant attention because of their biological significance and their intriguing dynamics. We constructed a synthetic gene-metabolic oscillator in Escherichia coli K12 using the glycolytic flux to drive transcriptional and metabolic oscillation through a signaling metabolite, acetyl phosphate. We showed that metabolic oscillation can be generated from metabolic flux imbalance (5-7).

Furthermore, single-cell characterization of the proteolysis system used in the metabolator showed zeroth-order kinetics. In addition, we showed that zeroth-order degradation kinetics, coupled with long-tailed initial protein distribution, will generate first-order population degradation kinetics. Through simulation, I demonstrated that zeroth-order kinetics can significantly enhance the robustness of the metabolator. This result highlights the importance of single-cell measurements in determining the dynamics of a process (8).

Currently at UCSF, I am employing the systems and synthetic approach to understand human T-cells signaling and create T-cells with novel function for therapeutic purpose. Adoptive transfer of T-cells expressing chimeric antigen receptors (a fusion of tumor specific single-chain antibodies and intracellular signaling domains from the T-cell receptor (TCR) signaling pathway) has shown promise as an anti-tumor therapy. However, their efficacy is limited because of limited in vivo activity, T-cell exhaustion after ex vivo expansion, low persistence inside the host, and potential safety concerns. In the Cell Propulsion Lab, we are applying synthetic biology approaches to engineer and optimize the behaviors of these tumor targeted T-cells. By engineering synthetic positive and negative feedback loops in T-cells, we show that we can precisely tune the amplitude and dynamics of the T-cell response. We have demonstrated that expression of bacterial effector proteins can be used to shut down and attenuate the response. Conversely, we have also shown that we can amplify output by expressing fusions of signaling proteins that mimic an activated signaling complex. The modularity of these circuits should allow us to construct and search for altered response profiles that improve the persistence and anti-tumor activity of engineered T-cells. Our approaches complement ongoing work on chimeric antigen receptor engineering. Furthermore, this work has yielded a new set of genetic tools to control the T-cell signaling dynamics.

1 Wong, W.W., Liao, J.C. The Synthetic Approach for Regulatory Circuits and Metabolism. (Systems Biology and Synthetic Biology, John Wiley & Sons)

2 Wong, W.W., Liao, J.C. A Synthetic Approach to Transcriptional Regulatory Engineering. (In Press, The Metabolic Pathway Engineering Handbook: Fundamentals, CRC Press)

3 Wong, W.W., Tran, L.M., Liao, J.C. 2009 A hidden square-root boundary between growth rate and biomass yield. Biotechnology and Bioengineering. 1(102): 73-80.

4 Wong, W.W., Tran, L.M., Liao, J.C. 2009 Microbial maximal specific growth rate as a square-root function of biomass yield and two kinetic parameters. Metabolic Engineering 11(6):409-14

5 Fung, E., Wong, W.W., Suen, J.K., Bulter, T., Lee, S., Liao, J.C. 2005. A synthetic gene-metabolic oscillator. Nature. 435(5): 118-122.

6 Bulter, T., Lee, S., Wong, W.W., Fung, E., Connor, M. and Liao, J.C. 2004. Design of artificial cell-cell communication for gene and metabolic circuits. Proceedings of the National Academy of Science, USA. 101(8): 2299-2304.

7 Wong, W.W., Liao, J.C. 2006. The design of intracellular oscillators that interact with metabolism. Cellular and Molecular Life Science. 63(11): 1215-1220.

8 Wong, W.W., Tsai T.Y., Liao, J.C. 2007. Single-cell zeroth-order protein degradation enhances the robustness of synthetic oscillator. Molecular Systems Biology. 3(130): 1-8.

9. Xie, X., Wong, W.W., Tang, Y. 2007. "Improving simvastatin bioconversion in Escherichia coli by deletion of bioH. Metabolic Engineering. 9(4): 379-386. 10. Wong, W.W., and Newman, J.S. 2002. Monte Carlo Simulation of the Open-Circuit Potential and the Entropy of Reaction in Lithium Manganese Oxide. Journal of the Electrochemical Society. 149(4):A493-A498.

11. Wang, J., Chun, H.J., Wong, W.W., Spencer, D.M., and Lenardo, M.J. 2001. Caspase-10 is an initiator caspase in death receptor signaling. Proceedings of the National Academy of Science, USA. 98 (24): 13884-13888.

12. Brynlidsen, M.P., Wong, W.W., Liao, J.C. 2005. Transcriptional regulation and metabolism. Biochemical Society Transactions. 33(6): 1423-1426.

See more of this Session: Meet the Faculty Candidate Poster Session
See more of this Group/Topical: Education

(4u) Electronic Tools for Student Engagement in Introductory Engineering

Joshua A. Enszer, Chemical and Biomolecular Engineering, University of Notre Dame, Notre Dame, IN

The primary goal of our undergraduate program is to produce engineers who are one step ahead of their peers, who have begun to prepare themselves for more than just their entry-level jobs. In order to accomplish this, we seek improvements to the educational process outside the classroom and ways to encourage students to have a more direct role in their own personal development. I present here overviews of two electronic projects designed to meet this goal.

An electronic portfolio is a collection of electronic files that is created and maintained by a user as evidence of professional development. As a portion of the academic advising experience, the Notre Dame e-Portfolio (NDeP) project [1,2] includes tools to document academic and professional progress in a way that allows for organization and easy presentation. The e-portfolio serves as a tool to make communication between students and advisers more efficient, and to help students formulate short- and long-term plans for their own career development. The NDeP project makes use of Google Apps to help students create their own e-portfolio site and manage their electronic documentation. This includes a set of surveys designed to help students consider their academic coursework, in which we also have students assess their own progress in the program. The survey addresses the six main tools every engineer in our program should have: analytical and problem solving skills, comprehension of engineering concepts, understanding of technical problems, communication skills, creativity and independence, and organizational skills. For each tool, students are asked to consider what deliverable proof exists that demonstrates their current proficiency. These artifacts can be electronically archived on their own NDeP site.

Educational literature on video games has discussed the idea that good games can promote beneficial learning opportunities [3]. Students are engaged in learning in such an environment, which promotes ideas of identity (a required personal commitment to achieve a goal), interaction (such as "just-in-time learning" through the game interface), production (control over the outcome of a scenario), and risk taking (but without the danger of failing a course!). I present here prototypical screenshots from a chemical engineering "game" that allows students to explore the basics of material and energy balances. The game is designed to increase in complexity and freedom through a series of levels that gradually drop assumptions and restrictions on models including systems of storage tanks, mixers, separators, and reactors.

I present here the current progress and future goals of the NDeP program, including information from the pilot group of students and plans to improve the implementation for this year's incoming students, making the process more interactive for both students and faculty. I also present the current and future work on the chemical engineering game, including plans to add material for higher level chemical engineering courses and making connections to common industrial simulation software.

[1] J. A. Kuczenski, J. A. Enszer, M. J. McCready, J. B. Brockman, X. Duan, and P. Turner. "Student Electronic Portfolios for Professional Development Using Google Apps." Presented at the 2009 AIChE Annual Meeting, Nashville, TN, 8-13 Nov 2009.

[2] J. A. Kuczenski, J. A. Enszer, M. J. McCready, and J. B. Brockman. "Student Electronic Portfolios for Professional Development Using Google Apps." Presented at the 2010 ASEE Annual Conference & Exposition, Louisville, KY, 20-23 June 2010.

[3] J. P. Gee. Good Video Games and Good Learning: Collected Essays on Video Games, Learning and Literacy (New Literacies and Digital Epistemologies). New York: Peter Lang Publishing, Inc. 2007.

See more of this Session: Meet the Faculty Candidate Poster Session
See more of this Group/Topical: Education

(4v) Evaluating Zero Valent Iron Nanoparticles with Acoustic Resonance Microbalance Techniques

Lauren F. Greenlee, Materials Reliability Division, National Institute of Standards and Technology, Boulder, CO

Zero valent iron (ZVI) nanoparticles are able to catalyze both oxidation and reduction reactions and therefore are of particular interest for the degradation of organic micropollutants during drinking water treatment. Many organic water contaminants have been recently identified in drinking water sources and in treated wastewater effluent. The persistence and toxicity of these compounds have motivated research in developing novel technologies and strategies to remove the compounds during drinking water and wastewater treatment. Using nanoparticles such as ZVI particles to remove micropollutants takes advantage of the large surface area to volume ratio of the particles and the resulting enhancement of surface properties that increases contaminant adsorption and/or degradation.

This research focuses on the design of bimetallic ZVI nanoparticles for the reduction of halogenated disinfection by-products (DBPs) and the use of acoustic resonance microbalance techniques to analyze the reactivity and lifetime of the particles. Acoustic resonance microbalances have sensitive mass detection and can detect mass changes of nanograms to picograms. A quartz crystal microbalance with dissipation detection (QCM-D) is used to analyze ZVI nanoparticles in real time under different experimental conditions, including water composition, pH, and organic contaminant type. ZVI nanoparticles coated with a secondary transition metal are compared; the catalytic lifetime of the particles is studied as a function of ZVI oxidation. A heated QCM technique (0 – 400 ^{o}C) is developed and used to analyze the extent of organic compound adsorption onto the particles, and thermogravimetric analysis (TGA) (0 – 900 ^{o}C) is used to validate results.

See more of this Session: Meet the Faculty Candidate Poster Session
See more of this Group/Topical: Education

(4w) Thermochemical Water Splitting Using Ionic Liquid Solvents

Nicholas AuYeung, Alex Yokochi, Victoria Johnson and Sardar Sardari, Chemical, Biological and Environmental Engineering, Oregon State University, Corvallis, OR

Thermochemical Water Splitting Using Ionic Liquid Solvents

Thermochemical water splitting routes such as the Sulfur-Iodine cycle may be greatly improved via the use of ionic liquids as reaction media. In order to enable large scale implementation of this cycle, approaches that simplify product stream management following the initial Bunsen reaction are particularly important, especially with regards to the management of excess water and excess iodine. Using ionic liquids as reaction media allows large amounts of I_2 and SO_2 to be dissolved while limiting the amount of water. Beyond examining the kinetics of the Bunsen reaction, we identified conditions under which the reaction system may be biased to intentionally produce a stream of H_2S based on a normally undesired side-reaction. This led us to the development of a new "Sulfur-Sulfur" Cycle, shown below:

$I_2 + 2H_2O + SO_2 = 2HI + H_2SO_4$ eq 1

$H_2SO_4 + 8HI = H_2S + 4I_2 + 4H_2O$ eq.2

$3H_2SO_4 = 3SO_2 + 3H_2O + 3/2\ O_2$ eq. 3

$H_2S + 2H_2O = 3H_2 + SO_2$ eq. 4

H_2S generation and steam reformation steps (equations 2 and 4 respectively) have been shown to be feasible[1], and presented here will be the parametric dependence of these two reactions on reactant concentrations, temperature, and residence time. Catalytic activity of several metals in the steam reformation of H_2S was also tested. An overall thermal efficiency assuming a steam reformation temperature of 1100 K is estimated to be 55%, and details of these calculations will be presented. The use of ionic liquids in the Hybrid Sulfur (also known as Westinghouse) Cycle will also be discussed.

---

[1] A. Yokochi, N. AuYeung, "*A Proposed New Sulfur-Sulfur Thermochemical Cycle*" *2009 Meeting of*

the AIChE.

See more of this Session: Meet the Faculty Candidate Poster Session
See more of this Group/Topical: Education

(4x) Quenched Electrostatic Assembly of Colloidal Trimers

Joseph J. McDermott, Neetu Chaturvedi and Darrell Velegol, Chemical Engineering, The Pennsylvania State University, University Park, PA

Controlling the kinetics of colloidal aggregation is an easy and effective method of fabricating colloidal assemblies. A new controlled assembly technique, 'Quenched Electrostatic Assembly' (QEA) has been developed that allows for the assembly of two- and three-component colloidal trimers in a specific order, but with random orientation. The QEA technique introduces oppositely charged larger core and smaller satellite particles into solution together. The satellites begin to electrostatically adhere to the core particles. After the desired number of satellites has adhered, the aggregation is quenched by the addition of polymeric nanoparticles with the same charge as the satellites. These nanoparticles quickly coat the remaining bare regions of each core, effectively blocking further satellites from adhering. Fusing the nanoparticle coating by heating the assembled trimers above the glass transition of the nanoparticles greatly increases the mechanical stability of the assemblies. This technique is used to produce trimers of three different colloidal particles that are mechanically and thermodynamically stable, with a maximum theoretical yield of 15%. Quenched Electrostatic Assembly has been used to form trimers out of particles of varying sizes and materials, including polymer, oxide, and metal particles. Separation options of the resulting trimers from their component particles and other, higher-order aggregates are discussed, as well as various functional assemblies that may be fabricated using this technique.

See more of this Session: Meet the Faculty Candidate Poster Session
See more of this Group/Topical: Education

(4y) Modeling and Simulation of Soft Materials for Energy and Biomimetic Applications

Pratyush Dayal, Department of Chemical and Petroleum Engineering, University of Pittsburgh, Pittsburgh, PA

Understanding systems far from equilibrium is critical for manipulating matter and can provide us with major strides in the fields of nanotechnology and materials science. My research objective is to use modeling and simulation to design systems encompassing soft materials and establish design rules to control their properties. In the proposed projects I will (1) design self-actuated cilia which autonomously propel particles or cells through microfluidic channels and thus provide "conveyor belt" functionality for lab-on-a-chip devices, (2) establish facile routes for fabricating defect-free polymeric nanocomposites which are critical for the design of efficient photovoltaics and other photonic devices, and (3) design micro/nano structured tapes and films using electrospinning technique. The primary focus of the first project is to impart synthetic materials with unique properties, such as autonomous sensing, actuation, and reconfiguration and thus, introduce new functionalities into micro-scale devices. The focus of the second project is to design materials that can potentially yield relatively inexpensive, environmentally benign sources of energy. In the third project, the objective is to facilitate the manufacture of diverse micro-structured materials that would have significant impact on industries of specialty polymers. In carrying out the above studies, I will be able to make significant contributions in the fields of active materials, nonlinear dynamics, polymer processing and biomimetics which are of particular interest to the chemical engineering, bioengineering and materials science communities.

See more of this Session: Meet the Faculty Candidate Poster Session
See more of this Group/Topical: Education

(4z) Engineering Biomaterials for Regenerative Medicine: Elucidation of Factors That Regulate Stem Cell Fate

Randolph Ashton, Department of Chemical Engineering, University of California Berkeley, Berkeley, CA

The development of human embryonic stem cell and induced pluripotent stem cell technology has created enormous potential for the field of regenerative medicine, but the use of these cells in translational tissue engineering applications remains limited because of the dearth in understanding of the biological mechanisms that regulate stem cell fate. My research is focused on elucidating these mechanisms by combining bioengineering and stem cell biology to develop novel tools and biomaterials for investigating and controlling the intracellular signaling pathways and extracellular microenviromental cues that regulate stem cell fate. Several research projects will be presented that center around defining the biological mechanisms that control neural stem cell fate at both the genetic and extracellular microenvironment levels. Approaches to discovery and nanoscale engineering of factors that regulate stem cell fate and methods for incorporation of such factors into biomaterials in an effort to instruct lineage specific differentiation of stem cells will be discussed.

See more of this Session: Meet the Faculty Candidate Poster Session
See more of this Group/Topical: Education

(45a) A Chemical Eng'g Undergratuate Educational and Outreach Program

Majid Salim, Chemical Eng'g, Trine University, Angola, IN

In 2002, an effort was taken by our sophmore Che students to educate high school junior/senior students providing informations on field of chemical and engieering in general to high school students. Our che students made presentations on high school sites as part of che measurement lab.course requirement. These presentations served two objectives.One for communication ability and social awareness.Another to provide understanding and importance and needs for engineers to high school students.

Originally, high school teachers were provided with evaluation forms assessing presentation content materials and quality of presentation delivery. Upon reviewing data for 6 years, in the 2008 ,the evaluation forms were revised to allow for teachers to also comment on student presentation strngths, weakeness,and suggestions for improvement. These additional comments resulted in focusing more in the area of communications that would benefit our sophmore the most.In addition data collected over last 8 years indicated more successful recruitng of high school studentds into eng'g field.

In summary,this program showed the increase in awareness and understanding, the importance and need to pursue eng'g degree by high school students.

See more of this Session: Poster Session On Engineering Education
See more of this Group/Topical: Education

(45b) Creating An Effective Outreach Program

Edward P. Gatzke, University of South Carolina, Dept. of Chemical Engineering, Columbia, SC

This presentation will describe efforts at the University of South Carolina to create a high-school outreach program with the intent to build interest in chemical engineering. This consisted of bringing groups of students onto campus from local high school chemistry classes for a sequence of interactive programs designed to get the student engaged in chemical engineering.

See more of this Session: Poster Session On Engineering Education
See more of this Group/Topical: Education

(45c) Software for Review of Mathematical Fundamentals

Edward P. Gatzke, University of South Carolina, Dept. of Chemical Engineering, Columbia, SC

Every engineering course requires mastery of a variety of mathematical topics. Devoting significant course time to basic mathematics review is time-consuming. Forcing students that have mastered topics to repeatedly drill is excessive and undesirable. An online solution was developed to mitigate the various issues related to review of mathematical fundamentals. This system allows students to practice online without grade penalty, while providing feedback and instruction on how to master the material. Ideally, this will allow courses to proceed more smoothly as all students will have masters the basic levels of mathematics required for the course.

See more of this Session: Poster Session On Engineering Education
See more of this Group/Topical: Education

(45d) On Brittle Fractures That Appeared in New York Times and Material Design

Kal Renganathan Sharma, Adjunct Professor, Department of Chemical Engineering, Prairie View A & M University, Prairie View, TX

Undergraduate students learn ductile and brittle fracture mechanisms during their introductory Materials Science course. The DBTT, ductile to brittle transition temperature phenomena can explain some of the disasters that made the evening news. Some examples are;

i) 2003 Columbia Disaster ii) 1986 Challenger Disaster iii) Collapse of the Silver Bridge at Point Pleasant, WV in Dec' 1967 iv) Great Boston Molasses Disaster in 1919 v) Sinking of Titanic in 1912

The 1986 challenger disaster can be attributed to the failure of the O rings made of fluorelastomers. The lower outdoor temperature during the luanch date of Jan-28-2010 compared with the design use temperature of the otherwise material lead to brittle fracture. The DBTT and the closely related Tg the glass transition temperature lies between the launch day weather temperature and design use temperature. The sinking of the ship Titanic in 1912 can be explained due to a change in the lattice solid structure of steel from FCC to BCC. The water temperature got colder than the steel was designed for leading to a brittle fracture. A huge tank 190 feet in diameter, 50 feet high fractured catostropically flooding the streets of Boston with 2 million gallons of molasses in 1919. The court appointed auditor handed down a decision that the disaster was the result of structural failure rather than explosion. The weather temperature rose that day. Was the brittle facture due to DBT phenomena or due to creep phenomena ? Stress corrosion cracking and long exposure to hydrogen sulfide is the attributed cause of failure of the silver bridge in West Virginia in 1967. Carbon-carbon composite failure has been sugested as a cause of the recent Columbia disaster in 2003. The students are motivated to the subject when these disasters are discussed in class. The course asessment reveals this clearly.

See more of this Session: Poster Session On Engineering Education
See more of this Group/Topical: Education

(45e) Teaching Renewable Energy in El Salvador

Richard A. Cairncross, Chemical and Biological Engineering, Drexel University, Philadelphia, PA

Professor Cairncross spent four months on a Fulbright Lectureship at the University of El Salvador between March to June 2010 to teach a course on Renewable Energy and to develop a graduate program on Renewable Energy. He has taught similar courses at Drexel University over the past three years. This presentation discusses the challenges and rewards of teaching courses on Renewable Energy in a private university in the USA and in an under-funded public university in a developing country. As in the United States, Salvadoran students are enthusiastic to learn more about renewable energy, but they are less motivated by worries about climate change than American students; in El Salvador the main motivations for renewable energy seems to be development of new jobs and reduction of pollution. Although El Salvador has an energy portfolio with approximately 50% renewable sources of energy, the remaining 50% is imported petroleum (with roughly equal amounts being used for electricity generation and transportation); so in some ways the technologies for developing renewable energy projects in El Salvador are similar to the United States. This presentation will review the syllabus and activities used in the course and critique them. The challenges of organizing a course in a foreign country along with the challenges of learning a new language will be discussed; although the style of delivery of the Renewable Energy Course in El Salvador evolved continually during the course, it was a valuable experience to delve deeper into the global aspects of Renewable Energy topics.

See more of this Session: Poster Session On Engineering Education
See more of this Group/Topical: Education

An Effective and Economical Flow Cell Photometer for Classroom Demonstrations and Laboratory Use

Butterfield, Anthony[a]; Grobstien, Stephen; Brau, Christopher; Johnson, James
[a]*Department of Chemical Engineering, University of Utah, Salt Lake City 84112, USA*

Introduction:

Understanding of and experience with a wide variety of analytical techniques is a necessary component of a quality undergraduate curriculum in Chemical Engineering. Students often gain such experience in some sort of unit operations course, using commercially manufactured analytical equipment. However, such equipment is costly and may often be regarded as merely a black box source of data by students, sometimes leading them to erroneous conclusions. Therefore, it is important to introduce students to economical and effective means of data collection that also highlight the electronics and basic physical principles influencing the path between the actuality of physical samples and the ultimate quantified data.

For our unit operations course, our students were in need of a real-time estimation of yeast concentration within a bioreactor. In the past, such estimation had been accomplished offline, using measurements of optical density from a spectrophotometer. We constructed a rudimentary flow cell and photometer using a LED light source and an amplified photodiode. Our design was run in series with a commercial spectrophotometer, worth several thousands of dollars. The constructed photometer cost approximately $8.00 and performed as well as the commercial option at estimating yeast cell concentration, and detecting the transition from the yeast's stationary phase to the exponential growth phase. Results were also validated using a fluorescent staining method within a commercial cell counter. By switching out LED's of different peak wavelengths, this simple, inexpensive and compact optical analytical device will be used on a variety of student projects

Methods:

OPT101 Photometer:

Figure 1A shows a photograph of our rudimentary spectrophotometer contained within an *Altoid* tin, and Figure 1B shows a schematic of the same apparatus. The flow cell was constructed out of a 1x1x4.5 cm disposable spectrophotometer cuvette (Fisher 14-955-125). Holes were drilled at the top and bottom of the cuvette and standard couplings for laboratory tubing were attached using a urethane adhesive. The cuvette was capped with a disposable polyethylene stopper (Fisher 14-385-999) and sealed with a silicone adhesive. A foot of 1 mm inner diameter black tubing (Master Flex, 06412-13) was attached to each port on the flow cell, and curved into an "S" shape within the *Altoid* tin in order to diminish the background effects of stray light.

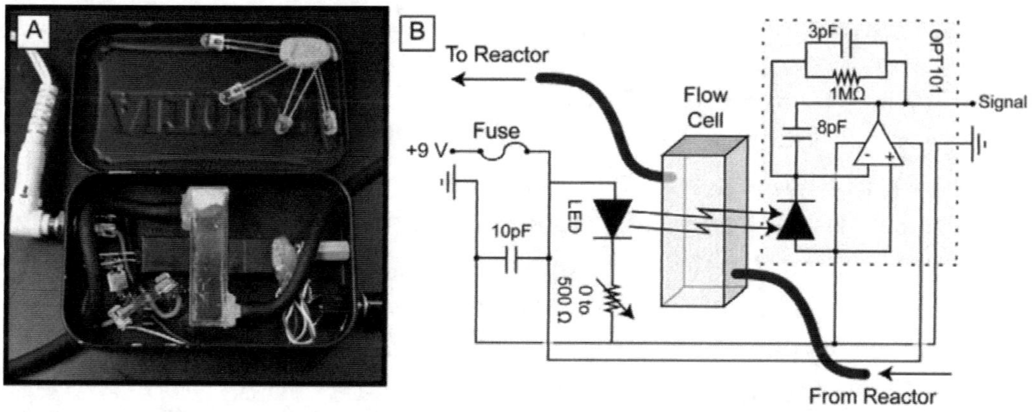

Figure 1: The OPT101 Device A) Photograph of the flow cell and accompanying electronics. B) Schematic of the device.

A separate cuvette was used to create a channel for the LED light and act as a positioner for both the LED and the photodiode. This cuvette was cut in half and its exterior was painted black. These haves were glued to the base of the container using a cyanoacrylate adhesive, at 1 cm apart so that the flow cell would fit snugly between them. A 7/16 inch hole was drilled in a disposable cuvette stopper, in order to hold the LED at one end of the channel. The photodiode and its associated circuitry fit snugly within the other half of the positioning cuvette.

A 9 V DC power supply was chosen to power the electronics. This supply could also be replaced with a 9V Battery. To protect against shock and damage to the electronics in case of a spill, a 65 mA fuse was used and the power was conditioned with a 10 pF capacitor. This power supply was used to power both the LED and the photodiode integrated circuit. To account for the various voltages needed to power various LEDs, and to control the photonic output of the LEDs, a 0 to 500 ohm trim pot was put in line with the light source.

To measure the transmitted light, a monolithic photodiode with a single supply transmittance amplifier was used (OPT101, Texas instruments; purchased from Digikey Corporation). The spectral responsivity of the OPT101 is shown in Figure 2 [1]. According to manufacturer, the signal voltage should increase linearly with light intensity from zero to the supply voltage minus 1.15V, where the signal levels off.

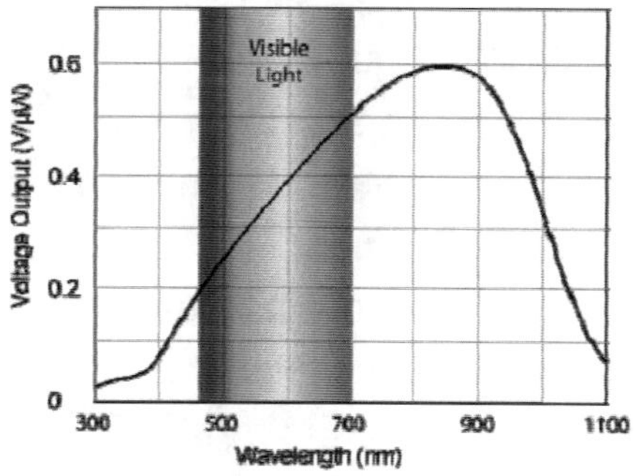

Figure 2: Response of the OPT101 to various wavelengths of light; adapted from [1].

Commercial Spectrophotometer and Cell Counter:

A Spectronic 21D spectrophotometer was put in line with the OPT101 device. Because this piece of equipment was to be dedicated to a separate experiment, it was included in this experiment as a temporary means of comparison, in the hopes that the OPT101 device could solely be used in future experiments. As an additional check on cell concentration, a Chemometec cell counter was also used (Chemometec, NeucleoCounter, YC-100). This device is an integrated fluorescence microscope which uses image analysis to determine the concentration of yeast cells in solution. Typically, samples had to be diluted by several orders of magnitude in order to obtain viable cell concentrations measurements from the YC-100 device, leading to additional sources of error in these measurements.

Bioreactor:

The yeast were grown in a 2 L commercial bioreactor, manufactured by New Brunswick Scientific Co. (Bio Flo 110). The growth media consisted of 10 g/L yeast extract (Sigma, Y1126-250G), 40 g/L glucose (Fluka, 50409047), and 20 g/L of bacto peptone (Sigma, P0556-250G). A 100 ml inoculum of yeast in media (Fleischmann's Yeast, 2192), was added to make the initial yeast concentration in the bioreactor 2 g/L and the total fluid volume 1.5 L. The temperature and pH in the reactor were set at 35° C and 6.5, respectively. Air was sparged through the stirred reactor at 120 cm^3/min. The system was monitored for approximately 8 hrs.

Results:

Figure 3 shows the calibration data for both the Spectronic 21D and the OPT101 device, plotted as the base 10 logarithm of percent transmittance verses the concentration of dried yeast in medium. According to Beer's law, the logarithm of transmittance should be proportional to the cell concentration multiplied by the path length of light and the molar absorbtivity of the cells. A linear fit and 95% confidence bounds are also plotted in Figure 3. As can be seen, the calibration slope for the OPT101 device is considerably steeper than that for the Spectronic 21 D, due to the increased path length in the flow cell. Is also evident that linear fit for the OPT101 device is slightly less precise than that for the Spectronic 21D. As Beer's law assumes the light travels uniformly in parallel

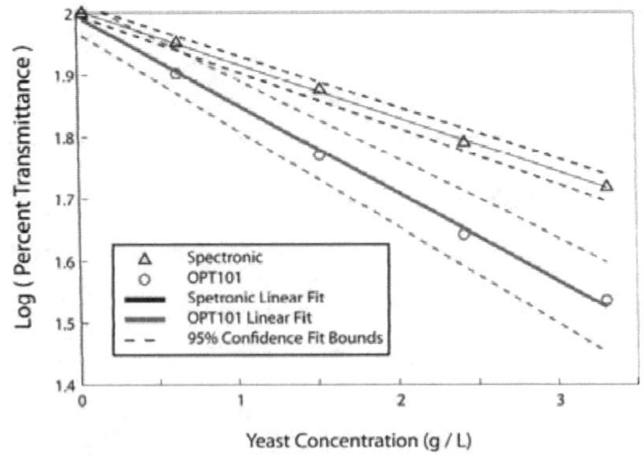

Figure 3: Calibration of the OPT101 and Spectronic 21D devices.

lines through the solution, and light from our LED will be more disperse, a degree of nonlinearity is to be expected. However, with an accurate calibration, the OPT101 device should be as effective as the commercial option at determining cell concentrations.

Figure 4 shows the response of the Spectronic spectrophotometer, the OPT 101 device, and cell counter over a single run of the Bioreactor. Both the signal from the Spectronic 21 D and the OPT101 device remain relatively stable during the lag phase of growth, as the yeast are acclimating (Figure 4A). Note however that the signal from the OPT 101 device was observed to have approximately two thirds the noise present in the Spectronic 21D, as can be seen in Figure 4A as a difference in line thickness. The flow cell in the Spectronic 21D is approximately 4 mm thick, compared to 1 cm in the OPT101 device. The thin flow cell in the Spectronic 21D lead to periodic entrapment of bubbles in the optical path, a likely source of the added noise. Furthermore, a thin flow cell tends to make the optical measurements more sensitive to entrapped solid particles and smaller bubbles, both of which were present in our bioreactor.

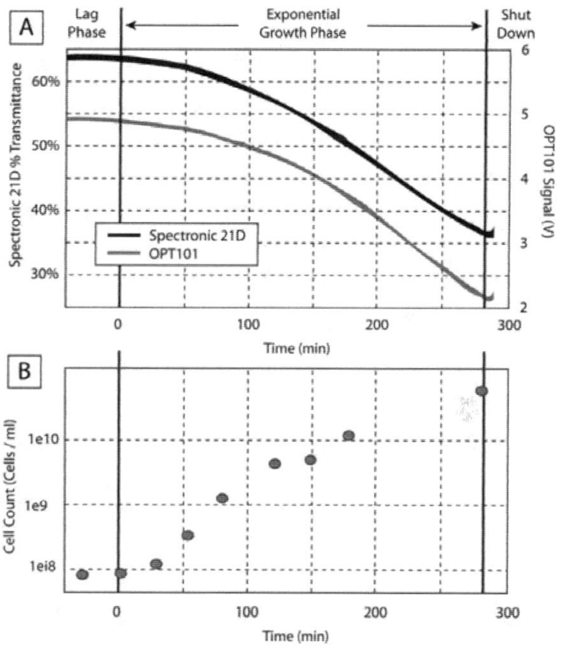

Figure 4: Measures of cell concentration over a bioreactor run. A) The response of the Spectronic and the OPT101 device. B) Cell concentrations measured offline with the YC-100 cell counter.

Both photometric devices were able to give the operator an adequate indication of the beginning of the exponential growth phase of the cells. These results were found to also agree with an offline census of cell population using the YC-100 device (Figure 4B). Because these offline measurements required repeated dilution to obtain an accurate cell count, they are deemed less precise than the calibrated spectrophotometers.

Conclusions:

The construction of this device and its comparison with more elaborate equipment provided a valuable education for our students regarding the electronics, optics, and costs that may be hidden within the physical measurements on which they will rely in their professional lives. Furthermore, such a device may serve as a compact and easily-understandable means of quantifying the results from classroom demonstrations of several chemical engineering topics, such as kinetics and mixing. Future plans include using this OPT101 device for in-class and outreach demonstrations in reaction kinetics using a NaOH/crystal violet alkali bleaching reaction, and in mass transfer using various dyes in a diffusion cell.

1. Monolythic Photodiode and Single-Supply Transimpedace Amplifier, Texas Instruments, http://focus.ti.com/lit/ds/sbbs002a/sbbs002a.pdf, 2003.

(52b) Use of a Scale Model Membrane Filtration System as a Small Scale Laboratory Demonstration

Jamie Hestekin, Ralph E. Martin Department of Chemical Engineering, University of Arkansas, Fayetteville, AR and Thomas M. Potts, Chemical Engineering, University of Arkansas, Fayettville, AR

The University of Arkansas Chemical Engineering department has provided local elementary schools, middle schools, Boy Scout groups, Girl Scout groups, and other groups with a demonstration of membrane separation technology. The technology is used to concentrate milk solids for cheese manufacture, to purify drinking water, to concentrate fruit juices, and many other applications. The unit used by the UARK presentation teams consists of a 125 mm square hydrophilic flat membrane housed in an acrylic holder with a feed side pressure gauge, a variable speed pump, a feed tank, and a collection cylinder. The feed is usually either milk or a colorful fruit juice. After a brief explanation of the parts of the system, the system is started by a volunteer who adjusts the feed side pressure to 20-30 psig. The permeate will consist of a colorless solution and is collected into the collection cylinder via clear silicone tubing, while the concentrate is returned to the feed tank via a different clear silicone tubing. While the system runs, the presentation team discusses the application of the technology to various separation problems. Children in K-12 seem to like the system because they can suggest separations and it is flexible enough to make these changes without destroying the membranes.

See more of this Session: Small Scale Laboratory and Software Demonstrations
See more of this Group/Topical: Education

AN UNLIKELY INTRODUCTION TO FICK'S SECOND LAW OF DIFFUSION: PVC AND TYGON TUBING

Asima Chakravorty, Daniel L. Johnson, Kenneth Wynne
Department of Chemical and Life Science Engineering,
Virginia Commonwealth University, Richmond, VA

Abstract

Polymer or "soft" surface science offers rich and diverse opportunities for education, but often experiments are quite complex or irrelevant to education in chemical engineering. Herein, we describe a laboratory experiment using a commonplace material, Tygon tubing, which is [poly(vinylchloride), PVC] plasticized with dioctylphthalate (DOP). A novel determination of the diffusion coefficient for hexane assisted DOP diffusion has been developed using Attenuated Total Reflectance Infrared (ATR-IR) spectroscopy. The procedure involves exposing sections of tubing to solvent for short time intervals. The area of the DOP carbonyl peak, which occurs in a PVC "IR window" is monitored vs. time. The peak areas are analyzed using Fick's second law of diffusion.

Introduction

Dioctylphthalate (DOP) is a typical PVC plasticizer added to obtain desired flexibility. The amount of plasticizer can reach 35-40 wt%. Depending on use conditions, plasticizer may migrate from the tubing to the environment.[1] Diffusion of plasticizer is a matter of concern for applications involving children's toys and medical devices.[2,3] This experiment addresses the phenomenon of DOP migration with the determination of the rate of solvent assisted diffusion. The rate of DOP diffusion is determined employing ATR-IR spectroscopy. Considering the common availability of Tygon tubing and ATR-IR spectroscopy, the procedure is a simple but novel experiment to measure the DOP diffusion coefficient and to introduce Fick's second law for data analysis.

Experimental

Materials

PVC – Tygon® tubing and hexane (≥95%) were sourced from Sigma Aldrich. A Nicolet iS10 FTIR with an ATR attachment was used for ATR-IR measurements.

Experiment

Eight pieces of Tygon® tubing, each 2 cm long, are cut in half lengthwise to make eight sets (1A, 1B, etc.). Each piece is placed on a hook (fishhooks work well and hold labels) and immersed in 400 ml hexane with continuous magnetic stirring. Samples are removed at 10, 20, 40, 60, 90, 120, 240, and 360 s. After removal, each sample set is dried for ~ 20 min at ambient temperature. The half-tubing samples are placed on the ATR-IR holder so that the outside of the tubing has optical contact with the Ge crystal. An ATR-IR spectrum is obtained (32 scans). DOP has a distinct carbonyl peak at 1750 cm^{-1} that is in an IR window for PVC.[4] After collecting 8 sets of spectra and saving as csv files, each spectrum is retrieved and the areas of the carbonyl absorptions are determined using the Nicolet data analysis program. Separately, a spectrum of "as received" Tygon® tubing (zero time) is obtained.

The reducing area of the carbonyl peak as a function of immersion time is used to determine the apparent diffusion constant. The diffusion coefficient is calculated using Fick's Law of Diffusion (Eq 1).

$$\frac{A_o - A_t}{A_o} = \frac{L}{\sqrt{(\pi D)}} t^{\frac{1}{2}} \qquad \text{Eq. 1.}$$

where,

A_o = initial peak area (area of "as received")

A_t = peak area at time t

D = diffusion coefficient

L = diffusion path length (IR penetration depth)

t = time immersed in hexane.

Results

The DOP diffusion experiment includes generating a spreadsheet with data suitably organized for analysis via Eq. 1. This includes an error analysis. Figure 1 shows a plot of $\frac{A_o - A_t}{A_o}$ vs. $t^{1/2}$ for a representative DOP diffusion experiment. The diffusion coefficient is 1.47E-10 cm^2/s. This determination compares favorably with the determination of diffusion of molecular species in polymeric solids.

Conclusion

With Tygon® tubing and ATR-IR spectroscopy, this experiment introduces the concept of Fick's Second Law of Diffusion with a practical example. Solvent use is minimal and solvent can be recycled. Thus, the experiment may qualify also as "green chemical engineering education".

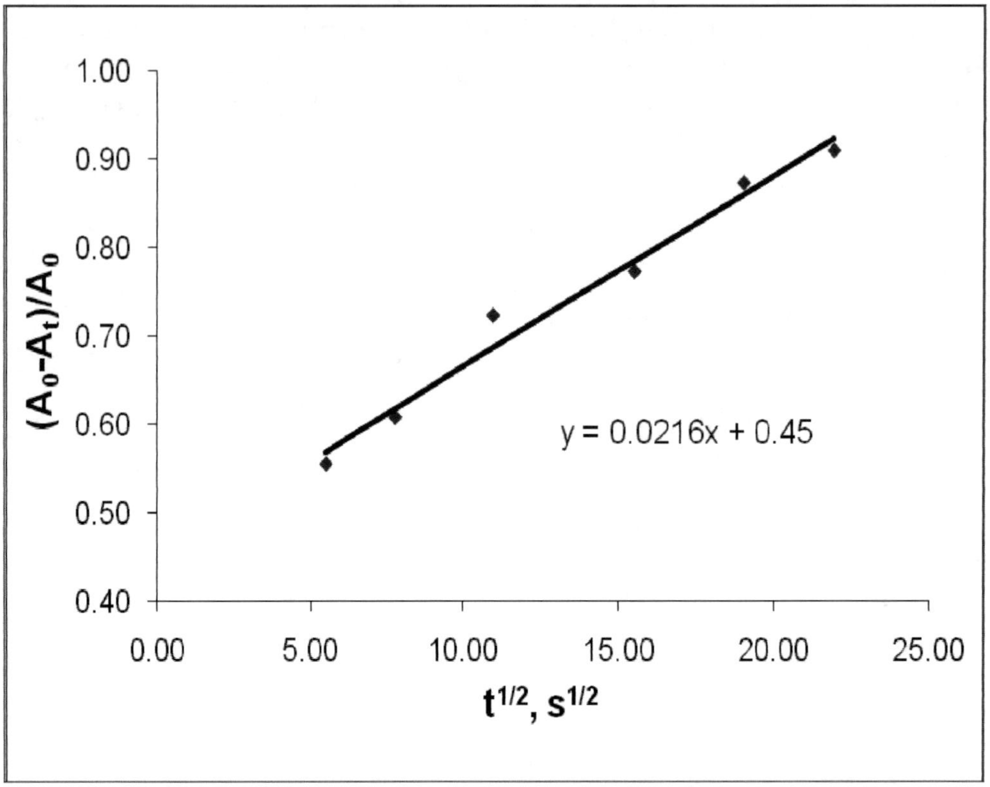

Figure 1. A typical graph for the determination of the diffusion coefficient for hexane assisted DOP diffusion.

References

1 Rodrigo Navarro, Monica Perez Perrino, Myriam Gomez Tardajos and Helmut Reinecke (2010) "Phthalate Plasticizers Covalently Bound to PVC: Plasticization with Suppressed Migration", *Macromolecules*, 43(5), 2377-2381.

2 Thomas, J.A. and Thomas, M.J. (1984) "Biological effects of Di-(2-ethylhexyl) phthalate and other phthalic acid esters", *CRC Critical Review of Toxicology,* 13, 283.

3 Jayakrishnan, K.; Lakshmi, S. (1998) "Immobile plasticizer in flexible PVC", *Nature*, *396*, 638-638.

4 Tabb, D. L.; Koenig, J. L. (1975) "Fourier transform infrared study of plasticized and unplasticized poly(vinyl chloride)", *Macromolecules* , 8(6), 929-34

Paper # 183864

Desktop Experiment: Pressurized Tank

Polly R. Piergiovanni
Department of Chemical and Biomolecular Engineering
Lafayette College, Easton, PA

Introduction: All first year engineering students at Lafayette College enroll in an "Introduction to Engineering" course that was developed to strengthen students' math skills. Laboratory exercises show applications of the math to engineering. For one laboratory, we constructed a simple pressurized tank that models the solution of a differential equation. In class, we show the students how many systems can be modeled by this equation, and mention several engineering applications. We also provide examples of pressurized tanks in chemical engineering. In the lab, students pressurize the tank, then open a valve with a small orifice (0.013" diameter) and note the pressure at intervals. Students in a Process Control course could repeat the experiment and develop a first order model of a pressurized tank, estimating the gain and time constant.

Description: The tank (see Figure 1) was constructed from 2" threaded galvanized steel pipe and threaded head caps. All parts to construct the vessel were purchased from McMaster-Carr (www.mcmaster.com) and a complete parts list is provided in Table 1. The total cost of each unit was $68.40. The vessel is mounted on an aluminum base (80/20, Transmission Engineering Company, www.tecoinc.com, $13.00). The students use a bicycle pump to pressurize the tank (to 30 psi, although it is capable of withstanding pressures of 100 psi). High quality bicycle pumps with longer hoses were purchased from McMaster-Carr (item 6028A23, $65.57) because the longer hose allowed the pump to be on the floor and reach the tank on the table.

Figure 1. The pressure vessel

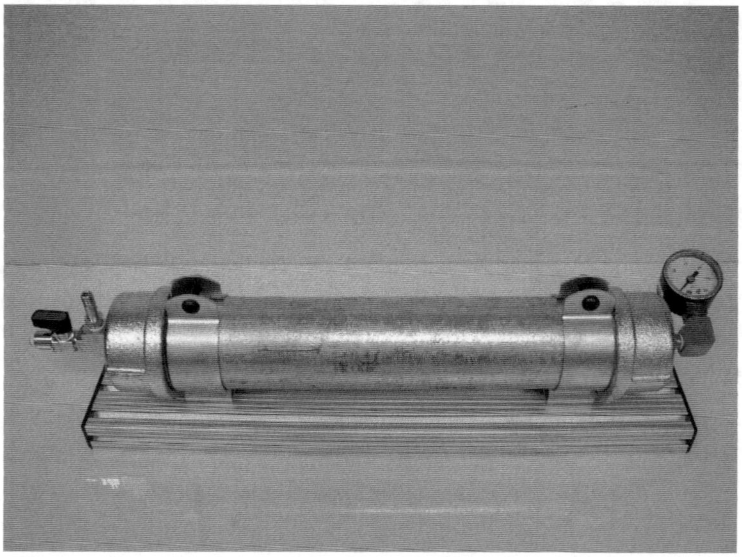

Table 1. Parts list for the Pressure Vessel.

Item	Quantity	Description	Unit Price	Total Price
4549K784	1	Std.Wall Galv Welded Steel Thrd Pipe Nipple 2"Pipe, 12" long, Thrded Ends, 3/4"L Thread, Sch 40	$17.05	$17.05
4638K259	2	Zinc-Pltd Iron, 2" Round Head Cap, 125 psi	$9.82	$19.64
3847K71	1	Multipurpose Gauge +/- Mid-Scale Accuracy 1 1/2" Dial, 1/8" NPT Male Bottom, 0 - 30 psi	$9.63	$9.63
8063K31	1	Brass Air Tank Valve 1/8" NPT, 1-5/16" Overall Length	$3.09	$3.09
50785K219	1	Med-Pressure Brass Thrd Pipe Fitting 1/8" Pipe Size, Female X Female X Male, Tee	$1.74	$1.74
50785K35	1	Med-Pressure Brass Thrd Pipe Fitting 1/8" Pipe Size, 90 Deg Elbow	$1.78	$1.78
50785K112	1	Med-Pressure Brass Thrd Pipe Fitting 1/8" Pipe Size, Hex Socket Plug, 3/16" Hex	$0.57	$0.57
50785K21	1	Med-Pressure Brass Thrd Pipe Fitting 1/8" Pipe Size, Hollow Hex-Head Plug	$0.61	$0.61
3006T75	1	Bottom-Load Clamping Hanger Zinc-Plated, W/O Bolt & Nut, for 2-3/8" OD, 100# WII, packs of 10	$9.18	$9.18
4912K71	1	Miniature Chrome-Plated Brass Ball Valve Wedge Handle, 1/8" NPT Female X Male Connections	$5.11	$5.11

TOTAL PARTS for Pressure Vessel: $68.40

Procedure: The mathematical model for the tank was provided to the students because many of them had just begun taking calculus. They were given the differential equation and its solution, $P(t) = P_0\, e^{kt}$, where P_0 represents the initial pressure in the tank, and P(t) is the pressure at any time t.

The students (in lab groups of two) pressurized the vessel to slightly over 30 psi and measured the pressure drop as a function of time. Originally, we gave the students a detailed procedure to collect the data using a stopwatch. However, we found it was much better to give the students the equipment, ask them to measure pressure drop as a function of time, and allow them to be creative. The students responded well, and many figured out a way to use their i-Phone and an app to collect the data.

Analysis: The students began by plotting their data on semilog paper. Many had not used any type of log paper, so we explained how the plot was set up, and how to interpret the slope.
Next, they used Excel to create a scatter plot of the data, and fit an exponential curve to the data. They found the values for P_0 and k from the equation. Then, they used Excel to calculate the natural log of the pressure for each point, and were asked to guess what type of curve ln (P) vs t would be. They fit a line to the data and noted if the parameters were the same as the first plot.

Results: Typical results are shown below in Figure 2. The students can see the relationship between the x-y plot with the exponential fit, and the semilog plot. They can also compare it with their paper plot on semilog paper.

Figure 2. Typical student data.

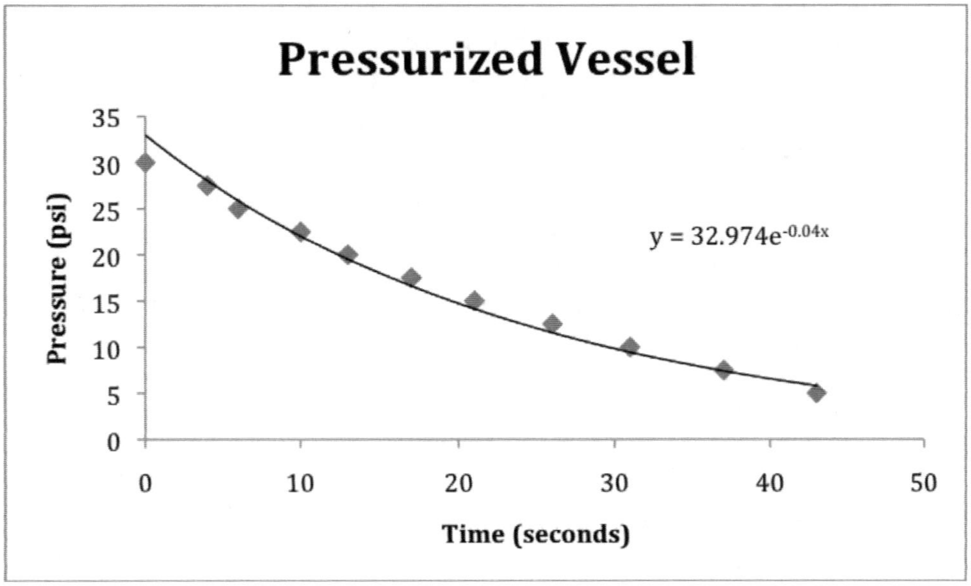

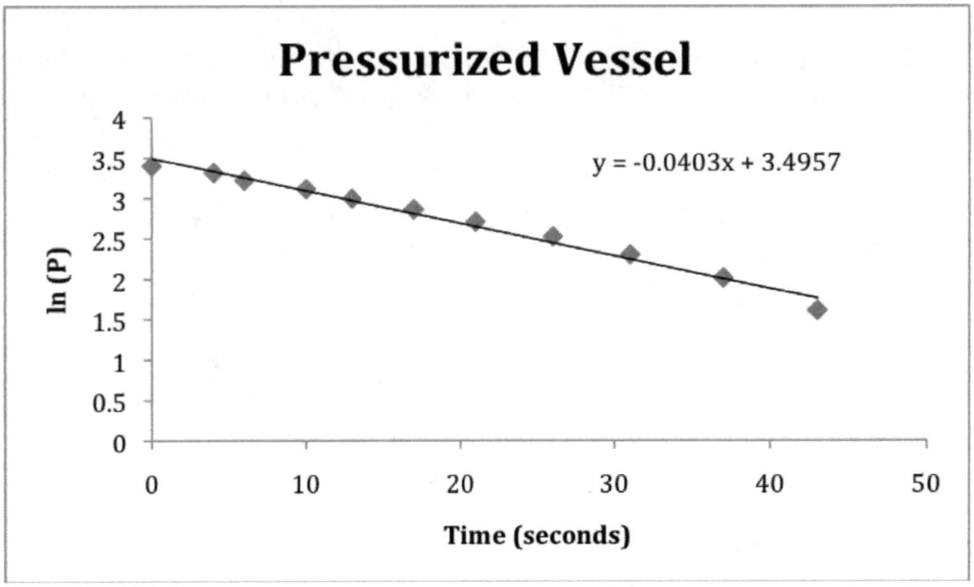

The lab has been completed by two groups of 14 students each. All students were able to complete the lab, analyze the data, and report the values of P_0 and k. They were able to repeat the analysis on an exam shortly after completing the experiment. Later during the Fall 2010 semester, 160 students will complete the experiment, and we will obtain more assessment data.

Conclusion: The lab was inexpensive to construct and operate for a large number of students. The results follow the model, and students learn to work with the exponential decay response that is common in many engineering applications.

(52f) Beginning of the End for the Handheld Calculator?

John Wagner[1], Allen Hersel[2] and Majid Salim[2], (1)Chemical Engineering, Trine University, Angola, IN, (2)ChE, Trine University, Angola, IN

The modern scientific electronic calculator has been at the side of engineering students for the past 40 years. It has the advantage of being portable, reliable, efficient and precise. However, today's multifunctional electronic devices may replace this old standby in the coming years. In particular, the iphone or ipod are common among today's students. Since 2008, when Apple allowed outside developers to create and sell applications (apps) for the iphone / ipod touch there has been an explosion in the availability of applications related to engineering. Now there are several good choices for scientific calculators available for the iphone. These app based calculators take advantage of the iphone's touch sensitive screen and user friendly interface. Some of these "calculators" provide capabilities such as solving multiple nonlinear equations, reverse polish notation, symbolic processing, and three dimensional high resolution color graphing. In addition, there are also apps that provide access to steam, psychometric and chemical safety data, quickly, without the need for web access. Specific engineering analysis applications are also being rapidly developed. For instance, in the area of fluid flow, there are applications that help size pipe for both liquids and gasses. There is even a simple process simulator that includes a rigorous distillation module. These apps are generally available at very reasonable prices. For these reasons, the McKetta Department of Chemical & Bioprocess Engineering at Trine University will start a pilot program of requiring the incoming freshman in 2010 to purchase an iphone, ipod touch or ipad. Engineering and Mathematics applications for the ipad will be demonstrated. These will include: ALPH, Spacetime, HiCalc, m48+, PCalc, Chemhazards, Steam HX, and the HVAC Toolkit.

See more of this Session: Small Scale Laboratory and Software Demonstrations
See more of this Group/Topical: Education

(52g) NSF Broader Impacts: K-12 Outreach

Priscilla J. Hill, Dave C. Swalm School of Chemical Engineering, Mississippi State University, Mississippi State, MS

The goal of the K-12 outreach is to develop simple hands-on laboratory activities and demonstrations to teach pre-college students about crystallization and particle technology. This is being done by having students at existing summer camps at Mississippi State University participate in these activities.

This presentation will discuss the activities and discussion with the K-12 students. Examples of activities will be given such as panning for gold by dry panning. The connection between the activities and NSF broader impacts will also be discussed.

See more of this Session: Small Scale Laboratory and Software Demonstrations
See more of this Group/Topical: Education

(68a) Chemical Product Design Course at Carnegie Mellon University

Nick Sahinidis and **Apurva Samudra**, Chemical Engineering, Carnegie Mellon University, Pittsburgh, PA

This presentation describes the chemical product design course at Carnegie Mellon University. The course introduces undergraduate students to the design of a chemical product. The lectures address aspects of the design of molecular structure, microstructure, and devices/processes that effect chemical change. Topics include market pull and technology push case studies; the design of refrigerants, polymers, and mixtures; techniques in the design of experiments, molecular property prediction, database searches; and intellectual property rights. This is a project-based course, for which an extensive report must be submitted. Recent course contents and projects worked on by students will be detailed in the presentation.

See more of this Session: Technology Transfer In (A) Product Design Via ChE Education and (B) Commercial Product Manufacturing
See more of this Group/Topical: Process Development Division

AIChE 2010 Annual Meeting, Paper 68b, November 8, 2010

A Product Design Capstone Course in the Chemical Engineering Undergraduate Curriculum at SUNY-Buffalo

Paschalis Alexandridis
Department of Chemical and Biological Engineering
University at Buffalo (UB) - The State University of New York (SUNY)
Buffalo, NY 14260-4200
palexand@buffalo.edu
http://www.cbe.buffalo.edu/alexandridis

Introduction

In addition to addressing the question "How should we make it?" (*process engineering*), several practicing chemical engineers are concerned nowadays with the question "What should we make?" (*product engineering*). Chemical engineers and the companies that employ them have traditionally focused in the efficient production of commodity chemicals and specialty chemicals, but are increasingly active in the development and production of (i) formulated products, (ii) consumer goods, (iii) bio-based concepts, and (iv) devices.

Formulated products are multi-component systems that have been designed and manufactured to meet end-use requirements. They are often multifunctional (i.e., they accomplish more than one function of value to the customer) and microstructured (i.e., their function and perceived value derives significantly from their internal structure in the range 0.1–100 μm). Formulated products, which include bulk and shaped solids, semisolids, liquids and gases, are also known as structured products, engineered products, dispersed systems, or chemical-based consumer products. Consumer goods of interest to chemical engineers offer functionality that is based on chemical/physical technology. Bio-based concepts include biomaterials, drugs, and technologies based on metabolic, cell, or tissue engineering. Devices include in their function physical or chemical transformations. (Costa et al., 2006)

The various chemical products may have little in common in terms of appearance or performance, however, they share common principles and practices in terms of development and manufacturing. Thus the need and the opportunity for a chemical product design course.

Product Design in the Chemical Engineering curriculum

The Product Design course has been developed for seniors majoring in chemical engineering, and is offered in the fall semester of the senior year. By that time, chemical engineering majors have completed successfully several math, physics, and chemistry courses, as well as a materials science and engineering course (which equips students with of knowledge of structure-processing-property for main classes of materials such as metals, ceramics and plastics). The students have also completed core courses in chemical engineering thermodynamics (including thermophysical properties, phase transitions), transport phenomena (including fluid mechanics, heat and mass transfer), and chemical kinetics / reactor design, and have some chemical engineering laboratory experience. Taken together, the above courses provide a broad foundation of knowledge that can be synthesized toward addressing problems of product design. The capstone Process Design course is offered to chemical engineering students in the spring semester of their senior year.

The Chemical Engineering Product Design course at SUNY-Buffalo encompasses the (1) general framework for product design and development (identify customer needs, convert needs to specifications, create ideas/concepts, select concept, formulate/test/manufacture product; also intellectual property, safety, environmental, marketing and financial considerations), and (2) structure-property relations that guide the search for materials with particular properties, as well as modifications of existing materials that would improve properties. These two main themes are reinforced by (3) case studies of successful products with well-documented history of invention, development, production, and marketing. The course material is integrated in (4) product design projects that synthesize (i) the market pull for innovations in the development of improved or new products, together with (ii) the technical push from advances in our fundamental understanding of structure-property relations at the atomic, molecular, nano-scale and continuum levels. The content covered in the course is rather broad, while the depth is limited by mostly time constrains.

Course organization

The class meets four hours a week throughout the semester, with the Instructor typically presenting new material or leading the discussion. Guest lecturers are invited to present information on their areas of expertise, e.g., library resources, intellectual property, product design practice. Several class periods are devoted to project presentations delivered by the students. Chemical engineering faculty members serve as consultants for the team projects. The teams meet with their consultants about once a week to review the progress and exchange ideas. Several books related to the course are placed on reserve for the students to use. Students have access to (through the university library) and are encouraged to utilize for their projects electronic books, journal articles, trade publications, and patents.

The course has been organized along several principles that are recognized as enhancing the student learning experience: (Davidson & Ambrose, 1994)

- Active learning is encouraged (the course material is relevant to every-day life and this makes it interesting to students; students are provided with real-life situations to analyze)
- Effective learning experiences are designed for students (problem-based learning is a significant component of the course; assignments apply class material to new contexts)
- Prompt feedback is provided (grading is done on the basis of detailed rubrics, and all results are communicated to the students; feedback is aimed to be corrective and supportive)
- Importance of time and effort spent learning is emphasized (instructor plans learning activities to permit best use of time; students are encouraged to make effective use of time)
- Student faculty-contact is encouraged (in and out of class; during meetings with consultants)
- Cooperation among students is encouraged (students work on several team projects throughout the semester; they also cooperate during in-class exercises; such cooperation can help students enhance self-esteem, collaborative skills, and personal responsibility)

• High expectations are communicated (the instructor sets high but attainable goals)
• Diverse talents and ways of thinking are respected (the course material is delivered in a way that varying learning styles are engaged; all students are encouraged to voice their views)

Assessment

Each student is assessed on the basis of quizzes and assignments that are completed individually, and projects that are carried out as team efforts (teams consist of preferably 3 members). Quizzes are held during class time and aim to test material presented in class and/or assigned for reading. Individual assignments include "TV/magazine Ad", "What's That Stuff?" and "Read the Label" (each of the these three assignments should address a different product). Team projects comprise "Project 1: Understanding Products", "Project 2: From Needs to Concepts" (problem statements for projects 1 and 2 are taken from the textbook, Wesselingh, Kiil, and Vigild: "Design & Development of Biological, Chemical, Food and Pharmaceutical Products"), and Project 3: "Find a Solution: Real-World Product Design and Development" (project statements have been obtained from www.ninesigma.com).

The overall course grade reflects performance in quizzes and in-class participation (counting for 10% of the overall grade), individual assignments (35% total: 10% for A.1, 10% for A.2, and 15% for A.3), and team projects (55% total: 20% for P.1, 10% for P.2, and 25% for P.3). Grading of assignments and projects is based on rubrics (that are provided to the students in advance). For each assignment/project, students submit a properly documented report document and a PowerPoint presentation (presented in class, as time permits).

The assignments/projects focus on synthesis (create, predict, design, improve, propose) and evaluation (select, decide, critique, justify, verify, debate, assess, recommend). For example, Project 3, that counts the most (25%) toward the course grade, is rather open-ended. Students are challenged with proposing a solution to a real-world problem articulated in a "Request for Proposals" (RFP) published by NineSigma, a company that facilitates solution-provider knowledge searches. RFPs identify critical innovation needs and seek a solution or access to knowledge and capabilities missing in the client's internal program. RFPs define the client's needs so that researchers ("providers") can understand what is needed and respond with the best possible solutions. Teams prepare a proposal for a solution to a NineSigma RFP using the RFP information as a springboard, and following the product design approach: (i) customer needs, (ii) product specifications, (iii) idea/concept generation, (iv) idea/concept selection, (v) product formulation and processing/manufacture, (vi) cost/market. The following RFP titles exemplify the types of problems the students have addressed: Development of Eco-Friendly Solvents; Enabling Freshness of Baked Cookie and Cracker Products; Peroxide-based Sterilants with High Sporicidal Activity; Seasonings or Ingredients to Impart the Taste of Fats to Non-Fried Potato Chips; Soil Conditioning Technologies to Enhance Nitrogen Use Efficiency; Natural Conditioners for Hair and Skin; Water Resistant, Recyclable Packaging Stock; Bio-degradable, Compostable Oxygen Barrier Film; Seeking Cleaning Formulations for Biofilm Removal from PMMA; Monomer Technologies for Imparting Strength in Hair; High Absorbent Desiccants for Coatings; Improving Gas Barrier Properties of Polylactic Acid.

Concluding remarks

A course on product design has been developed that is geared toward chemical engineering applications such as formulated, microstructured products and functional consumer goods. The course covers the general framework for product design and development (i.e., customer needs, product specifications, concept generation, concept selection, product formulation and manufacturing), and guides students to utilize structure-property relations in the search for new and improved products. Problem-based learning challenges students to analyze existing products and to propose solutions to real-world problems using their knowledge of chemical engineering principles. The students are supported in their projects by faculty consultants and are encouraged to utilize library and web-based resources. Students are assessed in multiple ways, based on work they do by themselves in and outside the class, and on team projects. The students appreciated the relevance of the course material to every-day life products and the opportunity for active learning and problem-based learning in a team setting. At the same time, the students viewed with some trepidation the open-ended nature of the course deliverables.

Bibliography

Journal articles advocating product design in chemical engineering:

- Do changes in the chemical industry imply changes in curriculum? Cussler, E. L. Chemical Engineering Education 1999, winter, 12-17.
- Molecular structure and property: Product engineering. Wei, J. Ind. Eng. Chem. Res. 2002, 41, 1917-1919.
- Chemical product engineering. Cussler, E. L.; Wei, J. AIChE Journal 2003, 49, 1072-1075.
- Chemical product engineering: An emerging paradigm within chemical engineering. Costa, R.; Moggridge, G. D.; Saraiva, P. M. AIChE Journal 2006, 52, 1976-1986.

Textbooks on product design with applications related to chemical engineering:

- Wesselingh, J. A.; Kiil, W.; Vigild, M. E. "Design & Development of Biological, Chemical, Food and Pharmaceutical Products"; J. Wiley & Sons, 2007 (ISBN 978-0-470-06154-1).
- Cussler, E. L.; Moggridge, G. D. "Chemical Product Design"; Cambridge University Press, 2001 (ISBN 0-521-79633-4).
- Wei, J. "Product Engineering: Molecular Structure and Properties"; Oxford University Press, 2007 (ISBN 9780195159172).
- Seider, W. D.; Seader, J. D.; Lewin, D. R.; Widagdo, S. "Product and Process Design Principles: Synthesis, Analysis, and Evaluation", 3rd ed.; J. Wiley & Sons, 2009 (ISBN 978-471-04895-5).

Textbooks on product design from the more common, mechanical engineering perspective:

- Ulrich, K. T.; Eppinger, S. D. "Product Design and Development", 4th ed.; McGraw-Hill, 2008 (ISBN 978-0-07-310142-2).
- Otto, K. N.; Wood, K. L. "Product Design (Techniques in Reverse Engineering and New Product Development)"; Prentice Hall, 2001 (ISBN 0-13-021271-7).

Textbooks on product design from a management/business perspective:
- Baker, M.; Hart, S. "Product Strategy and Management", 2nd ed.; Prentice Hall, 2007 (ISBN 978-0-273-69450-2).
- Rainey, D. L. "Product Innovation: Leading Change through Integrated Product Development"; Cambridge University Press, 2005 (ISBN 0-521-06601-8).
- Crawford, C. M.; Di Benedetto, C. A. "New Products Management", 9th ed.; McGraw-Hill, 2008 (ISBN 978-0-07-352988-2).

Books on teaching, helpful in developing a new course:
- Davidson, C. I.; Ambrose, S. A. "The New Professor's Handbook"; Anker Publishing Company Inc., 1994 (ISBN 1-882982-01-0).
- Wankat, P. C.; Oreovicz, F. S. "Teaching Engineering"; McGraw-Hill, 1993 (ISBN 0-07-068154-6).

(68c) Using Process Simulation for Technology Transfer and Process Facility Fit

Charles Siletti, INTELLIGEN, INC., Mount Laurel, NJ and Demetri Petrides, INTELLIGEN, INC., Scotch Plains, NJ

Technology transfer is, in large part, transfer of process knowledge from the development group to a manufacturing site. Because process simulation models represent a succinct, functional repository of process knowledge, they are a natural addition to the technology transfer process. This paper focuses on the capabilities and limitations of simulation models for technology transfer. Process-Facility fit is part of the technology-transfer process and is ultimately a determination of whether and how well a given process can be executed in a particular facility. Gap analysis is often the technique of choice for evaluating a facility. This presentation primarily focuses on the use of process simulation and scheduling tools to add quantitative information to gap analysis. A series of examples demonstrate how simulation can aid in determining gaps in primary equipment, buffer preparation equipment, purified water, utilities, labor and staffing. Techniques for managing uncertainties in process facility fit will also be covered.

See more of this Session: Technology Transfer In (A) Product Design Via ChE Education and (B) Commercial Product Manufacturing
See more of this Group/Topical: Process Development Division

(68d) Technology Transfer in Outsourcing Manufacturing of Specialty Polymers

Zhaoyang Ou, Zhen Lai and Chieh-Min Cheng, Xerox Corporation, Webster, NY

Outsourcing manufacturing of specialty chemicals has steadily grown over the years as a wider spectrum of business segments (biopharmaceutical, semiconductor) seek lower development and product costs, and faster product cycles. In contrast with commodity products with mature and widely applied technologies, specialty chemicals are often first developed and scaled-up internally in pilot scale before being transferred to external vendors for manufacturing. As such, it has presented unique challenges in technology transfer where not all critical manufacturing requirements are known and solutions are being developed alongside with vendor scale-up, which raises manufacturing risks and potential product costs. In this talk, some key lessons learned from the outsourcing manufacturing of specialty polymers will be discussed.

See more of this Session: Technology Transfer In (A) Product Design Via ChE Education and (B) Commercial Product Manufacturing
See more of this Group/Topical: Process Development Division

(68e) Technology Transfer of Pharmaceutical Freeze-Dried Sterile Liquid Manufacturing - A Case Study

John F. Peragine, Bristol-Myers Squibb Company, New Brunswick, NJ

A pharmaceutical freeze-dried sterile injectable liquid has been produced at the same manufacturing site for the past 40 years. This unique product has never been commercially manufactured at any other site since its inception. Since manufacturing at the current site is being discontinued, production needs to be transferred to a new site. Initial experimental attempts to transfer the current process to the new site have been unsuccessful. Although the manufacturing processes at both sites are very similar, transfer of the freeze-drying cycle has been problematic. The physical form of the freeze-dried product at the new site has not been comparable to the current site and has not met acceptance criteria. Since the new site will be the only site for worldwide production, the new manufacturing process needs to be harmonized (demonstrating that the new process yields final drug product equivalent to the currently marketed product). Process equivalence is also important to demonstrate for the required FDA and multinational filings before the new site can be approved. This paper discusses the key challenges of transferring the freeze-drying cycle and lessons learned during the technology transfer process.

See more of this Session: Technology Transfer In (A) Product Design Via ChE Education and (B) Commercial Product Manufacturing
See more of this Group/Topical: Process Development Division

(94a) Introducing Undergraduates to Pharmaceutical Technology through Problem Sets for a Material and Energy Balance Course

Stephanie Farrell, Mariano J. Savelski, C.Stewart Slater, Vladimir DeDelva, Keith McIver and Kathryn Whitaker, Chemical Engineering, Rowan Univeristy, Glassboro, NJ

Problem sets have been developed for the introductory chemical engineering course, material and energy balances, that expose students to the principles of pharmaceutical engineering. These include problem sets that convey essential concepts in pharmaceutical terminology, drug delivery, and manufacturing within the context of a material and energy balance calculation. For example, students learn the role of active pharmaceutical ingredient (API) and excipients (binders, filler, lubricants) in the formulation of various over the counter and prescription drugs through simple unit conversion and mass/mole/volume composition problems. Drug delivery problems emphasize the use of mass balances and physical property laws. API syntheses are the basis of problems related to reaction stoichiometry as well as those for heats of formation. Mass and energy balance calculation problems focus on pharmaceutical manufacturing unit operations such as blenders, mills, dryers, tablet presses, etc. Actual manufacturing procedures and bill of materials are used to present realistic industrial operations. Some of the more original problem sets deal with regulatory issues (FDA, health and safety) engaging students in problem solving within the context of an actual event. All of the modules include both problem statement and complete solution with accompanying figures which can be readily utilized by faculty in their courses. These are linked to the topics typically taught in an introductory material and energy balance course. Several examples of these problem sets will be presented in the paper.

This educational project is part of the outreach efforts of the National Science Foundation (NSF) Engineering Research Center for Structured Organic Particulate Systems (C-SOPS). The Center, led by Rutgers University, is focused on developing structured organic particulate systems used in pharmaceuticals and their manufacturing processes. Rowan University chemical engineering students and faculty are working in collaboration with the Center to create educational materials based on research related to pharmaceutical processing.

See more of this Session: Free Forum on Chemical Engineering Education (Part 1)
See more of this Group/Topical: Education

(94c) The "Single Pellet Reactor": A "Cool" Multiscale Problem, or a Useful Chemical Engineering POK for Learning of Mass Transfer?

Jennifer Anne Pascal[1], Seth Wynne[2], Vinten Diwakar[2] and Pedro Arce[2], (1)Chemical Engineering, Tennessee Tech University, Cookeville, TN, (2)Chemical Engineering, Tennessee Technological University, Cookeville, TN

In the field of Chemical Engineering, multiphase mass transport is one of the most important subjects for students to learn since, in general, it is an area of study that sets chemical engineers apart from all other engineering disciplines. Multiphase (and usually systems with mixtures of numerous species and reactions) bring a challenging learning scenario for the beginner, frequently the students! In the current chemical engineering curriculum, the "classical" textbooks used do not provide a systematic, didactic approach for learning mass transfer in systems with multiphase or transient conditions. Unfortunately, more than frequently, students are faced with formulations that seem to be crafted by the "Michelangelos" of the conservation laws rather than effective engineers using general and practical principles.

The "Single Pellet Reactor" offers a useful and practical device to effectively learn the conservation equations for the case of "multiphase" systems and under transient conditions. In this presentation, we will describe an effective approach based on macroscopic species conservation laws to derive the correct "engineering" equations and be able to identify assumptions needed to reach the "classical equations" found in the literature. The approach can be viewed as an effective POK (Arce, 2004) to study multiscale mass transfer in systems with more than one phase. Several observations useful for the learning of students will be drawn at the end of the discussion. Possible relevant scenarios for the experimental determination of parameters will be offered.

See more of this Session: Free Forum on Chemical Engineering Education (Part 1)
See more of this Group/Topical: Education

Teaching Chemical Engineering to Non-Chemical Engineers

Abstract
Non-chemical engineers, including other engineering disciplines and operating technicians and maintenance personnel need to work with degreed chemical engineers throughout the chemical industry. Frequently they have no background or understanding in some of the basic concepts of chemical engineering, including reactive chemicals concepts, thermodynamics, material and energy balances, heat and mass transfer, process control, solids handling fundamentals, and specialized unit operations such as chromatography and liquid-liquid extraction.

For over 20 years, AIChE has offered a 3 day public and on site courses entitled "Chemical Engineering for Non-Engineers". After code required courses, it is the most popular course offered by AIChE. This course has been attended by EPA lawyers, engineering managers without a chemical engineering background, chemists, patent attorneys, operating technicians, sales personnel, and other types of engineers.
This presentation will review the evolution of this course, its materials, the manner of teaching, and learnings from the classes over the years.

1. Background

For over 20 years, AIChE has had a public training course and accompanying teaching CD entitled "Chemical Engineering for Operators/Non-Engineers" and a slightly more advanced course, "Chemical Engineering for Non-Chemical Engineers". This presentation and summary highlights some of the learnings this author and instructor has experienced in teaching the first of these courses for approximately 10 years. Approximately 7 years ago, this course was expanded from 2 to 3 days to incorporate significant topical areas relating to safety, reactive chemicals, chromatography, solids handling, extraction, drying, and filtration. The course CD, first produced in 1998, is still waiting funding to be upgraded both in its operating system and its content. This course is offered approximately 3 times per year publicly and on site as requested, normally an additional 3 times per year. Recent on site presentations, customized to some degree, have been made for FMC, the Department of Homeland Security, Gelita, and Victoria, TX Community College (for the benefit of local chemical industry through a state of Texas grant). This course is also available in condensed format through the ASME on line educational series and is being adapted for on line use by AIChE's new E-Learning program. It will be piloted this fall.

2. Course Content

This course covers, in a non-mathematical manner, all basic unit operations of chemical engineering, each using from ½ hour to 3 hours of the course:

A. Basic principles of chemical engineering
B. Safety and reactive chemicals
C. Economics

D. Stochiometry and mass balances
E. Fluid flow and fluid handling equipment, pumps, fans, physical property effects such as viscosity and Reynolds number
F. Mass transfer and separations including distillation, absorption/stripping, extraction chromatography
G. Evaporation and crystallization
H. Heat transfer, heat exchangers, heat transfer coefficients, and cooling towers
I. Filtration, drying, and solids handling
J. Process control and dynamics
K. Tanks and reactor vessels

Each of these topics has illustrations of both equipment and important design concepts and the course covers the response of processes and process equipment to operational changes. Major class exercises include analysis of a brewery, a bromine reactive chemicals incident, coffee brewing, and an oil refinery review. We also discuss some current issues that contain chemical engineering issues such as corn based ethanol and aspects of the current energy policy. When taught on site, analysis of a client's actual process is used as part of the class.

3. Observations

Over the years, these are some of the important learnings I have seen in teaching this class to a wide variety of participants which has included Department of Homeland Security inspectors (typically ex-military or FBI agents with possibly some high school chemistry 10-20 years previously), newly appointed technical managers who have ChE's reporting to them, plant and pilot plant operators, chemists who are working with ChE's on new process and product development, EPA lawyers, and mechanical or electrical engineers:

<u>A. Chemical reaction concepts and principles</u>. The understanding of the differences between endothermic and exothermic reactions is not just an issue relating to heat release, energy requirements, but also related to equilibrium concepts and the factors that affect equilibrium and how to shift and control equilibrium.

<u>B. Reactive chemicals concepts.</u> The potential hazards associated with uncontrolled chemical reactions are important. When a small scale beaker overheats in the lab, it's a mess and may injure someone, but that same uncontrolled reaction on a larger scale may cause hundreds of millions of dollars of damage and injure or kill many people. We use both the BP Texas City and the Dixie Sugar explosion to illustrate these points as they relate to improper process control and solids handling.

<u>C. Scaling variables.</u> Design variables such as kinetic rates and heat transfer do not scale in the same way. Linear vs. logarithmic scaling issues are discussed. This is critical in the scale up of exothermic reactions whose rates are increasing exponentially with

temperature while heat flow out is increasing linearly. The idea that a liquid viscosity can change by a factor of ten over a range of operating temperatures is usually a significant "Ahha".

D. Separation concepts. The common principles behind distillation, absorption, stripping, and chromatography are shown together to demonstrate how physical property differences can be used to decide on separation processes and how these processes are affected by temperature, pressure, and quality needed.

E. Solids handling issues are always interesting to discuss and the many horror stories of solids not flowing from bins, unexpected solids explosions, and subtle differences in solids formation and crystal characteristics affecting dissolution rates and solids handling. Physical property characterizations such as angle of repose are usually unknown to course participants.

F. Economics are discussed in the context of a large scale ammonia plant so that participants can see how to use economics to assist in deciding what process research variables are important.

G. Process control dynamics and the concepts of proportion al, integral, and derivative control are usually new to the participants as well as the fact that there's no "right" answer to any particular process control scheme. There are plusses and minuses and these must be thought through carefully. The differences between control in the lab and control in a large scale plant are always interesting discussions.

At the end of these courses, the attendees have a better understanding of:

- Why chemical engineers are (or should!) be asking certain questions
- The impacts of changing temperatures, concentrations, and pressures in a chemical reaction system
- The variables that are important in deciding how best to separate mixtures of materials
- The importance of process control
- The many variables affecting the design of solids handling equipment.

They also understand how to interpret process behavior and changes better than they did prior to the course.

(94e) Screencasts in Chemical Engineering Courses

John L. Falconer, Department of Chemical and Biological Engineering, University of Colorado, Boulder, CO, J. Will Medlin, Chem. & Bio Engineering, University of Colorado, Boulder, CO, Janet deGrazia, Chem & Bio Engineering, University of Colorado, Boulder, CO and Garret Nicodemus, Chem * Bio Engineering, University of Colorado, Boulder, CO

Screencasts, which are screen captures of a computer monitor display plus narration, were piloted in six chemical engineering courses, from freshmen to graduate level. They were produced on Tablet PCs using Camtasia Studio software and Windows Journal as well as other software. They were posted on line for easy student access, and are available on a web site for use by chemical engineering faculty and students at other universities. These short videos (approximately 10 minutes) were used to: (1) present solutions to example problems, (2) explain how to use software, (3) offer mini-lectures, and (4) clarify concepts in more detail than class time allowed. Screencasts can be used to present material so that class time can be used for active-learning activities. They can also be used to explain a question that arises during office hours instead of repeatedly explaining it. They are not professional-quality videos and thus can be prepared relatively quickly. Students can watch them on their schedules and can stop and replay. The feedback from students in these courses was overwhelmingly positive, with 95% of students in some courses rating them useful or very useful. Students found them to be effective supplements to class time and textbooks, and students requested more of them.

See more of this Session: Free Forum on Chemical Engineering Education (Part 1)
See more of this Group/Topical: Education

(94f) The Faculty Perspective On Student Interaction with a Department Culture

Lisa G. Bullard, Chemical and Biomolecular Engineering, North Carolina State University, Raleigh, NC, Donald P. Visco, Chemical Engineering, Tennessee Technological University, Cookeville, TN, David L. Silverstein, University of Kentucky, Paducah, KY and Jason M. Keith, Department of Chemical Engineering, Michigan Technological University, Houghton, MI

Each engineering department has a distinct departmental culture based on its history, faculty makeup, and geography, among other factors. Some departments have a long history with an established culture and traditions, while newer programs need to create their culture from scratch. This is important as creating and sustaining a desirable culture can have a significant impact on recruiting, retention, and the general satisfaction of its undergraduate students. In this work, we will review the engineering literature and describe different models and activities which departments have successfully used in establishing a successful culture. This includes:

- freshman courses - professional societies - departmental traditions - professional development courses - intentional interactions between faculty and students

We note that these activities also shape students' expectations about what it means to be engineers.

See more of this Session: Free Forum on Chemical Engineering Education (Part 1)
See more of this Group/Topical: Education

(94g) The Graduate Student as Leader

Edward P. Gatzke, University of South Carolina, Dept. of Chemical Engineering, Columbia, SC and Jed Lyons, Mechanical Engineering, University of South Carolina, Columbia, SC

As part of NSF funded "Creating Industry-Ready PhD Graduates" a course was developed to enhance various skills not typically explicitly covered in a traditional graduate curriculum. The course, "Professional Development and Ethics in Engineering and Computing," included a variety of topics, including: mentoring, teamwork skills, organizational skills, managerial skills, conflict resolution, intellectual property, and research ethics.

See more of this Session: Free Forum on Chemical Engineering Education (Part 1)
See more of this Group/Topical: Education

(115a) NSF CBET Overview and Other NSF Programs

John McGrath, Division of Chemical, Bioengineering, Environmental and Transport Systems, National Science Foundation, Arlington, VA

An overview of CBET programs

See more of this Session: NSF Workshop 1
See more of this Group/Topical: Education

(115b) Highlights of CBET Cluster On Biomedical Engineering and Engineering Healthcare

Theresa A. Good, Division of Chemical, Bioengineering, Environmental and Transport Systems, The National Science Foundation, Arlington, VA

An overview of the CBET cluster on biomedical engineering and engineering healthcare

See more of this Session: NSF Workshop 1
See more of this Group/Topical: Education

(115c) Highlights of CBET Cluster On Chemical, Biochemical & Biotechnology Systems

George Antos, Chemical, Bioengineering, Environmental and Transport Systems, National Science Foundation, Arlington, VA

cluster on chemical, biochemical and biotechnology systems

See more of this Session: NSF Workshop 1
See more of this Group/Topical: Education

(115d) Highlights of CBET Cluster On Transport and Thermal Fluids Phenomena

Henning Winter, Division of Chemical, Bioengineering, Environmental and Transport Systems, National Science Foundation, Arlington, VA

CBET cluster on Transport and thermal fluids phenomena

See more of this Session: NSF Workshop 1
See more of this Group/Topical: Education

(115e) Highlights of CBET Cluster On Environmental Engineering & Sustainability

Gregory Rorrer, Chemical, Bioengineering, Environmental and Transport Systems, National Science Foundation, Arlington, VA

CBET cluster on environmental engineering and sustainability

See more of this Session: NSF Workshop 1
See more of this Group/Topical: Education

(115f) Interactive Question and Answer Session with NSF Program Directors

Robert M. Wellek, Division of Chemical, Bioengineering, Environmental and Transport Systems, The National Science Foundation, Arlington, VA

A discussion with NSF program directors

See more of this Session: NSF Workshop 1
See more of this Group/Topical: Education

Evolution of Teaching Transport Phenomena Courses in Chemical Engineering Curricula

Benito Serrano Rosales[1*], Luis Enrique García García[1], Jonathan Edwin Rodríguez Ibarra[1], Raúl Zambrano Rangel[1] and Jesús Moreira del Rio[2].

[1] *Unidad Académica de Ciencias Químicas, Universidad Autonoma de Zacatecas, Campus UAZ Siglo XXI, Edificio 6, Carr. a Guadalajara. Km 6, Zacatecas Zac. Mexico, 98160.* beniser@prodigy.net.mx *, Tel 52-492-925-6690-Ext 6137.*

[2] *Faculty of Engineering, Chemical Reactor Engineering Centre, The University of Western Ontario, London, ON, N6A5B8, Canada.*

Introduction

Transport Phenomena began as a discipline with the goal of systematizing the common principles in unit operations and chemical reactors. It started around 1960 with several publications by Byron Bird, Warren E. Stewart and Edwin N. Lightfoot which culminated with the well known book "Transport Phenomena".

Unit operations such as distillation, lixiviation, liquid -liquid extraction, etc, share the same physical principle. It would be easier to treat all these unit operations as a whole in an undergraduate course rather than covering one by one.

R. Bird presented and published a paper in AICHE 2008 in Philadelphia, USA about the very interesting story of the book of Bird Stewart and Lightfoot, 1960, with second edition in 2002. A brief summary is presented now.

During the period 1941 to 1951 the subject of transport phenomena was not taught to undergraduate as a unified subject in any university in the United States. Fluid Flow was covered including the Bernouli equation, viscosity, Piseuille law, turbulence, etc. The diffusion of chemical species was generally covered in physical chemistry, but the mass flux and the diffusion coefficients were poorly defined. After 1945, the physics departments focused their attention in nuclear physics. The two unit operations books used at that time were *Principles of Chemical Engineering* by Walker, Lewis, McAdams and Guilliland (1937), and *Elements of Chemical Engineering*, by Badger and McCabe (1936). These books were application oriented and make little use of the principles of transport phenomena.

Olaf Hougen realized that the subject of transport phenomena was of interest in the future development of many topics in chemical engineering. Transport phenomena, as a discipline, definitively bears a "made in Europe" label: Maxwell (in England), Boltzman (in Australia), Newton (in England), Fourier (in France), Fick, Plank, Prandtl, Schlichting, Nusselt, Stefan and others (in Germany).

In September 1957, R. Bird was asked to give a very important lecture to the deans of engineering. The following was emphasized: It is important that the students see in one course the connections between the transports of the three entities (momentum, energy, mass) and the transport at the three levels (molecular, microscopic and macroscopic).

During the first half of August 1957, R. Bird wrote the text for chapters 1 to 12 of the book. During the next semester, the activities were continued with text preparation, making up problems, preparing figures, etc. In the fall of 1960, the book "Transport Phenomena" was finally published, and it got mixed reception. Young faculty members were generally enthusiastic, whereas older teachers were rather antagonistic. The authors of this new book felt that the transport phenomena course was supposed to play the supporting role of an intermediate physics course. This would enable chemical engineering teachers to do a better job in teaching unit operations, process control, kinetics and catalysis and other applied subjects. The authors also felt that the course could provide useful background for teaching courses in such diverse areas as mechanical engineering, metallurgy, soil mechanics, meteorology, and analytical chemistry.

Several books appeared shortly after Bird, Stewart and Lightfoot Book.
Heat, mass and momentum transfer, Rohsenow and Choi, 1961.
Momentum, heat and mass transfer, Bennett and Myers, 1962.

R. Bird closes his paper answering with "yes" the question posed in the title: Was a transport phenomena course in chemical engineering curricula inevitable?.
And according to the experience and studies of the main author of this paper, the scientific and technological community of those years (1960s, 1970s, and even early 1980s), were reluctant to accept transport phenomena as a new and unified discipline subject in Chemical Engineering curricula. Therefore, it is reasonable to do some research about the evolution of transport phenomena courses in several national and international universities and other educational institutions. It is also interesting to know how the teaching process has evolved due to advances in computational methods.

Methodology
The programs of the undergraduate course "Transport Phenomena" taught in chemical engineering in several Mexican and foreign universities were looked for using internet. Information was also asked directly to colleagues in several universities. It was possible to consult the programs of the following Mexican universities: Universidad Autonoma de San Luis Potosi, Universidad Michoacana de San Nicolas de Hidalgo, Unversidad Autonoma Metropolitana Iztapalapa, Instituto Tecnologico de Aguascalientes, Instituto Tecnologico de Celaya, Universidad de Yucatan, Universidad Veracruzana, Instituto Tecnologico de Durango and others.
International Universities: University of Guelph (Canada), University of Iowa (USA), Universidad Nacional Patagonia (Argentina), San Juan Bosco (Chile), Universidad de Sevilla (España), Universidad de Santiago de Chile (Chile), Universidad de Western Ontario, (Canada), Universidad Nacional Experimental de Guayana, Universidad de Buenos Aires (Argentina), University of British Columbia (Canada), Queens University (Canada), Mc Master University (Canada) and Youngstown State University (USA). At least 60 internet pages were investigated, and all these materials gave us a panoramic view about how this course is currently taught.

Different books from the years 1950, 1960, 1970, etc., were revised in order to understand the Chemical engineering curricula in those times: Badger and Banchero (1954), Foust (1971), Hougen and Watson (1955), *Unit Operations in Organic Syntesis*, P. H. Groogins (1958), *Fundamentals of Chemical Reaction Engineering*, C. D. Holland, (1979), *The Phenomena of Fluid Flows,* R. S. Brodkey, (1967), *Design of Equilibrium Stage Processes,* B. D. Smith, (1963), *Fluid Bed Heat Transfer*, J. S. M. Botterill, (1975), *Transport Processes and Unit Operations*, C. J., Geankoplis, (1978), *Mathematical Methods in Chemical Engineering*, N.R. Amundson, (1966), *Mass Transfer Operations,* R. E. Treybal, (1955, 1968), among others.

Additionally, the progress of continued editions of transport phenomena and unit operations books was studied thoroughly for different authors, revising their different book editions: *Transport Processes and and Unit Operations*, C. J., Geankoplis, editions of 1978, 1983, 1993, 2005. *Unit Operation Principles*, Mc Cabe and Smith, editions of 1957, 1980, 2001, *Fundamentals of heat, mass and momentum transfer*, Welty, Wicks, W, editions of 1982, 2001, *Chemical Reactor Principles*, J. M Smith, editions of 1974, 1986. And others.

The opinion of retired professors from different universities was also considered as a valuable contribution to this study, particularly the opinion of professors who went through the transition of teaching transport phenomena as a full course. The principal author of this material also contributed to this study since he went through all the process of transport phenomena courses evolution while being a student and even as a professor in his home university.

Results and Discussion

The results of this search are presented, and the discussion is performed observing the general characteristics of different periods of time.

Before 1985

It was observed that in general, transport phenomena courses before 1985 started dealing mainly with momentum, heat and mass balances in laminar flow. These balances were performed by using the differential element as well as the general equations. Some progress in teaching processes in transient state was found until 1985.

During this same time, unit operation and chemical reactor books began to include topics of transport phenomena in their first sections. Consecutive sections in the books still dealt with unit operations in a traditional way. It was observed that there were not relationships among sections and little usage of transport phenomena concepts. This is perhaps the reason of the slowly introduction of transport processes in undergraduate Chemical Engineering curricula.

A lot of universities had mass, heat and momentum balances taught as separate courses, followed by unit operations applying these concepts. For example mass transport was included as the first unit in distillation courses, heat transport as the first unit in the courses for the design of heat exchangers, momentum transport as the first unit in fluid flow courses. In other words, the study of mass, heat and momentum transport was not offered in one or two unified courses, but each of them was studied in the courses where they were supposed to have application. Undoubtely, proceding in this way the discipline of transport of mass, heat and momentum lost its identity, and did not systematize their common principles.

Although the purpose of these actions was to include transport phenomena in parts to the curricula, including them in different courses, it failed to unify a single "transport phenomena" course because teaching these topics and concepts in a separate way never allowed its systematization. Some universities also mixed transport phenomena with unit operation courses. However, and in spite of the knowledge of diffusion, balances in the differential element, usage of the general equations, the design methods in some courses were based on semi-empirical correlations, usage of tables with information, avoiding mathematical models and their solutions, velocity profiles, concentration profiles, etc. These design methods were used under very specific conditions that introduced some level of uncertainty when used in different situations for which they were deducted.

After 1985, with the development of computers

After 1985, some progress was observed. Little by little this discipline was conceived as a scientific fundament to design the industrial pieces of equipment. More topics were added to these subjects, for example the study and deduction of correlations to calculate the interface parameters: friction factor, heat and mass transfer convective coefficients in order to design pieces of different equipment. Also, the study of turbulent flow was introduced. This new knowledge of transport phenomena was incorporated in different optional subjects, offered in chemical engineering, and in other careers, such as: bio engineering, food engineering, bio genetics, fabrication of medicines, etc. Obviously, different approaches were considered for different disciplines. Also, after 1985 a huge development of computation was registered, and this allowed having more tools to solve almost all kind of equations or systems of equations, be linear algebraic, non linear algebraic, ordinary differential, partial differential, integral- differential, etc. In this way it was possible to pose a physical problem, and after the consideration of the phenomena included, to achieve the mathematical model, to solve it, and then to get information about the behavior of the system under the change of the different variables.

Indeed, this was a huge progress, because it was possible to include the mathematical modeling as a part of the design of the pieces of equipment and with this modeling it was possible to reduce the experimental work. Undoubtedly, for an undergraduate student, to be able to handle a relatively complicated system, with two or three coordinates, transient, etc, and to observe the behavior under changes in variables once the solution of the mathematical model is available, is a big motivation to pursue graduate studies, to do inventions, to understand better the physical problem, etc.

The development in computation has had a big contribution in the usage of transport phenomena and it is very significant in graduate studies, where many theses are based in mathematical modeling taking advantage of the available softwares.
All of this indicates that transport phenomena teaching advances according to the needs of equipment design where the physical and chemical processes are developed and also according to the scientific needs for new mathematical models.

Currently...

Now, according to the results of the survey with the programs of transport phenomena courses taught in these days, a variety of tendencies are observed. In some universities the book of Bird et al is followed without variation. At the elemental level, the transport phenomena and unit operations are mixed, two courses of transport phenomena are offered, etc, but the general tendency is to teach this course in combination with the available software, so that the solution of the mathematical models is illustrated, and in this way, a better discussion of the behavior of the system is achieved. The scientific and fundamental discussion of the deduction of the mathematical models, their solution, the deduction of the interphase parameters based on the theory when it is possible, etc, let us to understand that each time this subject is conceived as the scientific fundament for the design of the pieces of equipment in chemical engineering, mechanical engineering, biomedical engineering, food engineering, biotechnology, human body understanding, etc.

Opinion of the author

After a conscious analysis of the results found in the surveys, as well as the information regarding old courses in transport phenomena, etc, the opinion of this author is that Transport Phenomena subject in the undergraduate program in Chemical Engineering should be taught in two semesters with the following contents: First course.- Calculations of molecular transport parameters (viscosity, mass diffusivity and thermal conductivity), study of problems in laminar flow in steady state, using the differential element and general equations, for momentm, heat and mass transport, problems of non steady state using some analytical thechnique such as the method of separation of variables, and also finite differences method. Second course.- macroscopic balances (for mass, linear momentum, moment of momentum, energy) turbulent flow (nature of the turbulent motion, universal profiles for fluid velocity, temperature and concentration), study of the heat and mass transfer coefficients, friction factors in the interface (deduction of Reynolds number, Nusselt number, Pi Buckingham method, boundary layer theory, Blasius solution, Von Karman Method, analogies among mass, momentum and heat transfer), Mass transfer in the interphase (double layer theory). Currently these two courses are taught in the undergraduate program of chemical engineering in the Universidad Autonoma de Zacatecas Mexico. It should be emphasized that these courses are a preparation for the design of the equipment, providing the scientific fundamentals inasmuch as possible, and combining the mathematical models with experimental data, to estimate parameters.

It is important to conclude this material by reminding that the objective of Transport Phenomena should be to present in a fundamental, systematized and unified way the common principles of unit operations and chemical reactors. This will allow establishing the mathematical models for laminar and turbulent flow systems where transfer of mass, momentum and heat is carried out, to find the velocity, temperature and concentration parameters, through the solution of the mathematical models. All of these should be done with the goal of obtaining correlations for the parameters of phase and interface, as a preparation and tool for the modeling and design of the required equipment to perform unit operations and chemical reactions. This for the chemical engineering undergraduate program, but this subject can be extended to biology, human body study, and other engineering careers.

References.

1.- R. B. Bird. “Was a transport phenomena course in chemical engineering curricula inevitable?”, Annual Meeting of American Institute of Chemical Engineers, Philadelphia PA, USA., November -----2008.

2.- R. Byron Bird, Warren E. Stewart, Edwin N. Lightfoot, “Transport Phenomena”, Second Edition, John Wiley & Sons, Inc, 2002, ISBN 0-471-41077-2

(147b) Mass Conservation Principles: Macro. Vs. Micro. A Powerful Learning Road Map in the Scaling of Transport Phenomena

Jennifer Pascal[1], **Parvin Golbayani**[2] and Pedro Arce[1], (1)Chemical Engineering, Tennessee Technological University, Cookeville, TN, (2)Chemical Student, Tennessee Technological University, Cookeville, TN

The concept of mass is introduced to students during the early stages of their school activities. For example, high school physics (i.e. mechanics) is a typical place where mass is analyzed for its role in inertial forces. Concepts of mass are also discussed when talking about properties of matter such impenetrability, density, etc. At the college level, and in engineering in particular, mass plays an important role in introducing the concept of continua (Arce et al., 2010) among other related aspects. In the field of chemical engineering, it is imperative that students are knowledgeable of the various conservation laws (e.g. total mass, momentum, energy). Since mass is a concept students are generally familiar with, it didactically makes sense to begin the process of learning the aforementioned conservation laws, with mass. Furthermore, in chemical engineering, "mass" concepts need to be carefully characterized before attempting to write and understand a conservation principle. Since we deal with chemical reacting systems, the distinction between "total" mass and "species" mass is quite useful. Unfortunately, the usual textbooks deal with these types of masses in a "casual" way largely contributing to the misunderstanding of the role that "mass" plays in continuum conservations laws. Once students are confident in their knowledge of the "mass" conservation law, they can move to the more complex, or less familiar cases, such as the conservation laws of linear momentum and energy, for example.

In this contribution, after carefully analyzing the different kinds of "mass" and introducing the different types of "control domains," an integral, or macroscopic level formulation is used to capture all aspects of mass conservation. The authors will present a method to introduce chemical engineering students to the conservation of total mass (first) through a fundamental approach that builds upon the previous knowledge of the students. This approach begins with fully understanding what mass is (i.e. total mass vs. component mass), defining total mass from a continuum point of view, systematically deriving the conservation of total mass from a continuum point of view for closed and open systems, and finally incorporating the idea of scaling into the analysis by converting the macroscopic equations into microscopic equations. One novel aspect is the introduction of the learning concept based on "switches" both in time and space so that "tools" (such as the divergence theorem and the Reynolds Transport Theorem) are needed to conduct the scaling. The authors have found that the approach is highly beneficial for the student learning and sets a "template" or POK (Arce, 1994) for other conservation laws.

Arce, P. Jr. of Sci. Educ. and Technology, 3, 145, (1994a). Arce, P.E., J.A. Pascal, C.M. Torres.

Chemical Engineering Education. In press.

See more of this Session: Free Forum on Chemical Engineering Education (Part 2)
See more of this Group/Topical: Education

(147c) Learning Separations

Kamalesh K. Sirkar, Chemical, Biological and Pharmaceutical Engineering, New Jersey Institute of Technology, Newark, NJ

I will describe a different way of learning separations. Separation of a mixture involves movement of a specific species to a specific region in the separation system which gets enriched in that species. This is caused by species-specific forces driving a species relative to other species. The same argument is valid for particle separation. The forces involved are primarily based on chemical potential gradient, electrical potential gradient, centrifugal force field, gravity and magnetic field. The physical systems involved can be two immiscible phases, membranes or single-phase systems; further the separation involved may include chemical separation or particle separation. There may be chemical reactions facilitating the development of species-specific regions. In most chemical engineering applications, the separation systems are open and there is bulk flow in and out of the separation system. The interactions of the direction of the bulk flow with the direction of the species-specific force(s) introduce four major types of separation device/operational configurations: bulk flow parallel to the force(s); feed-phase containing bulk flow perpendicular to the force(s); countercurrent, cocurrent or cross flow of two bulk phases/ regions perpendicular to the force(s); continuous stirred tank separator. Couple these with a few additional operational aspects such as reflux, recycle, time-dependent operation, cascade etc., bulk of the separation universe becomes quite clear and understandable.

See more of this Session: Free Forum on Chemical Engineering Education (Part 2)
See more of this Group/Topical: Education

THE ACCUMULATION TERM: PERSPECTIVES IN TEACHING AND LEARNING UNSTEADY-STATE MATERIAL AND ENERGY BALANCES

Whitney Schaper, Lale Yurttas

Abstract

This study seeks to understand and explain the underlying issues that cause students to have difficulty solving transient or unsteady-state problems. In a survey given to sophomore level undergraduate students within the Artie McFerrin Department of Chemical Engineering at Texas A&M University, eighty percent of the students indicated that they find unsteady-state problems more difficult to formulate. The most prevalent reasons given to explain why unsteady-state problems were more difficult were that the students have less experience with unsteady-state, they have difficulty setting up the differential balance, the problems are more difficult to visualize conceptually, and the application of differential equations makes them more complicated. Professors who were interviewed also indicated that their students have difficulties in solving unsteady-state systems. They were asked to discuss their thoughts on why students are often unable to solve unsteady-state problems and what could be done to minimize this issue. From these interviews, a list of recurring ideas was compiled. Sophomore chemical engineering students were given two unsteady-state material balance problems to solve in one occasion and one unsteady-state energy balance problem in a second occasion. They were asked to do so in a group but were not allowed to use textbooks. Half of the groups were provided with a step by step problem-solving strategy for unsteady-state problems. In addition, three of the groups were recorded as they attempted to solve these problems. For comparison, three senior chemical engineers were given the same set of problems and asked to solve them without the use of a textbook. The solutions were then reviewed and graded. Errors were grouped into several categories.

In this paper, the authors will describe their preliminary findings and future plans to expand their studies to understand the student discomfort with transient systems. A web learning module on conservation of mass and energy that was developed for both steady and unsteady state systems will also be presented. The authors hope that their findings will help them to develop effective tools and methodologies that can be integrated throughout the curriculum for teaching unsteady-state systems.

Introduction

The conservation laws of mass, energy, and momentum are the basis of many problems in chemical engineering. These laws state that mass, energy, and momentum can neither be created nor destroyed. It is essential that students are able to correctly apply conservation laws to solve problems. Both steady-state problems, in which no variable in the system changes with time, and unsteady-state problems, in which at least one system variable does change with time, are basic principles which are fundamental to chemical engineering.[1] Failure to understand and correctly apply conservation laws to unsteady-state systems is a common mistake among undergraduate chemical engineering students.

Unsteady-state, or transient, processes are prevalent in chemical engineering. Batch and semi-bath processes always operate in an unsteady-state condition, and start-up and shutdown of equipment is an unsteady-state operation as well. The procedure for deriving balances on transient processes is much the same as that for deriving balances on steady-state systems, however the accumulation term, which is zero for steady-state systems, is now nonzero, which results in a differential equation as opposed to an algebraic equation.[1]

Identifying and Defining the Problem

It was observed by faculty at Texas A&M University that undergraduate students seem to have more trouble with unsteady-state problems than with steady-state problems. To gain more insight into the issue, four professors of chemical engineering were interviewed and asked why they thought students had difficulty with unsteady-state problems and what they thought could be done differently in the future to minimize such issues.

During the professor interviews, there were several recurring ideas about why students have difficulty with unsteady-state problems. All of the professors interviewed agreed that a lack of understanding in fundamental areas of mathematics, particularly in the area of differential equations, is a contributing factor. It was mentioned that students are not familiar with dealing with differential equations, particularly partial differential equations. One professor believed that it is not always made clear what the application of differential equations are as they are presented in the math department and that students find it difficult to make the connection between math and chemical engineering. In addition, several professors mentioned that students often begin to work on problems without having a thorough understanding of the concepts behind the problem, which could be a reason that unsteady-state problems present a challenge.

The professors were also asked to provide ideas to help students better understand unsteady-state problems in the future. The professors mentioned that practice with unsteady-state problems early on and often in a student's education will help them to gain experience and become better at solving unsteady-state problems. For example, unsteady-state problems should be given in all classes, and students should be taught to always begin with a differential balance and cancel out any terms that do not apply for the particular system. It was also discussed that engineering accounting should be taught uniformly across classes and that professors should emphasize to students the importance of deriving and understanding equations as opposed to just being able to use them. Another suggestion was having labs earlier in the chemical engineering education to demonstrate conservation principles.

Researching the Problem

A study was conducted of a class of chemical engineering sophomores as well as three senior chemical engineering students. The students were given two unsteady-state material balance problems to solve on one occasion and one unsteady-state energy balance problem on a second occasion. On each occasion, the students were allotted fifty minutes, and they were not allowed to use any outside resource such as textbooks or the internet. The sophomore students were allowed to work in teams, whereas the senior students were asked to complete the problems without help. In addition to the problems, the students were asked to complete a short survey.

For the first problem, half of the sophomore students were given an unsteady-state problem solving procedure. It was observed during the study, however, that most of the students that were given the procedure did not appear to be using it to assist in solving the problems. The procedure that the students were given is shown below.

Unsteady-State Systems

Problem Solving Strategy

Here is a step by step plan for approaching unsteady-state problems.

1. Draw and label a flow chart and show all the known and unknown information, by assigning variables to the unknowns.
2. Define:
   a. *System,* show the system boundary.
   b. *Time period*. Time period may be
      1. differential (an instant of time), or
      2. finite (two instances of time, such as the time right after initial loading and the time before withdrawal for a batch system)
3. Define *extensive property(ies)* to be counted. Extensive properties may be total mass, total moles, mass of a component, moles of a component. Concentration is not an extensive property.
4. Write your general balance equation in words for the extensive property that you are counting.
5. Define a math problem. This may be
   a. An *integral balance* if you have defined a finite time period. Each term in the equation then is an amount of the balanced quantity.
   b. A differential balance if you have defined a differential time period. Each term in the equation then is a rate, balanced quantity/unit time. This will be an ordinary differential equation and you must provide an initial condition for its solution.
6. Solve math problem.
7. Reality check - does the solution make physical sense? Verify that
   a. The initial condition is satisfied.
   b. The differentiation of your solution yields to the original balance equation.

The previously determined steady-state solution (if one has been determined) is approached as t --→ ∞ in the transient solution.

Figure 1

On the first occasion, there were eleven teams of sophomore students. These students had already taken material and energy balances class[1] and were enrolled in thermodynamics. They had covered chapters 1-8, and a significant portion of chapter11 from Felder and Rousseau[1] in their first sophomore class. The first problem given to the students was a simple mass balance problem. The problem statement is given below.

- Methanol is added to a storage tank at a rate of 1200 kg/h and is simultaneously withdrawn at a rate $\dot{m}_W(t)$ (kg/h) that increases linearly with time. At $t = 0$ the tank contains 750 kg of the liquid and $\dot{m}_W = 750$ kg/h. Five hours later, $\dot{m}_W$ equals 1000 kg/h.
    a. Calculate an expression for $\dot{m}_W(t)$, letting $t = 0$ signify the time at which $\dot{m}_W = 750$ kg/h, and incorporate it into a differential methanol balance, letting M (kg) be the mass of the methanol in the tank at any time.
    b. Integrate the balance equation to obtain an expression for *M(t)*. For now, assume that the tank has an infinite capacity.
    c. Calculate how long it will take for the mass of the methanol in the tank to reach its maximum value, and calculate that value. Then calculate the time it will take to empty the tank.

Of the sophomore students, 82% answered the first problem correctly. Two teams that answered the problem incorrectly made errors in the integration that caused them to get the answer wrong. Even though these two teams were given the problem strategy in Figure 1 which clearly states that they should provide an initial condition, they both failed to follow this step as in the following sample taken from a sophomore team's paper:

$$\frac{dm}{dt} = rate\ in - rate\ out = 1200(kg/h) - (50t + 750)(kg/h)$$

$$\frac{dm}{dt} = 450 - 50t$$

$$\int dm = \int (450 - 50t)dt$$

$$m(t) = 450 + 25t^2$$

Many solutions even though lead to a correct answer lacked the steps in Figure 1, indicating lack of understanding of concepts as well as difficulty in relating math and chemical engineering. Expressions such as

$$m = rate\ in - rate\ out$$
$$dm = 450 - 50t$$

were common except in a few cases.

Two senior students completed the first problem. One of them got it entirely correct, while the other made a minor error in math in the last step of the problem that caused him to get the answer incorrect; however his overall approach to the problem was correct. All student work, senior or sophomore, showed that the students have difficulties in relating math to real variables such as concentration, mass, and others as can be seen from an example taken from one of the seniors' paper:

$$\int \frac{dCH_3OH}{dt} = \int 1200 - (750 + 50t)$$

The second problem given to the students on the first occasion was more complex and involved a gas mixture, thus variable composition, and the use of ideal gas law. These added dimensions lead to confusion among students. Most teams did not have enough time to complete the problem, and only one team was able to finish within the time allotted. This team although got the correct answer, they had set up the differential balance using concentration as their variable. Students usually have tendency not to think in terms of extensive variables when applying conservation laws and frequently apply these laws on intensive variables like composition, one example:

$$\frac{dx(t)}{dt} = input - output$$

On a separate occasion, another problem was given to the sophomore chemical engineering students. It combined principles of conservation of mass and energy and required the students to identify that mass of the system with respect to time was constant, whereas energy was not. To solve the problem, an energy balance had to be performed on the system. The problem statement is given below[2].

- The mixing tank shown below initially contains 50 kg of water at 25°C. Suddenly the two inlet valves and one outlet valve opened, so that two water streams, each with a flow rate of 5 kg/min flow into the tank, and a single exit stream with a flow rate of 10 kg/min leaves the tank. The temperature of one inlet stream is 80°C, and that of the other is 50°C. The tank is well mixed, so that the temperature of the outlet stream is always the same as the temperature of the water in the tank.
  (a) Calculate the steady-state temperature that will finally be obtained in the tank.
  (b) Develop an expression for the temperature of the fluid in the tank at any time.

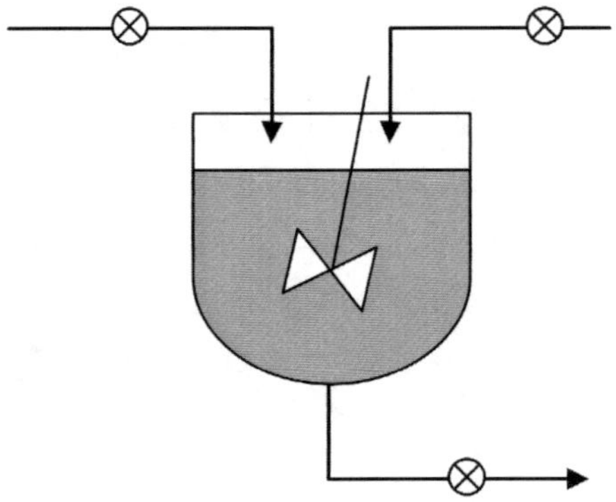

⇨Note that for a liquid C_P=C_V, and assume that heat capacity is not a function of temperature.

On this occasion, there were 13 teams of sophomore students. For this problem, 77% of the teams correctly answered part (a), while no team correctly answered part (b). The primary reason behind the incorrect results was that energy balances were not properly set up. Many of the teams started with an incorrect form of the energy balance. It was also noted that several of the teams did not start from the most general energy balance, and this caused the solutions to be incorrect. In addition, the students have been taught to follow a certain procedure to solve unsteady-state problems (shown above). The majority of the teams did not follow the steps given in the procedure. For example, only one team of students explicitly defined the system as they worked through the problem on paper. The final step to the unsteady-state problem solving procedure indicates that students should check to see if the solution makes sense and if the initial condition is satisfied. In this problem, since the steady-state solution was determined in part (a), it is also appropriate to check to see that the temperature found in part (a) is approached as time approaches infinity. Had the students checked their differential balance, they would have realized that their answers were incorrect.

The senior that completed the problem also got part (a) correct and part (b) incorrect. As was the case with many of the sophomores, the energy balance was not set up properly, no integration was performed, and the unsteady-state procedure was not followed.

In addition to the two problems, the students were also asked to complete a survey. They were asked which type of problem is more difficult for them to formulate, with the answer options being (a) Steady-state, (b) Unsteady-state, and (c) No preference. They were then asked why. Of the twenty surveys turned in, 80 percent of the students answered that they find unsteady-state problems more difficult to formulate, and 20 percent answered that they have no preference. The most prevalent reasons given to explain why unsteady-state problems were more difficult were that the students have less experience with unsteady-state, they have difficulty setting up the differential balance, the problems are more difficult to visualize conceptually, and the application of differential equations makes them more complicated. Three of the four students that answered no preference explained that both unsteady-state and steady-state are difficult for them.

Conservation of Mass Web Module

Although not as part of this study but as part of an NSF supported Departmental Level Reform Grant[3], but with the same motivation in mind as this study, Artie McFerrin Department of Chemical Engineering has developed web based modules called interlinked curriculum components, or ICCs. An ICC is a web-based resource for teaching and learning as well as a 'chunk' of material that is significantly smaller than a typical semester course. ICCs can be used and reused in multiple courses throughout the curriculum. ICC on conservation of mass in particular helps students to improve their understanding of the concepts and includes several elements in solving problems that deal with both:

- Steady-state and unsteady-state problems

- Non-reactive and reactive systems
- Single and multiple unit processes

Examples and exercises sections involve several unsteady-state and steady-state examples. Each example takes the user through the five-step problem-solving framework. To enhance learning, active graphics and video features provide visual input to the user. Examples are presented using a consistent problem-solving framework so that student, when they encounter a new problem, will have a strategy and set of questions that they can ask themselves if they do not immediately formulate an approach for solving the problem. Use of the five-step problem-solving framework is reinforced through its consistent presence and appearance throughout each example[4].

The module may be accessed at: http://alcheme.tamu.edu/

Conclusion

The results of the study show that undergraduate students are more comfortable working with steady-state systems, largely because they have more experience with these problems and because steady-state systems result in balances that require simpler math. In addition, the students that participated in the study performed better on the problem that required a mass balance than the problem that required an energy balance. This suggests that the students are more comfortable with material balances, and this is likely because they have more experience working with material balances than energy balances.

The results of this study are a step towards identifying the source of students' shortcomings when it comes to solving unsteady-state systems. Findings suggest that reinforcement of unsteady-state systems across all engineering classes may be a step towards improving student performance on such problems.

Limitations

There are several limitations to the study presented herein. One limitation is that only one chemical engineering senior completed both problems. In the next phase of this study, a larger number of chemical engineering seniors will be asked to participate for comparison to the sophomores. A second limitation is that on the first occasion that the problems were given to the sophomore students, half of them were provided with an unsteady-state problem solution procedure. In the future of this study, the researchers would like to first establish a base-line by offering the students no help whatsoever and then see how their performance improves when they have the procedure. Finally, the second problem that was given to the students on the first occasion of the study was decided to be too complex for studying students' understanding of unsteady-state problems. In the future, the problems will be more carefully selected so the students do not waste time solving a problem that ultimately cannot be used to deduce results.

Literature Cited

1. Felder, Richard M., and Ronald W. Rousseau. *Elementary Principles of Chemical Processes*. 3rd Edition, John Wiley and Sons, 2005.
2. Sandler, Stanley. Chemical and Engineering Thermodynamics. 3rd Edition, John Wiley and Sons, 1999.
3. "An Integrated Approach to Chemical Engineering Undergraduate Curriculum Reform," Yurttas, L, Pchenitchnaia L, Froyd, F, El-Halwagi, M, Glover, C, Osborne-Lee, I, Mills, P, Pelehvari, A, Proceedings of the 7th International Conference on the Foundations of Computer-Aided Process Design, FOCAPD 2009.
4. "A Web-based Approach to Teaching Conservation of Mass in a Chemical Engineering Curriculum," Lale Yurttas, Jeff Froyd, Mahmoud El-Halwagi, Charles Glover, Adam Shaw, Zachry Kraus, Jean Layne, 2007 Annual Conference, Honolulu, Hawaii.

(147e) Lessons and Discoveries From Teaching (nearly all) the Core Chemical Engineering Courses Over a Three Year Period

Eric L. Maase, Chemical and Biomolecular Engineering Department, Lafayette College, Easton, PA

Over the last three years I have had the unique experience of teaching nearly the entire core course sequence in chemical engineering. The courses taught during this period include: • Computational Methods and Computer Programming for Chemical Engineers • Thermodynamics I (General Principles) • Thermodynamics II (Theoretical and Applied Modeling of Equilibrium) • Process Control and Dynamics • Unit Operations I • Separations (sometimes called Rate Ops or Unit Ops II) • Mass Transport Laboratory • Unit Operations I Laboratory • Unit Operations II Laboratory For the program at Lafayette College the above list represents nine of the 14 core requirements and in two of the missing core courses I have had significant previous experience: Introduction to Material and Energy Balances (Intro ChemE) and Reaction Kinetics. The few contemporary core courses that in which I have not been the primary instructor include Transport and the Capstone Design sequence. In the process of what represents nearly redoing my original undergraduate degree I was presented with both unexpected challenges and also provided with an unusual opportunity to perceive our curriculum from a new perspective. My presentation aims to first outline our present curricula considering the goals of accreditation (ABET) alongside the objectives and requirements of specific and individual course objectives. Following this foundation approach the curricula will be discussion in regard to the issues of what students learn in comparison to what we teach – specifically do our course and degree objectives and the outcomes match? Several curricula improvement suggestions are proposed that range from the somewhat simple (communication) to the complex (integration) that may provide value to both students and to broader field of chemical engineering as we continue to move forward with new challenges.

See more of this Session: Free Forum on Chemical Engineering Education (Part 2)
See more of this Group/Topical: Education

(147f) Emphasizing the Biological Side: Graduate Certificates and Secondary Majors

John R. Schlup, Department of Chemical Engineering, Kansas State University, Manhattan, KS

In this day when biological science plays an increasing role in science, engineering, and every day life, the question of how to incorporate interdisciplinary topics in the usual university environment becomes important. Two approaches taken at Kansas State University to address this need, the graduate certificate and the secondary major, will be described. A graduate certificate in biobased products and biofuels has been initiated to recognize and encourage the interdisciplinary nature of this rapidly evolving field. More and more undergraduate students have an interest in biological topics, and applications are appearing in nearly every engineering discipline. In the College of Engineering, the secondary major has been adopted as one alternative to meeting the needs of students while allowing them to pursue the major of their choice.

See more of this Session: Free Forum on Chemical Engineering Education (Part 2)
See more of this Group/Topical: Education

(147g) Re-Envisioning Particle Technology: Filling the Void(age) in the Chemical Engineering Curriculum

Daniel Lepek, Department of Chemical Engineering, The Cooper Union, New York, NY

Although approximately 60-70% of all chemical engineering processes involve the solid phase in some step, whether as a raw material or a final product, many chemical engineering curriculums do not include a course on solids or particle technology. In this talk, the development of the Particle Technology course at The Cooper Union will be discussed.

This course, which consisted of a mixture of undergraduates and graduate students, introduced students to the role of particulates in the petrochemical, pharmaceutical, and other chemical process industries. In doing so, students were able to link previous coursework (fluid mechanics, heat transfer, materials science, numerical methods, mechanics) to problems and situations involving particles, either in processing, or in situations such as handling and storage. In order to further develop this course, computer-based examples using COMSOL, MATLAB, and FLUENT were introduced to provide insight into complex systems involving multiphase flow and equipment.

This talk will provide interested instructors with an introduction to the importance of a particle technology course in the chemical engineering curriculum.

See more of this Session: Free Forum on Chemical Engineering Education (Part 2)
See more of this Group/Topical: Education

(162a) Proposal Writing Tutorial

John R. Regalbuto, U. Illinois at Chicago, Chicago, IL and **Gregory Rorrer**, Chemical, Bioengineering, Environmental and Transport Systems, National Science Foundation, Arlington, VA

This will be a tutorial in proposal writing.

See more of this Session: NSF Workshop 2
See more of this Group/Topical: Education

(162b) Interactive Breakout Panels

Robert M. Wellek, Division of Chemical, Bioengineering, Environmental and Transport Systems, National Science Foundation, Arlington, VA

The programs at NSF will have interactive breakout panels.

See more of this Session: NSF Workshop 2
See more of this Group/Topical: Education

(192a) Program Preparation Prior to An ABET Visit

Gary K. Patterson, Department of Chemical Engineering, Missouri University of Science and Technology, Rolla, MO

Place holder for an invited talk on the title

See more of this Session: ABET Updates and Insights
See more of this Group/Topical: Education

(192b) Evaluator Preparation Prior to An ABET Visit

Edward Rosen, EMR Technology Group, St. Louis, MO

This is a place holder for the title.

See more of this Session: ABET Updates and Insights
See more of this Group/Topical: Education

(192c) A Typical ABET Visit Schedule

Jeffrey J. Siirola, Eastman Research Division, Eastman Chemical Company, Kingsport, TN

A placeholder for the title

See more of this Session: ABET Updates and Insights
See more of this Group/Topical: Education

(192d) Post ABET-Visit Interactions

Gary K. Patterson, Department of Chemical Engineering, Missouri University of Science and Technology, Rolla, MO

Place holder for the title

See more of this Session: ABET Updates and Insights
See more of this Group/Topical: Education

(192e) ABET Accreditation Issues

Edward Rosen, EMR Technology Group, St. Louis, MO

Place holder for title...

See more of this Session: ABET Updates and Insights
See more of this Group/Topical: Education

(192f) ABET Evaluator Expectations

Gary K. Patterson, Department of Chemical Engineering, Missouri University of Science and Technology, Rolla, MO

Place holder for title.

See more of this Session: ABET Updates and Insights
See more of this Group/Topical: Education

(192g) ABET Accreditation Resources

Jeffrey J. Siirola, Eastman Research Division, Eastman Chemical Company, Kingsport, TN

Place holder for title

See more of this Session: ABET Updates and Insights
See more of this Group/Topical: Education

(229a) Why University-Industry Partnerships Matter

Anthony M. Boccanfuso, The National Academies, University-Industry Demonstration Partnership, Washington, DC

Increasing attention is being paid to university-industry research partnerships since they are perceived to be critical to advancing the discovery to innovation process and competitiveness.

While several advantages of these type of collaborations are generally agreed upon, there are numerous challenges and barriers and this session will discuss a number of topics:

* The advantages of engaging in university-industry partnerships

* The challenges faced in developing effective collaborations

* The various perspectives of companies, universities and the faculty and industry scientists and engineers who pursue these type of partnerships

* Suggestions for identifying appropriate partners

* The impact of government agencies (state and federal) and foundations on these partnerships.

and finally, an update on the activities of the National Academies' University-Industry Demonstration Partnership and the projects being pursued to achieve its mission.

See more of this Session: Managing Technology Collaborations I - Academia, Industry and Government
See more of this Group/Topical: Management Division

(229b) IP Challenges in University-Industry Collaborations

Ken Horton, Gore School of Business, Westminster College, Salt Lake City, UT

Universities and their Industry partners generally have the same business goals when they collaborate on research and development. One of the exceptions to this general rule, though, arises when the partners discuss the topic of the intellectual property (IP), who owns it, and how it is used. Some of these challenges are a result of the traditional business models that are used in the collaboration. These challenges can be overcome by utilizing new business models in the collaboration that treat the IP--and how it is used--differently.

See more of this Session: Managing Technology Collaborations I - Academia, Industry and Government
See more of this Group/Topical: Management Division

(229c) Managing Multilateral Technology Collaborations at the Institute for Collaborative Biotechnologies

David H. Gay, Institute for Collaborative Biotechnologies, University of California, Santa Barbara, Santa Barbara, CA

The Institute for Collaborative Biotechnologies is a uniquely interdisciplinary research and development alliance. Led by the University of California, Santa Barbara, in collaboration with MIT and Caltech, together with partners from the Army and industry, the ICB transforms biological inspiration into technological innovation. The presentation will address opportunities, challenges, progress achieved and lessons learned in managing multilateral technology collaborations between the Army, industry and academia.

See more of this Session: Managing Technology Collaborations I - Academia, Industry and Government
See more of this Group/Topical: Management Division

Opening Stem Cell Research and Development: A Policy Proposal for the Management of Data, Intellectual Property, and Ethics

Krishanu Saha[1*], David E. Winickoff[2*], and Gregory D. Graff[3*]

[1] Whitehead Institute for Biomedical Research, Cambridge, MA 02142

[2] Department of Environmental Science, Policy, & Management, Univ. of California, Berkeley, Berkeley, CA 94720

[3] Department of Agricultural and Resource Economics, Colorado State Univ., Fort Collins, CO 80523

[*]Authors contributed equally to this work.

Abstract

The need to strike a better balance between free sharing and proprietary and regulatory restraint in the life sciences is becoming an important policy concern in the fields of health policy, law, and bioethics. Here we explore these issues in the field of stem cell research. Expanded funding for stem cell research and development holds unique promise for advancing medicine and human welfare. However, technical, proprietary, and regulatory conditions are not ideal: closed information, congested intellectual property, and regulatory complexity are likely to slow innovation, skew attention towards large markets, and prevent ethical settlement. Accordingly, we propose a new form of collective action among funding agencies, scientific journals, and research institutions to centralize data, cell lines, and technologies and to design and disseminate research tools. The goal is more efficient technology exchange within the research community, orientation of product development towards pressing public health needs, and promotion of 'best ethics' in research and development.

Three Pervasive Problems in Stem Cell R&D

Within each of three policy domains current conditions present serious problems for the pace of innovation, the distribution of resulting health benefits, and the public accountability of stem cell research. Moreover, these problems interact and are mutually compounding across domains.

Exchange of Data and Materials

Enhanced capacities to produce, manage, and disseminate data through new information technologies have fueled debates about the manner and timing of data release in the sciences [1]. Traditional norms around sharing research inputs have run headlong into the desire of scientists and their institutions to protect their interests in potentially valuable data, materials, and research results. Practices of delay and secrecy have resulted from the increased relevance of university life sciences to commercial markets and private investors [2, 3]. One mark of change is the proliferation of material transfer agreements between universities [2, 4]. Recent studies suggest that data withholding is becoming common and problematic [5].

Intellectual Property Rights

Following patenting trends elsewhere in the life sciences, scientists and their institutions have been claiming the essential building blocks of stem cell research as private assets. In the specific case of stem cell lines, the first and most obvious problem concerning materials arose from the combination of the Bush Administration's restrictive funding policies and the commanding patent position of the Wisconsin Alumni Research Foundation [6, 7]. Recent analyses show a significant rate of accumulation of new patents over stem cell technologies [8-11], with problematic implications for downstream innovation [12-14].

Bioethics and Regulations

As if technical and proprietary complexities were not enough, disagreement over ethics has given rise to two major problems for efficient and accountable governance of this work. First, in the U.S. federal moratorium on new human embryonic stem cell (hESC) lines has resulted in a vacuum both of funding and oversight. Even as private and state-funded research moves ahead, a national framework is lacking apart from some 2005 guidelines from the National Academy of Sciences. Second, the resulting state programs have proliferated a patchwork of state regulations that differ, sometimes only slightly, on a range of issues. To amplify matters, similar differences occur globally across national regulatory regimes.

Towards an Integrative Approach

Scientists and policymakers have, of course, identified each of these issues, and important initiatives in stem cell research have been undertaken. These efforts should be applauded, but, across the board, they remain inadequate on two counts. First, none of them goes far enough to solve the particular problem it targets. Second, by targeting technical, IP, or regulatory problems in isolation these efforts neglect the important interconnections between these domains. Overlooking interaction between the different domains involves both a conceptual and a practical error. The conceptual error is to ignore the profound ways in which these domains are mutually constitutive categories. Norms and practices of sharing data and materials are simultaneously issues of property and of ethics. Questions of sharing data and materials raise important questions of property which, in turn, raise questions of ethics. To separate these questions is to perform a conceptual purification that, at a minimum, prevents optimal solutions. Yet, the overlap is not merely conceptual, it is practical. As the recently encountered problems with some of the NIH hESC lines illustrate, ethical accountability can be promoted only to the extent that provenance characteristics and data about cell lines are well known and shared [15, 16]. Efficient progress of ethically accountable stem cell research requires considerations that span the complexities and bottlenecks across all three domains. Not only would an integrative approach be more efficient, given the interdependencies, an integrative approach may be the only way to solve some of the individual problems.

The Logic of Collective Action

The problems discussed above have not been adequately addressed largely because they are classic collective action problems, in which a diverse set of actors share common interests yet no one individual actor's incentives are sufficient to overcome the inertia of inaction. Mobilization requires leadership in the form of coordination and initial investment. Sufficient conditions for collective action have not quite yet emerged in any one of the three domains, let alone across them. And, given the currently decentralized pattern of funding and regulation in the U.S.—despite the articulation of clear interests among many parties—there is no clear leader. Under these circumstances, it is necessary to motivate *many, heterogeneous* potential actors through the use of various carrots and sticks. We argue that collective action with a push from major funding agencies can be a basis for opening up stem cell research in the face of multiple compounding constraints. It is precisely the major funding agencies, such as CIRM or the NIH, that have important roles to play in supporting and enhancing common-access resources in stem cell research. Such an "opening up" would result in a more efficient exchange of data, and with further action, sufficient access to research tools within the stem cell research community. We propose four specific actions for stem cell policy:

1. Lead the development of practical common norms

First, major funding agencies could convene interested segments of the research community to construct a practical normative framework for how stem cell materials and data should be managed and circulated within the community. Participation in the efforts above indicate that this group is heterogeneous and large—including scientists, lawyers, ethicists, state funding agencies, nonprofit agencies, universities, and industry. Leadership from major funding agencies is needed to structure this discussion in finding common ground. This commonly developed normative framework should seek to redress the negative effects spanning the technical, proprietary, and ethical domains.

2. Require access in research contracts

Second, major funding agencies should use funding contracts to require the sharing of research data and materials. By many estimates, stem cell lines and the fundamental protocols of cellular reprogramming and differentiation constitute common resources for the research community that are expected to drive broad progress in biomedical research. Thus, cell lines contain the hallmarks of what have previously been called "community resource projects" [17, 18]. Accordingly major funding agencies could classify the development of new human stem cell lines as a "community resource project". A key consequence of invoking this norm would be to include sharp stipulations in research contracts concerning access to stem cell lines and their associated technologies. The stipulations need not use a simple "open access" model, but could be structured in a variety of creative ways that would increase their flow while respecting the private rights and claims of inventors and their institutions [19].

3. Build a common data resource

Third, major funding agencies should support a common data resource that encompasses all of the necessary technical, IP, and ethics information that will be used by its funded scientists, and hopefully, by the community of stem cell scientists globally. The lack of success in this area to date suggests that generating and sustaining support and interest for such an initiative will require new carrots and sticks, and complementary efforts among major funders.

4. Foster further collection action

Fourth, major funding agencies should support further efforts to coordinate, design, build, and disseminate research tools and platform technologies to alleviate the most important technical, IP, and ethical bottlenecks facing the field. These first three actions can construct the foundation for further collective action in which research institutions themselves, and not the government, could to take the lead. With sufficient data available, consortia of interested research institutions could perform systematic analyses of key stem cell lines, research tools, or other platform technologies where some combination of technical, proprietary, and ethical bottlenecks have stymied work at the participating institutions and/or their commercial partners. Analysis of particular constraints for a particular cell line, research tool, or other platform technology is a time-intensive task that, if centralized and done once for many, would have significant efficiency gains for scientists, technology transfer offices, and ethics review boards alike. Informed by the results of those common analyses the consortia could go so far as to designing new technologies that work around the most important technical, IP, or ethical bottlenecks. These new technologies could then be developed and made available for both research use and commercial development of downstream applications through collective licensing or "protected commons" strategies—such as open source or patent pooling. The constituent components would be ethically vetted and chosen to meet regulatory standards across multiple jurisdictions. Such creative initiatives have taken root in the fields of agricultural biotechnology and synthetic biology [19-22], and could prove useful for stem cell research [23].

Conclusions

The field of stem cell research is full of promise. Yet, those who seek today to realize that promise must wade through a morass of data and material transfer issues, navigate or negotiate numerous intellectual property claims, and stand in peril of violating ethical or formal regulatory requirements. Major funding agencies can and should do more with its leadership than simply fund more research, taking a more active role in expanding the institutional underpinnings of a science commons that is under strain. Such leadership could catalyze fundamental normative and institutional changes that would expedite the exchange of technologies within the research community, orient investments over a wider range of public health needs, and promote a more ethically sound advance of stem cell R&D.

References

1. Rai, A.K. and R.S. Eisenberg, *Harnessing and Sharing the Benefits of State Sponsored Research.* Berkeley Technology Law Journal, 2006. **21**: p. 1187-1213.

2. Merrill, S.A. and A.M. Mazza, *Reaping the Benefits of Genomic and Proteomic Research: Intellectual Property Rights, Innovation, and Public Health.* 2006: National Academy Press.

3. Merges, R.P., *Property Rights Theory and the Commons: the Case of Scientific Research.* Social Philosophy and Policy, 1996. **13**: p. 145-167.

4. Ku, K. and J. Henderson, *The MTA?rip it up and start again?* Nature Biotechnology, 2007. **25**: p. 721-721.

5. Caulfield, T., et al., *Patents, commercialization and the Canadian stem cell research community.* Regen Med, 2008. **3**(4): p. 483-96.

6. Wadman, M., *Licensing fees slow advance of stem cells.* Nature, 2005. **435**: p. 272-273.

7. Loring, J.F. and C. Campbell, *Intellectual Property and Human Embryonic Stem Cell Research.* Science, 2006: p. 1716-1717.

8. Esmond, R.W., R.A. Schwartzman, and T.J. Ebersole, *Stem Cells: The Patent Landscape.* Intellectual Property & Technology Law Journal, 2006. **18**(1): p. 1-4.

9. Glanzel, W., *Stem Cells - Analysis of an emerging domain of scientific and technological endeavour.* Steunpunt O&O Statistieken, K.U. Leuven, 2004.

10. Campbell, D., M. Noiseux, and G. Cote, *Potential for Stem Cells Science and Technology in Canada: Great Promises and Challenges.* Science-Metrix report, 2004: p. 63-63.

11. Holden, C., *Wisconsin Stem Cell Patents Upheld.* Science, 2008. **319**(5870): p. 1602b-1603-1602b-1603.

12. Scheinfeld, R.C. and P.H. Bagley, *The current state of embryonic stem cell patents.* New York Law Journal, 2001. **226.**

13. O'Connor, S.M., *Intellectual Property Rights and Stem Cell Research: Who Owns the Medical Breakthroughs?* New England Law Review, 2005. **39**.

14. Simkin, M., *How will intellectual property factors affect the commercialization of stem cell therapeutics.* 2005: BioEurope.

15. Streiffer, R., *Informed Consent and Federal Funding for Stem Cell Research.* Hastings Center Report, 2008. **May-June 2008**: p. 40-47(8)-40-47(8).

16. Baker, M., *Consent issues restrict stem-cell use.* Nature, 2008. **454**(7204): p. 556-556.

17. Wellcome Trust. *Sharing data from large-scale biological research projects: a system of tripartite responsibility. Wellcome Trust Meeting Report, January 14-15, 2003, Fort Lauderdale, FL.* 2003. Fort Lauderdale, FL.

18. Human Genome Program. *Summary of principles agreed at the first international strategy meeting on human genome sequencing.* 1996. Bermuda.

19. Winickoff, D.E., K. Saha, and G. Graff, *Opening stem cell research and development: A policy proposal for the management of data, intellectual property, and ethics.* Yale Journal of Health Policy, Law, and Ethics, 2009. **IX**(Winter).

20. Atkinson, R.C., et al., *Intellectual property rights. Public sector collaboration for agricultural IP management.* Science, 2003. **301**(5630): p. 174-5.

21. Rai, A. and J. Boyle, *Synthetic Biology: Caught between Property Rights, the Public Domain, and the Commons.* PLoS Biol, 2007. **5**(3): p. e58-e58.

22. Porceddu, M.B.C., et al., *Constructive Approaches to Intellectual Property Complexity in Today?s Agricultural Technology World.* Plant Mol. Breed, 2007. **5**: p. 294-295.

23. Winickoff, D.E., K. Saha, and G. Graff, *Opening stem cell research and development: A policy proposal for the management of data, intellectual property, and ethics.* Yale Journal of Health Policy, Law, and Ethics, 2008. **VIII**(Winter).

(247a) Introducing Risk Analysis in a Design I Course

Dimitrios V. Papavassiliou, Chemical, Biological, and Materials Engineering, University of Oklahoma, Norman, OK, Margaret Freeman, Chemical, Biological and Materials Engineering, University of Oklahoma, Norman, OK and Georgia Kosmopoulou, Economics, University of Oklahoma, Norman, OK

A major challenge that Chemical Engineering graduates will face at the modern workplace is the management and operation of plants under conditions of uncertainty. Developments in the field of industrial organization of microeconomics offer tools to address this challenge with rather well developed concepts, such as decision theory and financial risk analysis. In this work we will describe our efforts to integrate modern economics concepts and techniques in the teaching of Engineering Design.

The Chemical Engineering curriculum at the University of Oklahoma includes three courses in Design: Design Lab (2 credit hours), Design I (3 credit hours) and Design II (3 credit hours). The capstone projects are part of Design II, while Design I is the course that introduces engineering economics to students, as well as concepts such as profitability and cost analysis. The majority of the students who enroll in Design I have no background in Economics, since introductory Economics courses are part of the general education electives (as are several other courses in the social sciences). The type of engineering economics taught to them focused on cost analysis and profitability estimation with limited references to uncertainty and financial risk. For the past several years, we have included material on risk analysis and decision making in the Design I class. Lately, through collaboration between Chemical Engineering and the Department of Economics, we have been developing classroom material to introduce risk analysis in the class. In a presentation in the 2009 AIChE meeting [1] we discussed the development of illustrative classroom experiments and case studies that showcased concepts such as utility and decision making under uncertainty in Design I. The discussion in this presentation will include the development and implementation of assignments that utilize the concepts of risk and utility, and the use of software (Crystall Ball by Oracle) to incorporate financial risk in engineering economics.

References

1.Papavassiliou, D.V., Jog, C., Le, P., Freeman, M., and G. Kosmopoulou, "Development of Classroom Experiments to Introduce Decision Making Under Uncertainty to Students in a Design I Course," paper 329b, American Institute of Chemical Engineers Annual Conference, Nashville, November, 2009.

See more of this Session: Steal This Course!
See more of this Group/Topical: Education

(247b) Incorporating the Concept of Financial Risk Into Instruction in Economic and Profitability Analysis for Chemical Engineering Design

Michael J. Solomon, Chemical Engineering, University of Michigan, Ann Arbor, Ann Arbor, MI

Economic and profitability analysis is a key component of the process and product design instruction in the discipline of chemical engineering. Likewise, risk analysis and management is a crucial activity and responsibility of firms in the chemical process industry. Financial (or economic) risk analysis, which is the quantification and management of uncertainty and variability in cash flows, is a subject not often addressed in the undergraduate chemical engineering curriculum. Yet, because systematic methods from finance exist to quantify economic risk, its inclusion in the curriculum is an effective way to introduce the concept of risk into chemical engineering design. Here we describe the organization and outcomes of a one-credit course in chemical engineering economic analysis that was recently established as a degree requirement for all undergraduates in chemical engineering at the University of Michigan. A key component of the course is instruction in economic risk analysis including discussion of uncertainty in cash flows, the concept of expected net present value, and the use and evaluation of real options in profitability analysis. Each of these ideas is incorporated into the economic and profitability analyses that are a typical part of design in most undergraduate chemical engineering programs.

See more of this Session: Steal This Course!
See more of this Group/Topical: Education

(247c) A Different Approach to Fluid Dynamics

Erin Jablonski, Bucknell University, Lewisburg, PA

Using the open-ended nature of group projects that have real world applications coupled with detailed daily in-class exercises and supporting laboratory experiments, a Fluid Dynamics course is offered that can take any new professor from start to finish. The key to success is strategic scheduling, frequent opportunities for students to demonstrate their knowledge, and a detailed syllabus that keeps everyone on track. The course organizational and instructional materials will be presented and offered to anyone who would like to "steal this course"!

See more of this Session: Steal This Course!
See more of this Group/Topical: Education

Teaching Thermodynamics: Designing a Course that Compensate, Support, and Challenge Students' Learning

Daniel Forciniti
Chemical and Biological Engineering Department
Missouri University of Science and Technology
Rolla MO 65409
Daniel Cernusca
Office of Global Learning
Missouri University of Science and Technology
Rolla MO 65409

Abstract

Chemical engineering thermodynamics is one of the driest topics in the chemical engineering curriculum. In most schools thermodynamics is taken by first or second semester sophomores, who undergo a cultural shock as they move away from their freshman engineering experience. The combination of a very abstract topic and very young students makes the teaching of thermodynamics a challenging experience. One of us (D.F.) has been teaching thermodynamics for the last twenty years. Approximately two years ago D.F. began transforming his class by introducing elements of cooperative learning and video clips. These changes were made with the help of an instructional designer (D.C.). In this presentation, we will offer the novice or prospective thermodynamics teacher enough elements to design a course which has, we believe, good chances to be a successful one. The availability of easy-to-use video capture and editing technology combined with increased capability to make them available online make small educational videos, often called courselets, a useful tool for instructors interested in expanding their classroom activities beyond the lecture time. This presentation includes such attempts made at Missouri S&T for an introductory undergraduate thermodynamics course. The convergent perspective of an instructional designer and of the instructor of this course will introduce: a) the learning needs addressed from compensating for prior knowledge to challenging high achievers; b) the technical and organizational strategies used to produce these courselets and c) the instructional tasks developed to effectively integrate this technology in the classroom. Finally both the feedback from students exposed to these small instructional videos, and instructor's perspective on the impact of this tool are discussed.

Introduction

We read in Wikipedia: "Thermodynamics is the study of energy conversion between heat and mechanical work, and subsequently the macroscopic variables such as temperature, volume and pressure". It has to be demoralizing for a layman to read this cryptic statement. Independently of how the science is formally defined, chemical engineers know that elements of thermodynamics are needed to understand the functioning of unit operations such as distillation towers, condensers, boilers, mixers, pumps, crystallizers, etc. Also, we use the Laws of Thermodynamics for the analysis of processes. Therefore, the thermodynamics that we teach to chemical engineering undergraduate students must prepare them to understand detailed descriptions of the aforementioned units later in the curriculum.

There are several excellent introductory textbooks that cover most of the topics the students need to master. In our department we have used over the years three different books:

1) "Introduction to Chemical Engineering Thermodynamics" by Smith, Van Ness and Abbott (7th edition)
2) "Chemical, Biochemical, and Engineering Thermodynamics" by Sandler
3) "Introductory Chemical Engineering Thermodynamics" by Elliott and Lira.

Most students find Smith, Van Ness and Abbott's book easy to read (although later versions are a little bit more complex). Some students find Elliott and Lira's spread sheet approach to solve thermodynamic problems very convenient. Other students like the companion CD in Sandler's book with its collection of MathCAD sheets and Basic programs. D.F. supplements the textbooks with his own notes. Therefore, the students use the textbook as a general reference. One semester we asked our bookstore to bring the three books and then we allowed the students to choose one (after giving them some guidelines). Most of them bought the book by Smith, Van Ness and Abbott. None of them liked the idea of having to choose a particular textbook in spite of the fact that we provided them with detailed reading guidelines for each one.

The students should come to an engineering thermodynamics class with some thermodynamics background beyond their high school experience. Unfortunately, very often their background is weak. In our university the students learn some thermodynamics in a freshman physics class. Then they take a Physical Chemistry class, whose main topic is thermodynamics, in their first semester junior year. Unfortunately, the Phys. Chem. Class is taken simultaneously with our second semester Thermo class. Sometimes the students' experience is limited to the use of "off-the-shelf" equations. In other words, their ability to set up a problem from scratch is quite limited. This weakness is often compounded by some limitations in their mathematical skills. When we add these deficiencies to a per se abstract and conceptually challenging topic we are at the verge of disaster.

One of us started his teaching career convinced that he needed to transform the students in his classes into experts in thermodynamics. The fact that he was a untenured professor whose teaching evaluations were well below the average convinced him that he needed to reach a compromise between his views and his future. Therefore, he began focusing his classes on a few topics. For example, in the first semester of thermodynamics he was content if the students were able to explain to a layman concepts like entropy, internal energy, free enthalpy and energy and phase equilibrium. In the second semester of thermodynamics he expected the students to master the principles of phase equilibrium to be able to select the proper model for fugacities (equations of state) or activity coefficients.

Chemical Engineering Thermodynamics is taught at Rolla in a two semesters sequence. In the first semester (2nd semester sophomores) we introduce the principles of the science and limit our discussions to single components. Emphasis is put on phase equilibrium but some time is spent in power generation and refrigeration cycles (this is instructor dependent). In the second semester (1st semester juniors) we extend the principles to mixtures and focus our discussion on phase equilibrium. In addition we cover chemical reaction equilibrium and mixing processes.

Figure 1 shows a typical need-based distribution of students. About 10% of them have deficiencies either in mathematics or in the basic sciences. About 10% are high achievers that need to be challenged beyond what an introductory thermodynamics book has to offer. Most of the students need to develop basic engineering skills or problem solving skills. We have developed instructional videos addressing these populations. The purpose of the compensating videos is to facilitate students to overcome obvious deficiencies. High achievers need to be challenged beyond the standard material to keep them engaged into the class. Short videos are an excellent means to achieve that. The main focus of the videos is, however, to reinforce concepts or strategies developed in the classroom. These videos affect most of the students in the class.

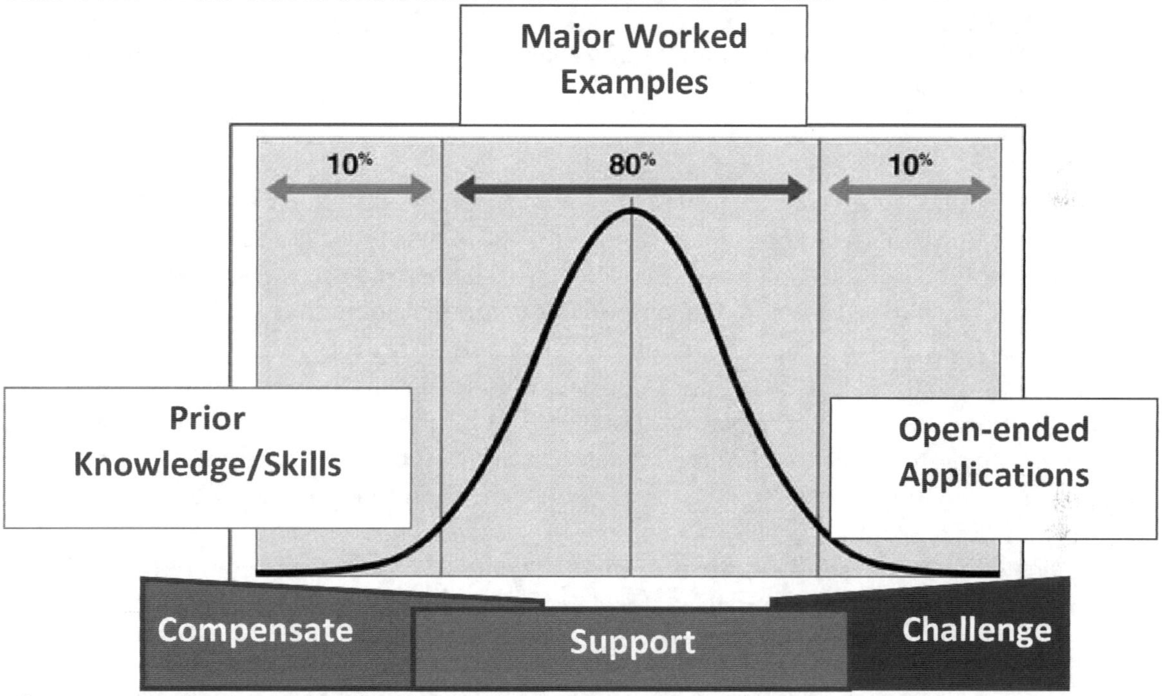

Figure 1. Typical student distribution in a Thermodynamics class according to their needs.

The following strategy was adopted to develop instructional videos. First we decided the focal topic and then selected the material to be included. The next step was to create a protocol to present the selected material and to convert the material in images. This was followed by audio recording with the selected material (although sometimes the audio was pre-recorded in the classroom). Images and audio were integrated (Camtasia Studio 6) and animated pointers and highlights were included to guide the viewer. The length of each video was set between 7 and 10 minutes. The instructor was responsible for the selection of the images, text and audio. The instructional designer put together the final product. The videos were posted in Blackboard for exclusive use of the students enrolled in the class. We found that the effectiveness of the video increased if we included a task that requires the students to use the video (Figure 2).

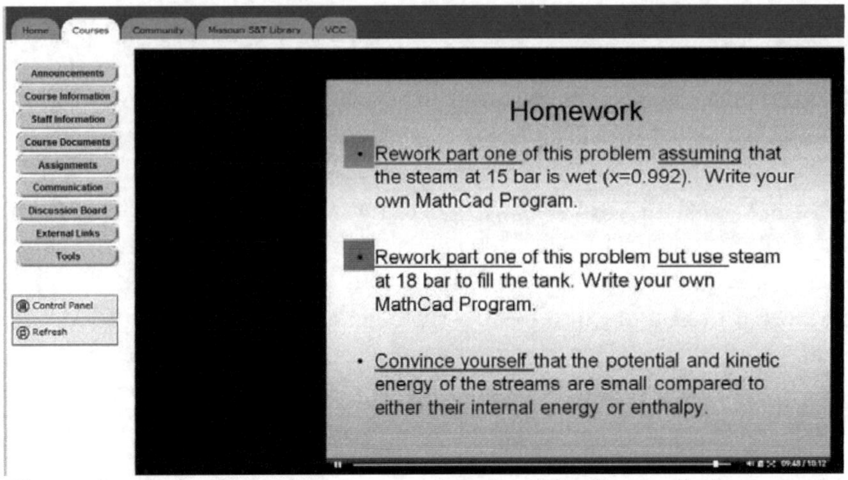

Figure 2. Snap shot of an instructional video that includes a series of tasks to re-enforce the material.

Three compensatory videos were created so far. The one covering basic calculus concepts was done from scratch. Most materials were handwritten. The other two are introductions to MathCAD. For one of them, the audio was recorded in the classroom and then combined with screenshots. The audio for the second one was recorded mostly outside the class. We have made one supporting video that includes a major example introduced in the class (a steam accumulator). The audio was recorded in the classroom for reference. The final audio was recorded outside the classroom. Slides were converted to images and overlapped on the audio. We found this approach to be the best one but, of course, it requires a higher time investment. We are currently generating a series of "challenging" videos that covers in detail the thermodynamics of elastomers (with experiments included).

The table below shows some preliminary analysis of two semesters before and after the introduction of the instructional videos. The first column is a list of standard questions included in our institution's teaching evaluation forms. The scores go from zero to four. Twelve students each semester completed the survey. Substantial improvement was observed in questions 3,4 and 7.

Table 1. Summary of teaching evaluations for the first thermodynamics class.

Questions (S&T official items)	FS08 (N=12) M (SD)	FS09 (N=12) M (SD)	t-test t(22)	p
1. Evaluate the course ,independent of the instructor's effectiveness, in terms of its educational value to you	3.17 (1.12)	2.83 (.72)	.87	.39
2. Rate the instructor's use of assignments and tests for facilitating your learning of the subject matter	1.83 (1.64)	1.58 (.90)	.46	.65
3. Rate the instructor's concern for your understanding of the material	**1.42** (1.24)	**2.42** (.90)	**-2.26**	**.03**
4. Rate the instructor's preparation for class	**1.42** (1.31)	**3.00** (1.04)	**-3.27**	**.003**
5. Rate the instructor's ability to communicate	1.50 (1.45)	1.83 (.58)	-.74	.47
6. Rate the instructor's ability to stimulate and motivate you	1.42 (1.31)	1.83 (.84)	-.93	.36
7. Rate the overall teaching effectiveness of this instructor	**1.25** (.97)	**1.92** (.67)	**-1.97**	**.06**

The generation of instructional videos requires a substantial time commitment. A lot of time is spent in preparing the material and in discussions between the instructor and the education specialist. The education specialist spent two semesters attending D.F. classes, observing both the instructor and the reaction of the students. We met periodically to exchange ideas. The overall teaching evaluation has not improved as much as we could like but we are convinced that the classes are better and that the students are better prepared for the challenges that they will encounter in their last three semesters at Missouri S&T.

(247e) Steal This Course: Chemical Engineering Thermodynamics

Margot Vigeant[1], Michael Prince[1] and Katharyn Nottis[2], (1)Chemical Engineering, Bucknell University, Lewisburg, PA, (2)Education, Bucknell University, Lewisburg, PA

New instructors can find a teaching assignment in Thermodynamics daunting, even at the undergraduate level. In this talk, we will share notes, quizzes, and two projects for a one semester undergraduate Thermodynamics course using Elliott and Lira as the text. The course is supplemented by "inquiry based activities" that help improve conceptual understanding in some of the trickiest areas.

See more of this Session: Steal This Course!
See more of this Group/Topical: Education

Steal This Course: Material and Energy Balances

Lisa G. Bullard and Michael D. Dickey
Department of Chemical and Biomolecular Engineering
North Carolina State University, Raleigh, NC, USA

Abstract

The material and energy balance course, sometimes referred to as the "stoichiometry course," is typically the gateway course in chemical engineering. It is the students' first exposure to the integration of chemistry, math, physics, and biology within the context of current chemical engineering practice, and their experience in the course often determines whether they will persist in the major. The authors have developed and compiled a comprehensive set of instructor resources that are available on a publicly accessible web site. Although most of these resources are linked to chapters in Felder and Rousseau's text, *Elementary Principles of Chemical Processes,* they are also useful for instructors using other texts. These resources include "lecture notes with gaps" (and instructor solutions, available from the authors) that can be used as a student coursepack or as an instructor reference. Ideas are compiled for each chapter of the text, including assignments, demonstrations, potential homework problems, illustrative videos links, and relevant analogies. Creative course ideas are included, as well as Engineering Jeopardy-based test reviews. Administrative materials include an illustrative course web site (with an example syllabus), guidelines for test proctors, boilerplate emails that are frequently needed, and a TA memo on grading and expectations. It is the authors' hope that these resources will improve both the instructor and the student experience in the material and energy balance course.

Introduction

Beginning in the late 1970s with the publication of Felder and Rousseau's *Elementary Principles of Chemical Processes* (1), a new approach to the stoichiometry course was adopted at N.C. State. Individual topics from physical chemistry were no longer presented on a stand-alone basis, but were introduced and applied in the context of chemical process engineering. The traditional homework problems related to chemical and petroleum processes were supplemented by problems from the growing range of fields that employ chemical engineers, including environmental engineering, biochemistry and biomedicine, and microelectronics. Most lectures included brief activities that provided practice and feedback in the methods that would be required on homework and tests. A 1990 paper by Felder outlined the new instructional approach and described the turnaround in student performance and evaluations that resulted from its adoption. (2)

The stoichiometry course has continued to evolve. Since the early 1990s, it has been taught using cooperative (team-based) learning, with measures being taken to hold all team members individually accountable for the entire content of team assignments. Instructional technology has played an increasingly important role in the course, with variety of software tools supplementing traditional instruction. Assignments include traditional closed-ended problems as well as open-ended problems that call for creative or critical thinking or both. A two-part paper written by Bullard and Felder in 2007 provides an update on the approach to teaching this class. The first paper outlines the course structure and policies, the preparation given to the teaching assistants who played an integral part in the course delivery, and the course assignments (3). The second paper summarizes the methods used in the course instruction and assessment (4).

The authors encourage readers to review the 2007 Bullard and Felder papers for direction as to course management and instruction. These include an example syllabus, course policies and procedures, teaching assistant preparation, first day activities, use of active and cooperative learning, assessment, and issues related to academic integrity. For a new instructor, in particular, these resources would be extremely helpful in approaching this course. However, the purpose of this paper is to describe instructor resources related to content that have been compiled on a publicly available web site, http://www.che.ncsu.edu/bullard//MEB_resources/MEB.htm. These include student handouts (with solutions for instructors), ideas for each chapter of the Felder and Rousseau text, fun and creative ideas, and a video illustrating how to use Solver within Excel.

Handouts (Course Notes)

We prepared a series of class handouts (i.e., course notes) that supplement the course textbook. In addition to covering the core content for the course, the handouts contain a number of questions and problems, with blank spaces for answers and solutions. In a typical lecture session, the class would work through part or all of a handout in a mixture of lecturing by the instructor, individual activities, and small-group activities focused on the questions and gaps in the handouts. The students would individually read a passage of text or part of a problem statement or solution and perhaps briefly discuss it in small groups to make sure they understood it. When they reached a gap, one of several different things might happen:

(a) the instructor might go through the solution at the board in traditional lecture format;
(b) the students might be given a short time (30 seconds–3 minutes) and asked to work individually or in small groups to try to fill in the gap; or
(c) the instructor might skip the gap and tell the students to be sure they knew what went in it before they got to the next exam.

To emphasize the importance of the handouts, the class was reminded frequently that some of the questions and problem segments in the handouts would show up on the exams, and they did.

When active learning (individual and group activities in class) was used, the instructor used a variety of formats. Sometimes students worked together in pairs or groups of three or four; sometimes they worked individually; and sometimes they worked individually first and then got into pairs, compared their solutions, and tried to reconcile any differences (*think-pair-share*). Occasionally they worked in pairs with one student doing the solving and explaining and the other asking questions and giving hints if necessary (*thinking-aloud pair problem solving*), with the roles reversing from one activity to the next. In all of these cases, when the instructor stopped an activity, he/she would call on several students for responses, ask for additional responses from volunteers, perhaps augment or elaborate on the responses, and proceed with the lesson.

The approach of using handouts and active learning exercises has several purposes. Students can read straightforward material much more rapidly than instructors can present it. Having them read prose descriptions, definitions of terms, and simple algebraic and arithmetic calculations saves an enormous amount of class time—enough to cover the difficult material in lectures and activities and still get through the complete course syllabus. In addition, people learn difficult material and develop skills through practice and feedback, not by being lectured on what they are supposed to know. We believe that more genuine learning resulted from those brief activities in class and problem sessions than from everything else we did in class. Feedback from students has been overwhelmingly positive regarding the usefulness of the handouts toward learning.

Students purchased a bound version of the handouts (a "coursepack") at the beginning of the semester and were instructed to bring the coursepack to class each day. At NC State, the AIChE student chapter prints and sells the coursepack notes ($20 per coursepack in 2010) as a fundraising activity. We used a tablet computer (a computer with a touch sensitive screen that responds to writing) along with a projector to display the coursepack file (as either a pdf file or Microsoft Journal Writer file). This approach facilitates writing directly on the coursepack page using the tablet computer as students work through "blanks" in the notes containing example problems or short answer questions in class.

Ideas for Each Chapter

The authors have both generated and collected a number of ideas to illustrate and augment the concepts contained within each chapter of the Felder and Rousseau text. These resources are compiled on the web site by chapter to allow instructors to preview the ideas for each chapter and select those which might supplement the lecture by illustrating particularly challenging concepts, showing applications, or giving students additional practice. Several chapters include links to potential homework problems using BioEMB (5) or the Fuel Cell Curriculum Project Web site (6), as well as the analogies of M. Fernandez-Torres (7). Examples of ideas from each include:

- Chapter 2: Ideas for initial or early assignments related to learning styles or chemical engineering as a profession; density illustrations using a lava lamp and a Galileo thermometer
- Chapter 3: Simple manometer demonstration, Youtube videos on effects of pressure
- Chapter 4: Simple demonstration to illustrate batch, semi-batch, and continuous; Youtube video on distillation as an example of a unit operation
- Chapter 5: Demonstrations using portable hand warmers, boiling water in a syringe; examples from cooking; Youtube videos on supercooled freezing and supercritical fluids
- Chapter 6: Fishbowl demonstration, weatherman's explanation of the dew point, colligative properties demonstration (making ice cream in a baggie)
- Chapter 7: Short answer conceptual questions; information literacy assignment
- Chapter 8: On-line psychrometric chart tutorial, illustrations using weather reports, sponge demonstration
- Chapter 9: Flame demonstration; article on using CO_2 scrubbing for environmental applications

Obviously the instructor cannot use all of the ideas given the time constraints of the course, but selective use of some of these may provide opportunities for active learning and help engage students in hands-on applications of the course material.

Fun and Creative Ideas

The authors have generated some fun and creative ideas which can be used in class or shared with students off-line. One goal of this content is to help students make connections between what they are learning in class and the real world. The "Fun Friday" PowerPoint slides briefly highlight everyday applications such as lava lamps, peeps, ice cream, Jell-O, sticky notes, and contact lenses, pointing out the engineering connections. The Chemical & Engineering News web site "What's That Stuff" (8) has an extensive collection of these types of applications. The "CHE Stars" PowerPoint file highlights the individuals for whom important equations and values in the textbook are named.

In addition to sharing creative material with the students, the instructors also solicit creative material from the students using an extra credit assignment on the final homework set:

> *You may earn up to 10 points of extra credit on your final homework assignment by submitting a creative expression of your experience in CHE 205. This might include a poem, song, puzzle, artwork — the sky's the limit! The only constraints are that your work must be original and your submission must be in good taste (something that could be shared with the rest of the class). You may work in groups if your idea requires multiple people to execute or is too big for one person to complete individually (but group projects will have a higher bar for grading).*

In the years in which we've given this assignment, we've gotten a remarkable collection of products, including crossword puzzles, cartoons, haiku and other poems, paintings, a murder mystery, a crocheted doll holding a tiny model of the textbook, a music video, and a live rap song with costumes and choreography. Some of the students kept doing the same sort of thing in other courses: two who did a music video in the stoichiometry course went on to do sequels in subsequent courses on process simulation and thermodynamics, and another student who submitted a personal course journal in CHE 205 continued the journal through her senior year, documenting her entire experience in chemical engineering. We spend the last day of class allowing students to share their contributions, which are generally received with lots of laughter and cheering. It's a great way to end the semester, and we have received enthusiastic student participation despite the fact that it has little bearing on the final grade.

Video Demonstration of Solver in Excel

An important aspect of this course involves solving multi-variable systems of equations. The 2005 edition of the Felder and Rousseau text contains an instructional CD with the easy-to-use equation solving package, EZ Solve. Unfortunately, due to operating system requirements, EZ Solve will not work on current Windows operating systems (a problem that will be addressed in an upcoming edition of the text). Alternatively, Excel's Solver is a tool that can be used to solve multi-variable systems of equations. Although many of these systems can be solved tediously by hand, the reoccurrence of these types of systems in the course offers a valuable opportunity to expose students to the power of the Solver function in Excel.

Hands-on instruction for Solver is an effective way for students to learn since demonstrating it in a lecture format can be difficult for students to follow, tends to be boring, and will not challenge those students who have previously used Excel. The instructors have created a video demonstration (which can be accessed via the webpage) that allows students to learn on their own computer at their own pace. The video can be accessed as many times as needed and is broken up into short video segments that allow students to easily navigate the content based on their competency level. The video starts with segments on Excel basics and ends with a chemical engineering-related demonstration on the utility of Solver.

Conclusion

In Felder's article, "The Alumni Speak," "...many [of those surveyed] cited the value of the stoichiometry course in their academic and/or professional careers, stressing its importance in the advice they gave to beginning chemical engineering students, while no other core chemical engineering course was cited by more than two respondents." One respondent admonishes that the stoichiometry course "...is the most important course you can take—the first step in any engineering calculation is a material/energy balance." (9) Clearly, this is a critically important course. Many students will decide whether to persist in chemical engineering on the basis of their experience and their success in the material and energy balance course. For many students, this may be the first technical course of their college career which contains truly new material (as opposed to their experience of having seen physics, chemistry, and calculus before); in addition, homework is in the form of the dreaded "word problem" and integrates chemistry, physics, and math to a degree not encountered before. While the material is not as difficult conceptually as in later courses such as thermodynamics and transport phenomena, it is critical that the instructor be well organized, clearly convey the concepts, set appropriate expectations, help students make connections between course content and real life applications, and provide additional resources to help students be successful. It is the desire of the authors that the material provided on the web site and referenced papers will assist instructors in accomplishing these goals.

References

1. R.M. Felder and R.W. Rousseau, *Elementary Principles of Chemical Processes,* 2005 Update Edition, New York, John Wiley & Sons (2005).
2. R.M. Felder, "Stoichiometry Without Tears," *Chemical Engineering* Education, 24(4), 188 (1990).
3. L.G. Bullard and R.M. Felder, "A Student-Centered Approach to Teaching Material and Energy Balances: 1. Course Design," *Chem. Engr. Education,* 41(2), 93 – 100 (Spring 2007).
4. L.G. Bullard and R.M. Felder, "A Student-Centered Approach to Teaching Material and Energy Balances: 2. Course Delivery and Assessment," *Chem. Engr. Education,* 41(3), 167-176 (Summer 2007)
5. C. Komives, R. Beitle, and E. Fernandez, BioEMB Problem Bank, http://www.engr.sjsu.edu/~bioemb/index.php.

6. S. Fogler, J. Keith, D. Chmielewski, M. Gross, and D. Allen, The Fuel Cell Curriculum Project, http://www.chem.mtu.edu/~jmkeith/fuel_cell_curriculum/.
7. M.J. Fernandez-Torres, "Analogies: Those Little Tricks That Help Students to Understand Basic Concepts in Chemical Engineering," *Chem. Engr. Education*, 39(4), (Fall 2005).
8. What's That Stuff, *Chemical & Engineering News*, http://pubs.acs.org/cen/whatstuff/stuff.html
9. R.M. Felder, "The Alumni Speak," *Chem. Engr. Education, 34*(3), 238–239 (Summer 2000).

(247g) Chemical Plant Design: Scope, Organization and Evaluation of Semester-Long Projects

Kevin D. Dahm, Chemical Engineering, Rowan University, Glassboro, NJ

This paper will outline the approach to teaching capstone design that has been employed at Rowan University for the past five years.

At Rowan, Chemical Plant Design is the second course in a two-course sequence, following a course on design of individual chemical process components. The Chemical Plant Design course is structured around a single design project completed in teams of 4-5. There are five graded deliverables: three progress reports, a final design report and a final presentation. A different project is used every year, but the scope and expectations for each of the five deliverables are applicable to most any process design project and are the same every year.

This presentation will discuss the attributes of a good project, the specific expectations for each deliverable, and the rubrics used to grade the projects.

See more of this Session: Steal This Course!
See more of this Group/Topical: Education

(247h) Ethics Education for Chemical Engineering Undergrads

Claire F. Komives, Department of Chemical and Materials Engineering, San Jose State University, San Jose, CA

Ethics education is often based on a utilitarian approach, namely – the best possible outcome for the most people. However, this does not address the need for making ethical decisions that impact individuals or conflicts in professional situations. An ethics course for undergrad chemical engineers has been developed that uses Aristotelian principles, laying a foundation for decision making based on the equality of humans based on their common nature. This presentation will introduce the ethics components for the Process Safety and Ethics course. Lecture notes and references for the lectures will be provided to all interested participants.

See more of this Session: Steal This Course!
See more of this Group/Topical: Education

(254a) ABET through the Looking Glass

Daina Briedis, Chemical Engineering and Materials Science, Michigan State University, East Lansing, MI

This presentation is a pseudo-insider's overview of the best practices and available resources for two audiences: programs that are seeking to develop or re-design sustainable improvement processes, and programs preparing for an ABET accreditation visit. The presentation will highlight recent changes at ABET including harmonizing criteria, expectations for compliance with criteria, and insights about how new program evaluators are trained and how this will influence what is considered to be "best practices."

See more of this Session: ABET Resources and Best Practices
See more of this Group/Topical: Education

Feedback Control Loop Guides ABET Process

Polly R. Piergiovanni
Department of Chemical and Biomolecular Engineering
Lafayette College, Easton, PA

Introduction:

The Department of Chemical and Biomolecular Engineering at Lafayette College was most recently evaluated by ABET in October 2008. The year before our visit was a transitional year, so the department re-evaluated its ABET preparation procedures during that time. The most helpful result was a feedback loop diagram, explaining the "Process to Establish and Revise Educational Objectives and Program Outcomes". From this diagram, the ABET responsibilities of the faculty, the department and the advisory board are clear. After the diagram was created, the department standardized assessment tools and procedures and a calendar for the various tasks.

Procedure:

Faculty generally care about how much students learn in their courses, and many make notes in the margins of their notes and handouts about how to improve each lecture for the next offering. This helps improve the courses. We decided to use a similar process to improve our curriculum and meet the ABET outcomes.

Processes:

1. Created: ABET Feedback Loop Diagram, "Process to Establish and Revise Educational Objectives and Program Outcomes". This was the most successful change we made, and was helpful in guiding the assessment process and writing the self-study document. Features of the loop:
    - Inner loop: annually adjust courses
    - Every two years, see if outcomes (matrix) needs to be adjusted
    - Every six years (ABET visit) – see if program outcomes should be adjusted
    - Every six years also look at Program Educational Objectives
    - Allows changes made in the inner loops to filter through before we make additional changes
    - Lets all faculty see what changes are being made to the new "incoming students"
    - Clearly shows the inputs to the process, and the outcomes from the process
    - Chemical Engineers like this kind of diagram

2. Created: Course assessment sheet. Faculty complete the sheet for each ChE course they teach. The front page takes minutes to complete, and this is the

minimum requirement. Failure to complete it affects merit raises. Additional information is added, but each faculty member determines how and what to present. Many create a standard final exam that is given each year, so student achievements can be tracked yearly. Some choose a different problem from each exam that illustrates the objectives for each course. Some provide statistical information from each exam during the semester, along with the material the exam covered. The main benefit of the form is that instructors know the objectives for each course, year by year. As different faculty teach the core courses, it is helpful to know what was done before.

3. Student interviews: At the end of each year, eight students from each class are invited to an "end of year" interview. The department provides pizza, and the students are asked a standard list of questions (primarily about course order, their teamwork and writing skills, etc.). The faculty who ask the questions have not taught the students during that year, so the students are candid in their responses.

4. Advisory Board input: For the two years preceding the ABET visit, we invited the Industrial Advisory Board to campus. They met with the faculty and discussed an important issue to the department, such as the structure of the laboratory sequence. The IAB (ten chemical engineers from a variety of industries, all but one Lafayette alumni) provided input. Many of the members recruit at the college, and could tell us what they were looking for, and how we could better prepare our students.
   For the first two years after our ABET visit, the department did not ask the IAB to come to campus, but instead conversed with them via e-mail. Every May, the department head sent a letter to the IAB, summarizing the conversation and changes made to the curriculum as a result. As the next ABET visit draws near, we will again invite the IAB to campus so they can observe the changes first hand.

5. Department Retreat: Every November the department holds its annual ABET retreat. At this meeting, the faculty members review the matrix of courses/outcomes, the results from the student interview, the advice from the IAB and a summary of the information from the course assessment sheets. We review how the changes implemented in the past few years have progressed through the loop, and determine if the changes have had the desired effect. We discuss changes such as course order (for the students), but larger curricular changes are discussed at a separate meeting.

6. Idea that didn't work: Typically, we collected samples of student work during the year before the ABET visit. In the 2009-10 academic year, we purchased a high speed scanner. We decided to have the students scan in all their work after it had been graded and returned, so we would have a permanent record of all student work. The professors attached a penalty to students who did not complete the task. While the majority of students

complied, a few did not. The professors decided it was more work to track down the students who did not turn in their work than the value of having the scanned homework.

Currently, students are not required to scan their graded homework. We are working on a new policy. Likely, we will require a few students to scan each graded homework and exam.

Lab reports and design projects have always been collected electronically, so we have a complete record.

Conclusion:

As a result, data is collected annually for each course, with little extra time required by faculty. If changes have been made for a course, the data can show if student learning improved. In addition, transfer of courses between faculty occurs more smoothly. With the review schedule, the department can observe trends and make adjustments at the correct time (allowing time for changes to flow through the loop). The feedback loop diagram, the tools and the procedures will be shared in the presentation, as well as initial results from the process.

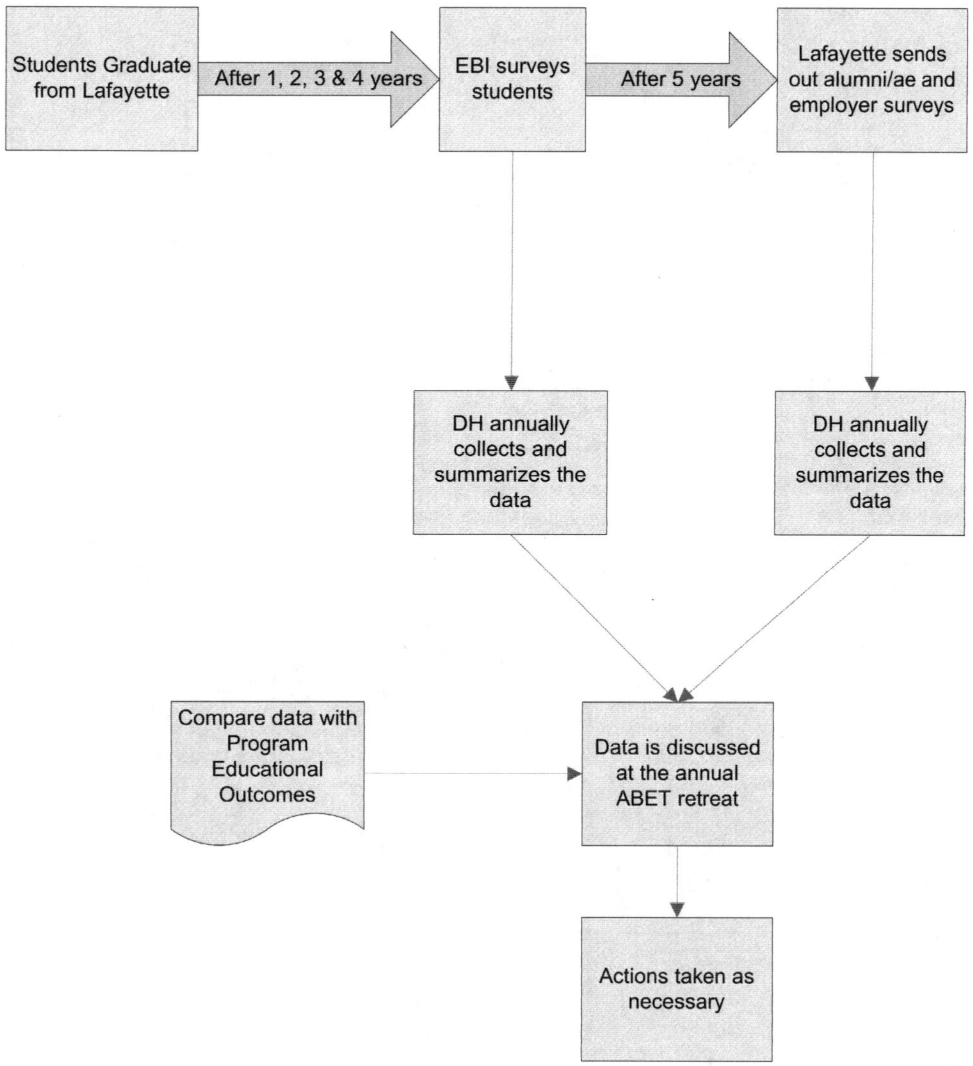
Students Graduate from Lafayette
After 1, 2, 3 & 4 years
EBI surveys students
After 5 years
Lafayette sends out alumni/ae and employer surveys
DH annually collects and summarizes the data
DH annually collects and summarizes the data
Compare data with Program Educational Outcomes
Data is discussed at the annual ABET retreat
Actions taken as necessary

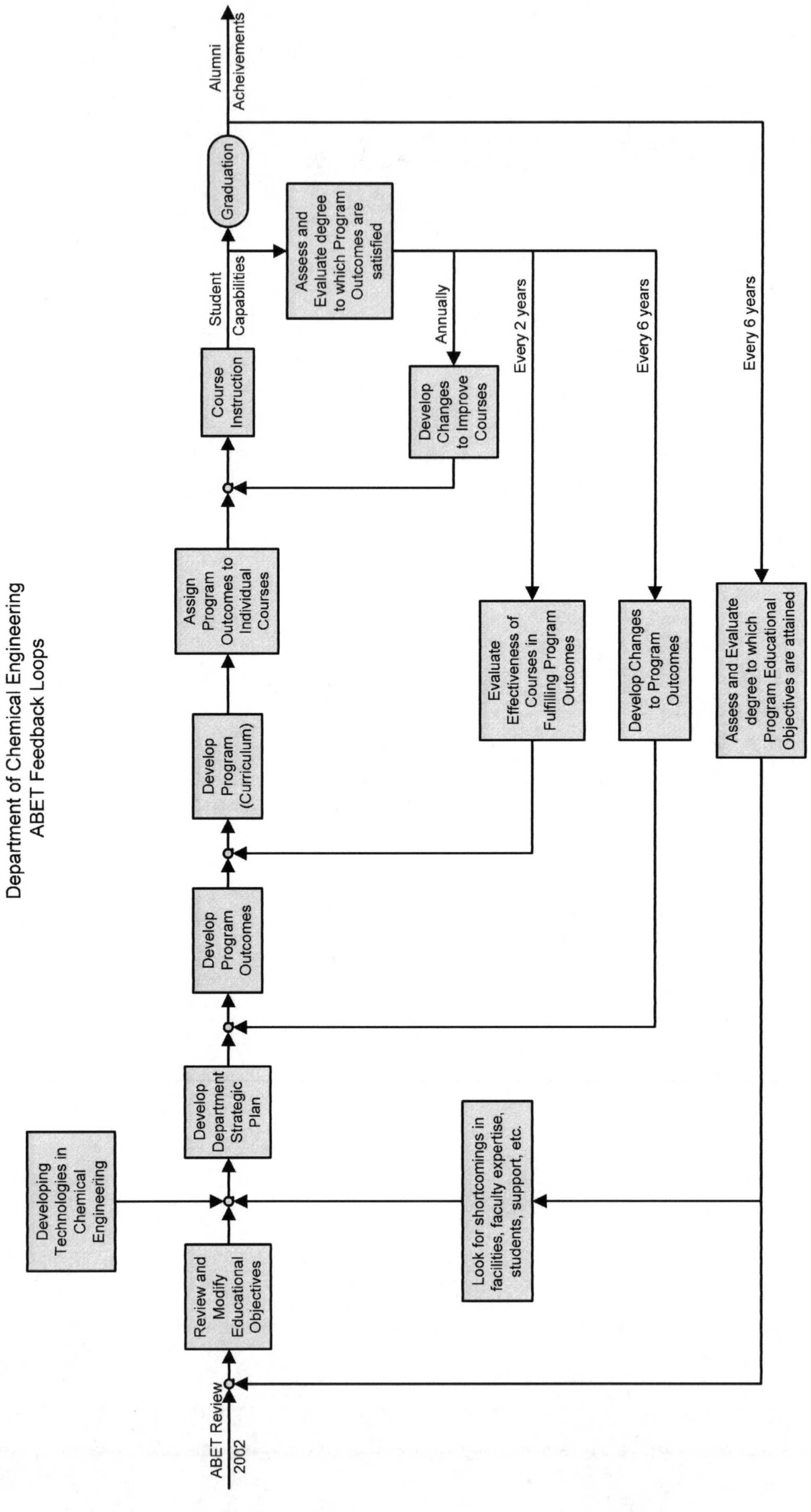
Department of Chemical Engineering
ABET Feedback Loops
ABET Review
2002
Review and Modify Educational Objectives
Developing Technologies in Chemical Engineering
Develop Department Strategic Plan
Develop Program Outcomes
Develop Program (Curriculum)
Assign Program Outcomes to Individual Courses
Course Instruction
Student Capabilities
Graduation
Alumni Acheivements
Assess and Evaluate degree to which Program Outcomes are satisfied
Annually
Develop Changes to Improve Courses
Every 2 years
Evaluate Effectiveness of Courses in Fulfilling Program Outcomes
Every 6 years
Develop Changes to Program Outcomes
Every 6 years
Assess and Evaluate degree to which Program Educational Objectives are attained
Look for shortcomings in facilities, faculty expertise, students, support, etc.

Assessment Sheet for Chemical Engineering Courses

Course: ____________________

Instructor: ____________________

Semester: __________

Program Outcomes to be satisfied by this course:

____ (a) Ability to apply knowledge of mathematics, science and engineering
____ (b) Ability to design/conduct experiments and analyze/interpret data
____ (c) Ability to design a system, component or processes to meet desired needs
____ (d) Ability to function in multi-disciplinary teams
____ (e) Ability to identify, formulate, and solve engineering problems
____ (f) Understanding of professional and ethical responsibility
____ (g) Ability to communicate effectively
____ (h) Understanding the impact of engineering solutions in a global and societal context
____ (i) Recognition of the need for, and the ability to engage in life-long learning
____ (j) Knowledge of contemporary issues
____ (k) Ability to use techniques, skills and modern engineering tools necessary for engineering practice
____ (l) Understanding of safety and environmental factors
____ (m) Understanding of engineering economics
____ (n) Understanding of theoretical concepts

ANALYSIS OF COURSE OUTCOMES

Objective # 1:

Method(s) of Assessment:

Degree to which students met this outcome:

__ very satisfactory ___ satisfactory ___ unsatisfactory

Response to previous year's recommendations:

Recommendations for future improvement:

Objective # 2:

Method(s) of Assessment:

Degree to which students met this outcome:

_ _ very satisfactory ___ satisfactory ___ unsatisfactory

Response to previous year's recommendations:

Recommendations for future improvement:

(254c) Coordination, Cooperation, and Data Analysis in the Assessment Process

Mark R. Anklam, Chemical Engineering Department, Rose-Hulman Institute of Technology, Terre Haute, IN and Julia Williams, Institutional Research, Planning and Assessment, Rose-Hulman Institute of Technology, Terre Haute, IN

In the Rose-Hulman ABET assessment processes there is coordination that exists between the engineering programs and the Institute. This coordination is centered primarily in the fact that each engineering program maps specific outcomes to the Institute outcomes. As a result, collection of student artifacts occurs in engineering courses through the use of our Learning Management System (Angel) while the evaluation of the artifacts occurs in an Institute-level process managed by the Office of Institutional Research, Planning, and Assessment (IRPA) with faculty evaluators drawn from many different departments. Reports on student performance are created by IRPA and then delivered to engineering departments for analysis and improvement decisions. The chemical engineering faculty uses these data along with data from rubrics collected in various chemical engineering courses to assess program outcomes. The engineering program/Institute partnership helps reduce the workload of the departments while still providing rich assessment data.

See more of this Session: ABET Resources and Best Practices
See more of this Group/Topical: Education

(254d) Nuts and Bolts of ABET Assessment at a Small Program with a Complementary Research Mission

John C. Prindle, Chemical & Biomolecular Engineering, Tulane University, New Orleans, LA and Vijay T. John, Department of Chemical and Biomolecular Engineering, Tulane University, New Orleans, LA

We describe assessment procedures at Tulane. This is a small program but one which has a mission to integrate education with research. A novel UO laboratory experience coupled with a long standing practice school and service learning are distinctive aspects of the program.

See more of this Session: ABET Resources and Best Practices
See more of this Group/Topical: Education

(254e) University of Maine's Chemical and Biological Engineering Program Assessment Process

John J. Hwalek, Douglas Bousfield, Albert Co, Darrell Donahue, Paul Millard and M. Clayton Wheeler, Chemical and Biological Engineering, University of Maine, Orono, ME

An overview of the assessment process used in the Chemical and Biological Engineering Department at the University of Maine will begin with a discussion of how the department developed Program Outcomes in relation to the ABET 3a-k criteria. Then, the unique features of the assessment process will be discussed which include: Online Faculty Course Assessment Reports, targeted input from the Industrial Advisory Board, feedback from coop and intern mentors and a database for tracking program changes. Finally, a few examples will be presented which illustrate how the assessment results have been integrated and used to close the assessment loop.

See more of this Session: ABET Resources and Best Practices
See more of this Group/Topical: Education

Evolution of Course and Curriculum Assessment at Ohio University

Michael E. Prudich, Chemical and Biomolecular Engineering, Ohio University, Athens, OH

Developing a comprehensive, integrated system to assess chemical engineering program outcomes can be a difficult and time-consuming process. Maintaining and continuing the assessment system, once it has been developed, can also be problematic. This presentation describes the curriculum and course assessment system currently used by the Department of Chemical and Biomolecular Engineering at Ohio University. This system consists of a set of individual course-level assessments which integrate together to constitute the backbone of a program assessment. Senior exit interviews and periodic surveys of recent graduates and alumni round out the assessment portfolio. Review of this complete package by a "committee of the whole" encourages faculty buy-in and the use of the assessment results in "closing the loop."

Evolution of our assessment process through two rounds of ABET visits will be discussed. The main features of the evolution have focused on attaining increased efficiency in the assessment process. The assessment process originally adopted early in the decade consisted of a set of educational objectives and outcomes (Table 1) that mapped into 47 separate assessment elements (Table 2) that, in composite, were assessed 252 times (the 'X's in Table 2 indicate assessment responsibilities) annually across the chemical engineering curriculum. Experience gained from approximately 8 years of operation with this system has enabled us to identify efficiencies that have recently been implemented into our assessment process. The primary source of efficiency was the adoption of a more concise set of education objectives and outcomes (Table 3) that has been mapped into 22 separate assessment elements (Table 4). These new assessment elements are now distributed throughout the chemical engineering curriculum in one of three categories: 'i' - introduce, 'P' - practice, and 'A' - assess. Only those elements with an assigned category of 'A' require explicit, course-level assessment. Courses that we consider to be capstone courses (process design and unit operations laboratory) are assessed annually, while assessments made in other courses are performed on a biannual basis. This results in course-level responsibility for only 64 assessments (approximately 45 annually), versus the 252 annual assessment points that were included in our initial process.

Table 1. "Original" Program Educational Objectives – Program Outcomes

Program Educational Objective 1: Graduates will have a strong foundation in chemical engineering theory and practice.

Program Outcomes for Program Educational Objective 1: Students will demonstrate the ability to:

a. apply knowledge to chemical engineering problems from subjects including mathematics, chemistry, physics, biology, and other engineering disciplines;

b. apply knowledge of chemical engineering fundamentals including material balances, energy balances, thermodynamics, momentum transfer and fluid flow, heat transfer, mass transfer, chemical reaction engineering, and bioengineering;

c. apply knowledge of chemical engineering unit operations such as heat exchangers, continuous contacting equipment, staged separation processes, chemical reactors, and mass transfer equipment;

d. complete experimental studies including designing and conducting experiments, formulating mathematical models, and analyzing and interpreting results using statistical tools;

e. solve engineering problems including identifying the problem to be solved, determining what data is and isn't needed, identifying probable causes and potential solutions, identifying applicable theory and constructing modeling equations, articulating underlying assumptions in the theory, identifying the type of math problem and appropriate solution techniques, solving several steps in sequence, and critically evaluating the solution for reasonableness;

f. and design chemical processes, using current engineering tools and considering controllability, product quality, economics, safety, and environmental concerns.

Program Educational Objective 2: Graduates will have communication and interpersonal skills needed to succeed in a professional environment.

Program Outcomes for Program Educational Objective 2. Students will demonstrate the ability to:

a. participate effectively in a team through leadership, individual contributions, and multidisciplinary interactions;

b. and communicate in oral, written, and graphical form.

Program Educational Objective 3: Graduates will be scholars and professionals and dedicated to the betterment of themselves and society.

Program Outcomes for Program Educational Objective 3. Students will demonstrate the ability to:

a. articulate the responsibilities of engineering practice including professional responsibilities and ethical responsibilities;

b. articulate the interaction between engineering solutions, contemporary issues, and cultural perspectives;

c. and engage in life-long learning by learning independently and articulating the importance of independent learning for future professional development.

Table 2. "Old" Matrix of Course and Program Outcomes - page 1/2

Table Key: U = Unacceptable, S = Satisfactory, E = Exceeds Expectations, N = Not Assessed. Revised 10/02/08		100	101	200	201	305-6	307	308	345	346	347	400	404	408	415-6	442-17	443-4	448	481	499	EXIT
Program Outcomes: Students will demonstrate the ability to																					
1a	apply knowledge to chemical engineering problems from subjects including																				
1a1	mathematics,			A	A	A	A	A		A	A	A		A		A		A	A		
1a2	chemistry,			A	A	A	A	A				A			A		A		A		
1a3	physics,					A	A														X
1a4	biology,						A						A						A		
1a5	and other engineering disciplines;		A	A													A				
1b	apply knowledge of chemical engineering fundamentals including																				
1b1	material balances,	A		A	A	A	A	A	A		A	A	A		A			A	A		
1b2	energy balances,				A	A	A	A	A	A					A						
1b3	thermodynamics,			A	A	A			A	A	A				A		A	A			
1b4	momentum transfer and fluid flow,								A	A					A						X
1b5	heat transfer,									A					A						
1b6	mass transfer,						A	A			A				A						
1b7	chemical reaction engineering,						A	A							A				A		
1b8	and bioengineering												A						A		
1c	apply knowledge of chemical engineering unit operations including																				
1c1	heat exchangers,				A					A							A				
1c2	continuous contacting equipment,										A				A						
1c3	staged separation processes,										A				A						X
1c4	chemical reactors,			A	A		A	A							A				A		
1c5	and mass transfer equipment;										A				A		A				
1d	complete experimental studies including																				
1d1	designing and conducting experiments,							A		A	A			A	A	A			A		
1d2	formulating mathematical models,							A		A	A			A	A	A			A		X
1d3	and analyzing and interpreting results using statistical tools;									A	A			A	A	A					
1e	solve engineering problems including																				
1e1	identifying the problem to be solved,								A	A		A		A	A		A	A	A		
1e2	determining what data is and isn't needed,	A			A	A			A	A		A			A		A				
1e3	identifiying probable causes and potential solutions,								A					A			A	A			
1e4	identifying applicable theory and constructing modelin equations,					A			A			A			A	A	A		A		
1e5	articulating underlying assumptions in the theory,					A			A	A		A			A		A				X
1e6	identifying the type of math problem and appropriate solution techniques,			A	A	A				A	A	A			A						
1e7	solving several steps in sequence,		A	A	A	A			A	A	A	A			A		A		A		
1e8	and critically evaluating the solution for reasonableness.		A	A	A	A				A	A	A		A	A		A		A		

Table 2. "Old' Matrix of Course and Program Outcomes - page 2/2																					
Table Key: U = Unacceptable, S = Satisfactory, E = Exceeds Expectations, N = Not Assessed. Revised 09/25/08		100	101	200	201	305-6	307	308	345	346	347	400	404	408	415-6	417	443-4	448	481	499	EXIT
Program Outcomes: Students will demonstrate the ability to																					
1f	design chemical processes																				
1f1	using current engineering tools				A					A	A					A	A				X
1f2	and considering controllability,															A		A			
1f3	product quality,							A									A				
1f4	economics,				A			A		A							A				
1f5	safety,				A											A	A	A			
1f6	and environmental concerns;				A												A	A			
2a	participate effectively in a team through																				
2a1	leadership,																				X
2a2	individual contributions,																				
2a3	and multidisciplinary interactions;																				
2b	communicate in																				
2b1	oral				A								A		A		A			A	X
2b2	written,			A	A	A		A	A	A		A	A	A	A	A	A	A	A	A	
2b3	and graphical form;		A		A				A	A		A	A	A	A	A	A				
3a	articulate the responsibilities of engineering practice including																				
3a1	professional responsibilities												A	A						A	X
3a2	and ethical responsibilities;												A							A	
3b	articulate the interaction between engineering solutions,																				
3b1	contemporary issues,				A								A							A	X
3b2	and cultural perspectives;				A								A							A	
3c	and engage in life-long learning by																				
3c1	learning independently												A		A		A		A	A	X
3c2	and articulating the importance of independent learning for future professional development.																			A	

Table 3. Ohio University Department of Chemical and Biomolecular Engineering

Program Educational Objectives – Program Outcomes

Objective 1: Graduates will have a strong foundation in chemical and biomolecular engineering theory and practice.

Outcomes for Objective 1: Students will demonstrate the ability to:

a. Apply knowledge of mathematics, basic sciences, and computational tools to the solution of chemical and biomolecular engineering problems.

b. Apply knowledge of chemical and biomolecular engineering theory and science including:

- b.1. Mass and energy balances;
- b.2. Thermodynamics (both mechanical and chemical);
- b.3. Transport phenomena (momentum, heat, and mass);
- b.4. Chemical kinetics and reactions (both homogenous and heterogeneous).

c. Apply knowledge of chemical and biomolecular engineering practice and application including:

- c.1. Material handling (fluid flow, e.g. piping systems, pumps, compressors);
- c.2. Heat transfer devices (e.g. heat exchangers);
- c.3. Separation systems (e.g. distillation, absorption);
- c.4. Chemical reactors (e.g. batch, CSTR, PFR, dispersion).

d. Complete experimental studies including:

- d.1. Designing and conducting experiments;
- d.2. Formulating mathematical models;
- d.3. Interpreting results using statistical tools.

e. Identify, formulate, and solve chemical and biomolecular engineering problems.

f. Design chemical processes including:

- f.1. Using process simulators;
- f.2. Considering controllability;
- f.3. Considering product quality and economics;
- f.4. Considering safety and environmental concerns.

Objective 2: Graduates will have the communication and interpersonal skills needed to succeed in a professional environment.

Outcomes for Objective 2: Students will demonstrate the ability to:

a. Participate effectively in a team through leadership, individual contributions, and multidisciplinary interactions.

b. Communicate in oral, written, and graphical form.

Objective 3: Graduates will be scholars and professionals and dedicated to the betterment of themselves and society.

Outcomes for Objective 3: Students will demonstrate the ability to:

a. Articulate the responsibilities of engineering practice including professional responsibilities and ethical responsibilities;

b. Articulate the interactions between engineering solutions, contemporary issues, and cultural perspectives;

c. Engage in life-long learning by learning independently and articulating the importance of independent learning for future professional development.

Table 4. "New and Improved" Matrix of Course and Program Outcomes																					
Table Key: i = related content must be introduced, P = related content must be practiced; A = student learning related to this outcome must be assessed. Revised 10/15/09		100	101	200	201	305-6	307	308	345	346	347	400	404 / 481	408	415-6	442-417	443-4	448	499	EXIT	
Program Outcomes: Students will demonstrate the ability to		0A-0P-5i	1A-3P-2i	2A-4P-5i	1A-5P-4i	1A-5P	3A-2P	4A-7P	3A-4P	5A-6P	3A-10P	3A-3P	1A-4P	4A-1P	12A-2P	4A-1P	12A-2P	1A-2P	3A-0P		
1a	apply knowledge of mathematics, basic sciences, and computation tools to the solution of chemical and biomolecular engineering problems		A	A	P	P	A	P	P	P	P	A	A	A	P		P				6A-8P
1b	apply knowledge of chemical and biomolecular engineering theory and science including:																				
1b1	material and energy balances,			A	A	P	P	P	P	A	A	A	P		A		P				6A-6P
1b2	thermodynamics,		i	P	P	A			P	P	P				P		A				2A-6P-i
1b3	transport phenomena,							P	A	A	A				A		A				5A-2P
1b6	chemical kinetics and reactions;						A	A				A			A						4A-0P
1c	apply knowledge of chemical and biomolecular engineering practice and application including:																				
1c1	material handling,			i					A								A				2A-0P-i
1c2	heat transfer equipment,				i					A							A				2A-0P-i
1c3	separation systems,										A				A		A				3A-0P
1c4	chemical reactors;			i	i		A	A				P			A						3A-1P-ii
1d	complete experimental studies including:																				
1d1	designing and conducting experiments,		P					A		P	P			A	A						3A-3P
1d2	formulating mathematical models,							A	A	P	P	P	P	A	A						4A-4P
1d3	interpreting results using statistical tools;										P			A	A	A					3A-1P
1e	identify, formulate, and solve chemical and biomolecular engineering problems;		i	P	P	P	P	P	P	P	P	P			A		A				2A-9P-i
1f	design chemical processes																				
1f1	using process simulators,			i		P					P						A				1A-2P-i
1f2	considering controllability,															A					1A-0P
1f3	considering product quality and economics,			i	i			P		P							A				1A-2P-ii
1f4	considering safety and environmental issues;			i	i						P					A	A	A			3A-1P-ii
2a	participate effectively in a team through leadership, individual contributions, and multidisciplinary interactions;	i	P	P	P			P		A	P	A			A	P	A				4A-6P-i
2b	communicate in oral, written, and graphical form;	i	P	P	P	P		P		A	P			P	A	A	A				4A-7P-i
3a	articulate the responsibilities of engineering practice including professional and ethical responsibilities;	i											P				P	P	A		1A-3P-i
3b	articulate the interactions between engineering solutions, contemporary issues, and cultural perspectives;	i											P				P	P	A		1A-3P-i
3c	engage in life-long learning by learning independently and articulating the importance of independent learning for future professional development.	i													A		A		A		3A-0P-i

(254g) A "Seven Trait Writing Tool" for Assessment of Technical Writing

John R. Schlup, Department of Chemical Engineering, Kansas State University, Manhattan, KS

It is often difficult to assess coursework in engineering classes that are writing intensive. It would be desirable to integrate assessment of both the student's communication skills and their technical content. An added benefit is that students can begin to see the interrelatedness of various learning and program outcomes. Over the past several years a coursework assessment tool has been developed for use within an undergraduate chemical engineering laboratory. Considerations of technical content, idea development, and student communication skills have been integrated into one rubric. These data then have been employed to assess three program outcomes. The basis for this approach will be discussed, and the rubric with performance criteria will be described. Example outputs and their application to assessment of program outcomes will be considered.

See more of this Session: ABET Resources and Best Practices
See more of this Group/Topical: Education

(339a) CSB Recommendations From the T2 Laboratories Inc. Investigation

Rachael T. Gunaratnam, Recommendations, U.S. Chemical Safety and Hazard Investigation Board, Washington, DC

The U.S. Chemical Safety Board investigated an explosion and fire at the T2 Laboratories, Inc. in Jacksonville, Florida on December 19, 2007. The incident killed four employees and injured 32 others, including employees and members of the public in nearby businesses. In the course of producing a gasoline additive, a runaway exothermic reaction led to an uncontrollable pressure and temperature rise, causing the reactor to burst, ignite and explode. The CSB concluded that the cooling and relief systems were inadequate to control the runaway reaction. The investigation found that the owners of T2 had no prior reactive chemistry experience, though one was a chemist and other a chemical engineer. Further investigation revealed that most baccalaureate chemical engineering curricula in the U.S. do not specifically address reactive hazard recognition or management. The CSB recommended that ABET and AIChE collaborate to require that reactive hazard awareness become part of baccalaureate chemical engineering curricula. This paper will discuss an overview of the ABET process; what changes AIChe has proposed for the CSB recommendations; and provide an update on the current status of these recommendations.

See more of this Session: Integrating New Safety Requirements Into the Chemical Engineering Curriculum
See more of this Group/Topical: Education

(339b) The Integration of Process Safety Into a Chemical Reaction Engineering Course: The Review of the T-2 Incident

Ronald J. Willey, Department of Chemical Engineering, Northeastern University, Boston, MA, H. Scott Fogler, Chemical Engineering, The University of Michigan, Ann Arbor, MI and Michael B. Cutlip, University of Connecticut, Storrs, CT

The explosion and subsequent death of four people at the T2 Laboratories chemical facility in Florida in 2007 has resulted in the CSB finding that undergraduate chemical engineering students do not receive knowledge in the hazards associated with chemical processing. This presentation will review the events that led up to the T-2 tragedy. Afterwards, an analysis of the event will be presented that can be used in a chemical reaction engineering classroom to show the hazards involve when dealing with exothermic reactions. This event can be integrated into the learning experience of chemical engineering students learning about runaway reactions

See more of this Session: Integrating New Safety Requirements Into the Chemical Engineering Curriculum
See more of this Group/Topical: Education

A SAChE Module

Designed to Bridge Process Safety's Troubled Waters:

Meeting the New Academic Process Safety Requirements

Bruce K. Vaughen
Cabot Corporation
700 E US Highway 36
Tuscola, IL 61953

ABSTRACT

Fatalities, such as those which occurred at the T2 Laboratory in 2007, could have been avoided with better understanding and implementation of the elements of Process Safety Management (PSM). Applications using the PSM elements have made significant advances in industry over the past three decades which help engineers reduce process safety risks with better tools for understanding, analyzing, designing and controlling hazardous processes. The concepts of conservation of mass and energy have been complemented with the process safety management concept of conservation of life. This paper describes a new SaChE module which has been created to help bridge part of the process safety-related awareness and education gap between academia and industry. The module provides a framework for organizing the SaChE products, combining the requirements of the senior design project with process hazards analysis principles which drive successful industrial PSM programs. Instructors and students can use this module to improve their awareness of the PSM-related resources and techniques, with particular emphasis on locating the SaChE products that have been designed to be integrated within the chemical engineering curriculum. Use of this module will help ensure that students graduate with, at minimum, an awareness of different hazards analysis techniques that must be used when they design their risk reduction controls, preventing irreversible life-changing process-related incidents from affecting them or their co-workers.

1- INTRODUCTION

The fatal explosion at the T2 Laboratories, Inc., in 2007 triggered an investigation by the US Chemical Safety Board (CSB) and resulted in several findings [1]. One of these findings included the following statement:

> Chemical Engineer (part owner) killed because of lack of proper understanding of reactive chemistry hazards and process safety & risk reduction design.

As a result of their investigation, the CSB recommended to the AIChE:

> "Work with the Accreditation Board for Engineering and Technology, Inc. (ABET) to add reactive hazard awareness to baccalaureate chemical engineering curricula requirements."

Since the report, AIChE has developed and proposed guidelines to ABET for teaching safety to undergraduates. These guidelines are shown in Table 1.

The "**S**afety **a**nd **Ch**emical **E**ngineering Education (SaChE)" Program, formed in 1992 as AIChE's link between industrial process safety experts in the Center for Chemical Process Safety (CCPS) and academe, has developed more than 100 process safety-related products to assist instructors [2]. SaChE program members also organize faculty workshops to help share best industrial practices with academe [3, 4, 5, and 6].

This paper presents the basic concepts for a SAChE Product that provides guidance to engineering design teams to help them meet the combined process safety guidelines of both academia and industry [7].

2- PROJECT ROAD MAP

This section describes the basis for the module and the framework (the "Road Map") using a block flow diagram.

The SaChE module combines of the steps in a senior design project (the basis) with the AIChE Process Safety Guidelines and the principles of Process Safety Management (PSM). As is shown in Figure 1, the design team's project has ten basic steps:

1 Define Project
2 Research Technologies
3 Understand Risk
4 Understand Process Safety
5 Understand Risk Reduction Strategies
6 Hazards Evaluation
7 Evaluate Options
8 Select Final Design
9 Understand Facility Siting
10 Write Final Report

The flow of the blocks noted in Figure 1 helps the design team organize their efforts to meet their academic and industrial process safety requirements, including references to the industry's basic principles of Process Safety Management (PSM).

The connection between the process safety Guidelines listed in Table 1 and the Road Map in Figure 1 are shown in Table 2. The steps shown in Figure 1 that are focused on safety include steps 3, 4, 5, 6, and 9. These process safety-related steps are described in more detail in the next section.

3 – THE SACHE PRODUCT

The SACHE Product provides guidance for engineering design teams to help them meet the combined process safety requirements of academe and industry [7]. It can be used by professors, industrial trainers, and students who are working on a process design project, as well.

The Product includes:

1. An Overview document written to assist the instructor on how to use the module,
2. A PowerPoint presentation (the module), and
3. Handouts that are used with the PowerPoint presentation.

The presentation begins with a brief description of the T2 accident that was the genesis of the CSB recommendation to AICHE and ABET. The design project "Road Map" is presented next, helping the design team visualize how the elements of Process Safety Management (PSM) used in industry are incorporated into their project.

The handouts include four tables. Table 1 describes the AICHE/SACHE/ABET proposed process safety Guidelines; Table 2 compares the steps in a design project that meet part of the process safety Guidelines. Note that the team's final written design report helps departments document that their students meet the AIChE/SACHE process safety guidelines. Specific SACHE products and SACHE Safety Certificates are referenced in Tables 3 and 4 to point the designers to available resources for learning about, understanding how and then developing a safe design.

The module helps describe the process safety-related steps shown Figure 1:

Step 3: Understanding Risk

The module starts with a definition of risk and includes an example of a risk matrix. The matrix consists of a Frequency term and a Consequence term. The term "unacceptable risk," with its high frequency and high consequence is compared to the term "acceptable risk," with its correspondingly low frequency and low consequence. Each team must establish their own risk tolerance levels before proceeding with their analysis – just like any business must do before allocating resources to reduce their process safety-related risks.

Step 4: Understanding Process Safety

The types of process safety hazards are: Fires, Explosions and Toxic Releases. A simplified PSM definition is included in the charts; PSM's goal is to reduce the process safety risk to people, the environment and the business.

Step 5: Risk Reduction Strategy

The risk reduction strategies include lowering the expected frequency with engineering or administrative controls and/or lowering the consequence with inherently safer designs. Engineering design considerations include alarms and interlocks (Safety Instrumented Systems, SIS). Administrative controls include written operating procedures. Inherently Safer Design considerations (especially crucial at this point!) include, for example, less hazardous materials or reduced processing temperature or pressure extremes. Consequence reduction strategies could include quick detection of the potential event and a quick emergency response, such as smoke detectors, sprinklers and then immediate action by an emergency response team.

NOTE: The IMPORTANT Operational Discipline (OD) concepts must be considered when applying risk evaluations in industry. For example: If people do not follow the process safety "rules" (i.e., operating procedures, equipment codes or standards, or maintenance and reliability programs, etc.), then their risk is larger since their OD is less than 1 (where 1 equals perfect compliance to OD). This is shown in a "perceived risk" versus "actual risk" chart when there is poor OD.

Step 6: The Hazards Evaluation

The elements of Process Safety Management (PSM) are briefly discussed in the module, with particular focus on reinforcing the importance of the Process Safety Information (PSI) and Process Hazards Analysis (PHA) elements and their considerations in the design process.

Step 9: Facility Siting.

Please refer to more details of Facility Siting, which is industrially significant, in the notes written in Table 2, Step 9.

The module concludes by reminding the students of the process safety-related steps in the design team's project Road Map chart. The final two charts were influenced by the Deepwater Horizon disaster in early 2010 (the time when this Product was being prepared). Hopefully the students will understand: although we have made progress in process safety, we still have work to do.

4 - SUMMARY

The SACHE Product helps bridge a gap between academia and industry in process safety-related awareness and education. The module provides a Road Map (a framework) for organizing the existing SaChE products, combining the requirements of the design project with PSM principles. The module can be used to improve the awareness of the PSM-related resources and techniques, with particular emphasis on locating the SaChE products that have been designed to be integrated within the chemical engineering curriculum. Using this module will help ensure that students graduate with, at minimum, an awareness of different hazards analysis techniques that must be used when they design their risk reduction controls.

REFERENCES

1 csb.gov

The U.S. Chemical Safety and Hazard Investigation Board (CSB) Investigation Report: "T2 Laboratories Inc. Reactive Chemical Explosion, Jacksonville, FL, Dec. 19, 2007."

http://www.csb.gov/investigations/detail.aspx?SID=8&Type=2&pg=1&F_InvestigationId=8

2 Safety and Chemical Engineering Education Program (SaChE), sache.org

3 Schneider, R., "SAChE: What exactly is it and how can it help you?," SAChE Faculty Workshop, Baytown, TX, August 17, 2010.

4 Cox, K. R., "Process Safety in the Classroom," SAChE Faculty Workshop, Baytown, TX, August 16, 2010.

5 Chosnek, J., "Process Safety and Chemical Engineering Design," SACHE Faculty Workshop, Baytown, TX, August 17, 2010.

6 Edwards, V. E., "Develop and Design Inherently Safer Process Plants," SACHE Faculty Workshop, Baytown, TX, August 17, 2010.

7 Vaughen, B. K., "Safety Guidance for Design Projects," Safety and Chemical Engineering Education Program (SaChE), Product issued in 2011, SACHE.org.

http://www.sache.org/products.asp.

Figure 1: The Design Project Road Map

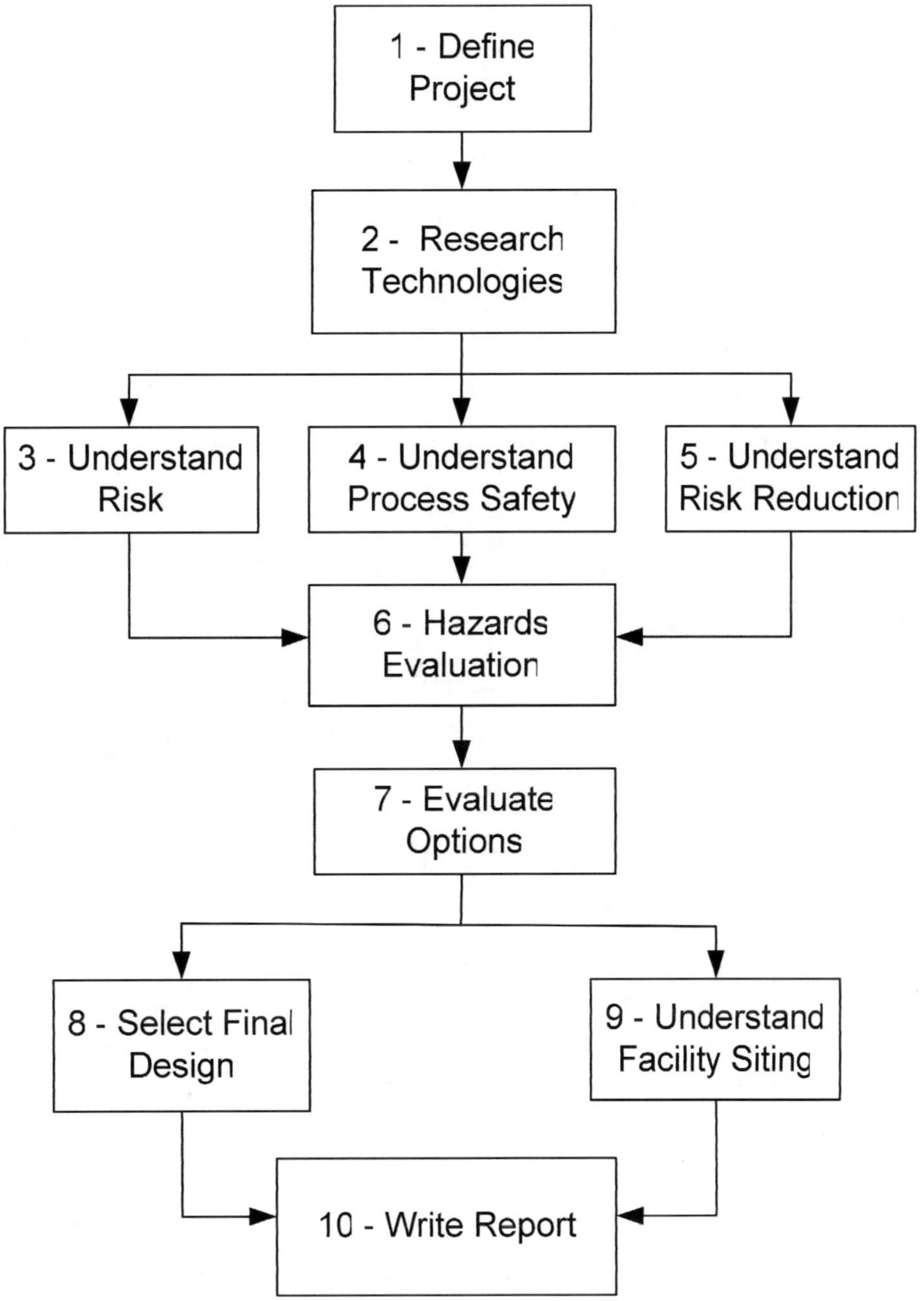

Table 1

AIChE/SACHE Guidelines for Teaching Safety and Design

1	The graduate must understand the importance of process safety and the resources and commitment required. This should include the important incidents that define process safety, and how these incidents affected the practice of chemical engineering.
2	The graduate must be able to characterize the hazards associated with chemicals and other agents. This must include toxic, flammable, and reactive hazards.
3	The graduate must understand and be able to apply concepts of inherently safer design.
4	The graduate must understand how to control and mitigate hazards to prevent accidents. This should include generally accepted management systems, plant procedures and designs to prevent accidents.
5	The graduate should be familiar with the major regulations that impact the safety of chemical plants.
6	The graduate should understand the consequences of chemical plant incidents due to acute and chronic chemical releases and exposures.
7	The graduate should be reasonably proficient with at least one hazard identification procedure.
8	The graduate should have an introduction to the process for hazard evaluations and risk assessments.

Note: The AIChE and ABET guidelines are an abbreviated version of this table.

Table 2

The Design Project Steps Tied to the AIChE/SACHE/ABET Guidelines

Design Project Steps (Road Map)		Description of Project Steps [with referenced AICHE/SACHE/ABET Guidelines – see Table 1]
1	Define Project	Project Proposal - must have clear understanding of objectives and business case (what is the benefit of making this product?) Will it make a profit? Must be clearly written and understood by all Team members.
2	Research Technologies	Research and locate potential technology options that will meet project goals. (Literature review) Establish high-cut types of Process Safety Information (see step 4 below)
3	Understand Risk	The graduate should have an introduction to the process for hazard evaluations and risk assessments [Guideline 8].
4	Understand Process Safety	a. The graduate must be able to characterize the hazards associated with chemicals and other agents. This must include toxic, flammable, and reactive hazards [Guideline 2]. Understand Process Safety Information (PSI) required to manufacture product. Includes basic description of Hazards of Materials, Process Design Technology and Equipment Design Technology [Guideline 2]. b. The graduate should understand the consequences of chemical plant incidents due to acute and chronic chemical releases and exposures [Guideline 6]. c. The graduate must understand the importance of process safety and the resources and commitment required. This should include the important incidents that define process safety, and how these incidents affected the practice of chemical engineering [Guideline 1]. d. The graduate should be familiar with the major regulations that impact the safety of chemical plants [Guideline 5].

5	Understand Risk Reduction Strategies a. Understand Engineering and Administrative Controls b. Understand Inherently Safer Design c. Understand Human Factors	a. The graduate must understand how to control and mitigate hazards to prevent accidents. This should include generally accepted management systems, plant procedures and designs to prevent accidents [Guideline 4]. b. Reduce the Frequency as much as is practical. Note for US - Required for OSHA PSM [Guidelines 4 and 5]. c. The graduate must understand and be able to apply concepts of inherently safer design [Guideline 3]. d. Improve the "Human Factors" term in the risk equation ("Operating Discipline," "Conduct of Operations," etc.) - improvement in human factors will reduce risk. Note for US - Required for OSHA PSM [Guideline 5].
6	Hazards Evaluation	The graduate should be reasonably proficient with at least one hazard identification procedure [Guideline 7].
7	Evaluate Options	Evaluate overall Risk associated with chosen technologies (effect of safety, health, environmental and business) to select final design project. Perform Risk Gap Analysis.
8	Select Final Design	Must define final Process Safety Information (PSI) required to manufacture product. Includes Hazards of Materials, Process Design Technology and Equipment Design Technology. Analyze Cost / Benefit for Final Design.
9	Understand Facility Siting	Must understand impact on personnel and communities. The difficulty for a design project is that design concepts do not have a site layout. An option for Design Team could be to identify a part of their hazardous process (if any), propose a site layout and then define their worst case scenario using a Facility Siting hazards analysis checklist. Note for US - This analysis is required for OSHA's PSM standard (29 CFR 1910.119) and is essential for the EPA's Risk Management Plan (RMP) [Guideline 5].

10	Write Final Report	Use Report Template. Note that a chapter is included to address findings discovered late that affect the final choice (hence it is titled "Directions for Future Efforts" - a very important chapter when deadlines approach!)

Table 3

SACHE Products Tied to Design Project Steps (Online via SACHE.org)

Design Project Step	**SACHE Product (Development year via SACHE.org)**
1. Define Project	See Table 3
2. Research Technologies	
3. Understand Risk	a. Risk Assessment (2008) b. Green Engineering Tutorial (2004) c. Project Risk Analysis (2009) d. Understanding Atmospheric Dispersions of Accidental Releases - CCPS book (2010)
4. Understand Process Safety	a. Chemical Process Safety (2006) b. Seminar on Fires (2009) c. Explosions (2009) d. Dust Explosion Control (2006) e. Chemical Reactivity Hazards (2005) f. Properties of Materials (2007) g. Fire Protection Concepts (2010) h. Introduction to Biosafety (2005)
5. Understand Risk Reduction Strategies	a. Runaway Reactions – Experimental Characterizations (2005) b. Design for Overpressure and Underpressure Protection (2006) c. Safety Valves: Practical Design Practices for Relief Valve Sizing (2003) d. Compressible and Two-Phase Flow with Applications Including Relief System Sizing (2011) e. Inherent Safety (2006)

6. Hazards Evaluation	a. Layers of Protection Analysis (2011) b. Static Electricity as an Ignition Source (2008) c. Process Hazard Analysis (2009)
7. Evaluate Options	Inherently Safer Design Conflicts and Decisions (2008)
8. Select Final Design – Risk Based	
9. Understand Facility Siting	
10. Write Final Report	Improving Communication Skills (2004) Includes Oral Presentations

Notice: See SACHE.org for Supporting Case Histories

Table 4

SACHE Safety Certificates Tied to Design Project Steps (Online via SACHE.org)

Design Project Step	SACHE Product (Year via SACHE.org)
3. Understand Risk	Risk Assessment (2008)
4. Understand Process Safety	a. Chemical Process Safety (2008) b. Chemical Reactivity Hazards (2008) c. Runaway Reactions (2008) d. Dust Explosion Control (2010)
5. Understand Risk Reduction Strategies	Inherent Safety (2009)

(339d) The CHEM-E Car Competition: Furthering Undergraduate Education in Process Safety

Tara Henriksen, CASE Forensics Corp, Seattle, WA and Daniel Crowl, Chemical Engineering, Michigan Technological University, Houghton, MI

The CHEM-E car competition is a tool to assist educators in teaching undergraduates the importance of process safety in the chemical engineering industry. The competition has two competition dates, a regional competition each spring semester and a national competition each fall semester. Student teams are required to complete an engineering design package with a focus on the safety standards set forth by the safety committee. The engineering documentation package requires students to consider process safety as they design, analyze, and create processes to control the operation of a chemically powered car which must carry an initially unknown load a fixed distance. The students participating in the program are expected to become familiar with management of change documentation, the appropriate level and use of PPE, the creation of standard and emergency operating procedures, and the design of independent protection layers.

See more of this Session: Integrating New Safety Requirements Into the Chemical Engineering Curriculum
See more of this Group/Topical: Education

(339e) Department-Based Lab Safety

Laura P. Ford and Christi Patton Luks, Chemical Engineering, University of Tulsa, Tulsa, OK

Some universities are fortunate to have a Health, Safety, and Environment office on campus to help with lab inspections, safety training, and hazardous waste disposal, but many smaller schools do not have this institutional support. The University of Tulsa's HS&E officer is the VP for Human Resources and Risk Management and has a background in benefits, not safety. There is little university support for lab safety beyond inspection of fire extinguishers, a 3-hour safety training for faculty every three years, and a 2-hour lab safety video for students working in research labs. Our department has pulled resources from several different places to put together a safety program for our chemical engineering labs.

1. Basic safety training: the rules of lab and an overview of material safety data sheets. 2. Lab inspection: Students complete a 3-page lab inspection which covers housekeeping, safety equipment, fume hoods, and hazards of electricity, chemicals, and compressed gases. The students test the safety showers and eye washes – the only time this equipment is tested. Obvious hazards, such as an electrical cord running through a sink, are planted around the lab by the faculty to overcome the hesitancy of students to report hazards. 3. SACHE certificate program: The students are required to complete the Safety in the Chemical Process Industries (Lab I) and Chemical Reactivity Hazards (Lab II) certificates offered by SACHE and AIChE. 4. Job safety analysis (JSA) form: Students complete a JSA as part of the prelab report for each experiment. This form is actually a university document. The form requires the students to list the potential hazards and preventative/remedial actions for each step of the lab procedure. Personal hazards (chemical contact) as well as equipment hazards (running a pump dry) must be included. 5. Material safety data sheets (MSDS): MSDS are required as part of the prelab report for all chemicals other than water and air. 6. Chemical Reactivity Worksheet: Students are required to consider the reactivity of all chemicals that they are mixing during labs in Lab II using the Chemical Reactivity Worksheet. 7. Chemical Process Safety by Crowl and Louvar: the prelab report must contain a statistical risk analysis to determine the probability of an equipment failure as describe by Crowl and Louvar.

Safety is emphasized throughout each lab in the prelab report, video, and formal or oral report. Through these materials the students learn to systematically apply personal and process safety.

See more of this Session: Integrating New Safety Requirements Into the Chemical Engineering Curriculum
See more of this Group/Topical: Education

Developing Process Safety Capsules for the Chemical Engineering Classroom

Tuesday, November 9, 2010: 4:45 PM

251 B Room (Salt Palace Convention Center)

Delmar R. Morrison and Russell A. Ogle, Exponent, Lisle, IL

The news media present us with continued reminders that we have a long ways to go towards eliminating chemical process accidents. An important strategy for reducing the number and severity of chemical process accidents is to educate our future chemical engineers in process safety principles. Some chemical engineering departments may find it difficult to add new course requirements for process safety when the chemical engineering curriculum is already so crowded. Most faculty members have also moved through the academic ranks without the benefit of industrial experience and daily exposure to chemical process safety applications. One solution to bridge the gap in experience and the loaded course schedule is to integrate process safety principles into the chemical engineering core curriculum. This can be accomplished in a very natural way by incorporating process safety capsules into the curriculum.

Process safety capsules are anecdotes, example problems, homework problems, laboratory assignments, or group projects that illustrate process safety principles. First we outline the key process safety principles that should be integrated into the chemical engineering undergraduate education. Next we demonstrate the relationship of these process safety principles to the chemical engineering core curriculum. Then we illustrate how to package process safety principles into capsules that can be incorporated into the chemical engineering core curriculum. Finally, we identify reference sources for these capsules.

Introduction

We believe that chemical process safety should be pervasive throughout the chemical engineering curriculum. Our objective is to make fire, explosion, and hazardous release awareness second nature to the graduating chemical engineer. With this foundation, new chemical engineers will be well prepared to understand process safety management objectives once they enter the workforce. The best way to build intuition towards chemical process hazards is to incorporate them into the traditional chemical engineering curriculum. This can be done without disruption in a very convenient way by sprinkling process safety capsules throughout the core chemical engineering courses.

A process safety capsule is essentially a homework problem or lecture example problem that uses engineering fundamentals to characterize some aspect of fire, explosion, or toxic release phenomena. The capsule consists of four elements: identification of the hazard, description of a realistic scenario, engineering calculations(s), and a final decision or conclusion. Ideally, we could envision one or more such process safety capsules being incorporated into the weekly lesson plan. The following discussion will elaborate on each of the elements.

Chemical Process Hazards

Chemical process hazards can be categorized as fires, explosions, and hazardous releases that have the potential to cause harm or damage to personnel, equipment, and the environment. To understand the characteristics of these phenomena requires a solid foundation in mass and energy balances, thermodynamics, fluid mechanics, heat transfer, mass transfer, chemical kinetics, and process dynamics and control. These domains of knowledge are, of course, the foundation of chemical engineering science. Many engineering professors are career academics with little formal exposure to practical management of chemical process safety; thus, they may not readily correlate fundamental textbook problems with realistic process hazards.

Fires may involve solid, liquid, or gaseous fuels and are the result of exothermic oxidation processes between the fuel and oxidizer reactants to yield gaseous products. Combustion problems involve stoichiometry, kinetics, and thermodynamics.

Explosions can be caused by both chemical reactions (e.g., TNT) and mechanical energy (e.g., overpressure). Explosion problems can involve energy balances, pressure-temperature-volume relationships, statics and dynamics, and fluid dynamics for flow through pressure relief valves.

Hazardous releases may take many forms including toxic exposures for personnel, pressurized releases, and toxic releases to the environment. Examples could include diffusion problems, gas and liquid jet calculations, or Henry's Law calculations for water contaminants.

Accident Scenario

Many accidents in the chemical process plant are the result of a complicated series of events and may involve multiple phenomena; however, they can usually be reduced to one or two primary calculations to illustrate the phenomena. To engage the student, an interesting and realistic scenario should be constructed such that the student can more easily understand and learn the significance of the hazard and utility of the calculations.

Consider, for example, a spill of flammable liquid. A drum of acetone is accidentally spilled inside a warehouse. Such a spill may be easily ignited to cause a pool fire that could injure personnel or lead to a larger conflagration that destroys the building.

- Assume firefighters enter the warehouse to suppress the fire. What is the radiant heat flux to which the firefighters are exposed? This is a heat transfer problem.
- Can a sufficiently large amount of ventilation prevent the formation of an explosive vapor concentration inside the warehouse? This is a fluid mechanics and mass transfer problem that may also be approximated as a well-mixed reactor to bound the problem.
- Instead of one drum of acetone, assume that a valve is opened on a transfer line charged by a centrifugal pump. This is a mechanical energy balance problem, which can be compared to the results of the small pool fire example.
- What if a drum of a second liquid such as propanol was also spilled? Will the liquids mix or form an immiscible layer? What will be the resulting vapor pressure? These are problems in phase equilibria.

One of the first steps in preventing or mitigating accidental fires, explosions, or toxic releases is to gauge the severity of their consequences. The chemical engineering sciences are essential to this task.

Linking Chemical Process Hazards to the Chemical Engineering Core

We define the chemical engineering core as mass and energy balances, thermodynamics, fluid mechanics, heat transfer, mass transfer, chemical kinetics, and process dynamics and control. The linkage between chemical process hazards and the core curriculum is illustrated in Table 1. Our proposal is to demonstrate the analysis of process hazards by incorporating them into the homework problems assigned in the core curriculum.

Table 1. Example matrix between Cores and Hazards.

Core Subject	Fire	Explosion	Toxic release
Mass and Energy Balances	Combustion stoichiometry	Inert gas addition required to achieve minimum oxygen concentration	Quantity of toxic gas generated by mixing incompatible chemicals
Thermodynamics	Adiabatic flame temperature	Phase equilibria of a boiling liquid, expanding vapor explosion (BLEVE)	Chemical equilibrium calculations of incompatible chemical mixures
Fluid Mechanics	Discharge of flammable liquid through an orifice	Detonation of ethylene oxide-air mixture	Dense gas dispersion
Heat Transfer	Fireball radiation	Steam explosion	BLEVE of chlorine tank car subjected to an external fire
Mass Transfer	Flashpoint modification by addition of contaminant	Binary flash calculation for failure of a storage tank heated by an external fire to the point of tank failure	Buoyant gas dispersion
Chemical Kinetics	Critical temperature determination for a rail tank car filled with a thermally unstable chemical	Runaway polymerization kinetics due to loss of coolant	Toxic gas generation by thermal decomposition of unstable fuel-oxidizer mixture
Process Dynamics and Control	Concentration history of flammable gas during vessel purging with inert gas	Temperature dynamics during thermal runaway in a chemical reactor	Toxic gas concentration history in a confined space as a function ventilation rate

The problems suggested in the table are an illustration only. There are numerous textbooks and monographs that instructors can turn to for inspiration for this purpose. A brief bibliography is provided below. The presentation will provide several specific examples to illustrate the relationship between "typical" textbook problems and chemical process safety problems.

Suggested Bibliography

Lee's Loss Prevention in the Process Industries, Edited by Sam Mannan

Chemical Process Safety: Fundamentals with Applications, Daniel A. Crowl and Joseph F. Louvar

Understanding Explosions, Daniel A. Crowl.

Chemical Process Safety: Learning from Case Histories, Roy E. Sanders.

What Went Wrong?: Case Histories of Process Plant Disasters, Trevor Kletz

Incidents That Define Process Safety, Center for Chemical Process Safety.

http://www.csb.gov, The U.S. Chemical Safety Board, completed accident reports in addition to illustrative video animations.

http://psc.tamu.edu/, The Mary Kay O'Connor Process Safety Center

http://www.hse.gov.uk/, The United Kingdom Health and Safety Executive

Process Safety Progress Journal

Fire Safety Journal

Journal of Chemical Health and Safety

Journal of Hazardous Materials

Journal of Loss Prevention in the Process Industries

Proceedings of the Loss Prevention Symposium of the American Institute of Chemical Engineers

Proceedings of the Process Plant Safety Symposium of the American Institute of Chemical Engineers

Proceedings of the Center for Chemical Process Safety Symposium of AIChE

(339g) Teaching Inherently Safer Design Concepts

Gavin P. Towler, UOP, Des Plaines, IL

Inherently safer design (ISD) is receiving increased attention from activists, legislators and regulators, who see the value of ISD techniques in reducing the potential impact of chemical plant accidents. The application of ISD methods in process design and analysis is usually not straightforward, and even experienced practitioners encounter difficulties in setting the right boundaries and basis for comparison in ISD studies.

While it is difficult to give undergraduate students more than a superficial awareness of ISD methods, it is important that they should be introduced to ISD concepts during their senior design classes. The paper will give a brief outline of ISD methodology and present some examples of design projects that highlight the need for ISD approaches.

See more of this Session: Integrating New Safety Requirements Into the Chemical Engineering Curriculum
See more of this Group/Topical: Education

(339h) Conservation of Life as a Concept for Chemical Engineering Education

James A. Klein, DuPont, Minneapolis, MN and Richard A. Davis, Department of Chemical Engineering, University of Minnesota Duluth, Duluth, MN

In 1918, Lewis DeBlois, DuPont's first corporate safety manager, wrote:

"Conservation of life should surely not be rated below the conservation of energy. Yet, few of our technical schools and universities offer instruction in this subject, and the graduates go out to their profession with only vague surmises on "what all this talk on safety is about."

Over 90 years later, a Mary Kay O'Connor Process Safety Center survey showed that only 11% of 180 universities required process safety education and that only 13% offered an elective course in process safety. The US Chemical Safety Board also has recommended that ABET accreditation require more awareness of process safety in chemical engineering programs.

This paper explores the use of "conservation of life" as a concept and unifying theme for increasing awareness, application, and integration of process safety in chemical engineering education. Students need to think of conservation of mass, conservation of energy, and conservation of life as equally important in engineering design and analysis. By giving them appropriate tools for evaluating and implementing conservation of life principles, we can help them to better understand "what all this safety talk is about," and what their role is in contributing to process safety in the chemical engineering field.

See more of this Session: Integrating New Safety Requirements Into the Chemical Engineering Curriculum
See more of this Group/Topical: Education

(360a) Active and Cooperative Learning Strategies for Introducing Biomaterials to Undergraduates

Jennifer Vernengo, Department of Chemical Engineering, Rowan University, Glassboro, NJ and Jennifer Kadlowec, Mechanical Engineering, Rowan University, Glassboro, NJ

This paper describes a Rowan University initiative to incorporate Biomaterials education into its undergraduate curriculum. Biomaterials is a relatively new interdisciplinary field within Chemical Engineering. Because of the increasing prevalence of musculoskeletal, cardiovascular and neurodegenerative disorders in today's modern society, there is a necessity to engineer biomaterials that improve the quality of life for people with painful and debilitating diseases. The objective of this initiative is therefore to teach students engineering concepts behind prominent biomaterials. Students will apply the engineering concepts to designing improved medical devices and also learn "soft skills" such as multidisciplinary teamwork, effective communication, and critical thinking. These lessons will be introduced to students within the framework of several active and cooperative learning activities that are utilized in the classroom setting and in hands-on laboratory experiences. In the classroom, students learn through interactive demonstrations, medical case studies, and student and instructor led discussions on primary literature. Juniors and seniors also participate in semester-long team based Biomaterials related projects in Rowan University's Clinic Program. Here, students work in interdisciplinary teams to gain hands-on experience in copolymer synthesis, hydrogel preparation, microencapsulation, and mechanical characterization of biomaterials. Students are given ample opportunities to sharpen their communication skills with presentations on proposed Biomaterials innovations and analysis of their laboratory results. The active and cooperative activities described in this paper can also be adapted for K-12 levels.

See more of this Session: Recent Applications of Active and Cooperative Learning in the Classroom
See more of this Group/Topical: Education

(360b) An Interdisciplinary Minor in Hydrogen Technology at Michigan Technological University

Jason M. Keith[1], Daniel Crowl[1], David Caspary[1], Jeffrey Allen[2], Jeff Naber[2], Dennis Meng[2], Abhijit Mukherjee[2], John Lukowski[3], Jay Meldrum[4] and Barry Solomon[5], (1)Department of Chemical Engineering, Michigan Technological University, Houghton, MI, (2)Department of Mechanical Engineering - Engineering Mechanics, Michigan Technological University, Houghton, MI, (3) Department of Electrical and Computer Engineering, Michigan Technological University, Houghton, MI, (4)Keweenaw Research Center, Michigan Technological University, Houghton, MI, (5)Social Sciences, Michigan Technological University, Houghton, MI

In this work we describe our efforts in using hands-on learning through in-class problems and team projects within the context of a new minor in hydrogen and fuel cell technology at Michigan Technological University. The minor can be obtained from students in any department, but targeted programs are Chemical Engineering, Mechanical Engineering, Electrical Engineering, Materials Science and Engineering, Mechanical Engineering Technology, and Electrical Engineering Technology.

To receive the 16 credit minor, students are required to satisfy requirements in four areas, which are: • Participation in multiple semesters of the Alternative Fuels Group Enterprise, where students work on hands-on integration, design, and/or research projects in hydrogen and fuel cells • Taking a fuel cell course which includes in-class problem solving • Taking a lecture or laboratory course on hydrogen energy, which includes in-class lab exercises or problem solving • Taking discipline-specific elective courses, which include in-class modules and team projects

In addition to describing the minor, the paper will also describe the content of the Alternative Fuels Group Enterprise as well as the fuel cell and hydrogen energy courses.

See more of this Session: Recent Applications of Active and Cooperative Learning in the Classroom
See more of this Group/Topical: Education

(360c) Use of Asynchronous Media to Facilitate Active Learning

Paul Blowers, Chemical and Environmental Engineering, The University of Arizona, Tucson, AZ and Jane Hunter, Systems and Industrial Engineering, The University of Arizona, Tucson, AZ

Many papers discussing learning styles and active learning techniques have been published in the literature and presented at discipline-specific, as well as ASEE, conferences. In general, use of active learning techniques is strongly promoted. However, several difficulties can arise when using active learning techniques. Some problems that faculty encounter during the class meeting time include student resistance to active learning techniques, students being unprepared, and poor classroom design. In addition, students who cannot attend class traditionally miss out on the enriched learning environment that active learning enables. This work investigates how electronic media that are readily available on many campuses can be used to enable active learning gains asynchronously.

The tools were used in a sophomore Chemical Engineering materials and energy balances course that routinely enrolls approximately 100 students per semester. Typically, of those enrolled, 35-50% are not Chemical Engineering majors that may have different backgrounds from the Chemical Engineering students enrolled in the course. The specific tools used to facilitate active learning asynchronously include Desire2Learn (D2L), a course management software application, and OneNote, a program available in Microsoft Office 2007. OneNote allows faculty with a PC Tablet to write on their computer screen and develop real-time notes during a discussion or lecture while also archiving recorded sound from either the laptop microphone or a portable lanyard that is relatively inexpensive to purchase. The combination of the two tools allows students to have 24/7 access to class materials that replay the active learning portions of the course. While students cannot ask live questions when reviewing the archived materials, email enables them to build understanding about topics after the lectures are archived. Fortunately, D2L has functions that allow course instructors to monitor the level of individual student activity on course materials, which was used to evaluated the effectiveness of the media.

The following hypotheses are investigated:

1) Students who miss class will access materials statistically more than those who attend lectures

2) Students who have English as a Second Language (ESL) will more heavily use the archived materials to enable them to translate more carefully the materials covered

3) Students who consistently review content in timely manners perform better than those who have inconsistent access patterns

4) Some patterns of use will be clear indicators of very weak learners that could be coached on how to be better students using quantifiable data of access of other students.

We found that students accessed the archived materials for lectures they missed significantly more often during semesters in which they were required to download the OneNote tool as compared to semesters in which download of OneNote was optional. In addition, we found that ESL students used the archived materials more heavily and a positive impact on their grades was demonstrated. Patterns of use showed that timely use of materials indicated mature students who were ready to move on, while other patterns indicated very weak students who needed coaching on how to use their resources more effectively.

This work shows how the synergies of online materials that are not directly active can help support active learning methods in courses.

See more of this Session: Recent Applications of Active and Cooperative Learning in the Classroom
See more of this Group/Topical: Education

(360d) Designing, Building and Operating Process Control Systems in Unit Operation Labs

Eric L. Maase, Chemical and Biomolecular Engineering Department, Lafayette College, Easton, PA

A small design and construction was undertaken to investigate the integration of process control theory with experimental practice into the undergraduate laboratory curricula. A small continuously stirred tank reactor vessel (CSTR), two source tanks each with a feed pump, and level (pressure), temperature, and flow sensors has been developed. With the use a LEGO NXT brick and LabView™ data acquisition and control system has been interfaced to provide control of the system. The system is designed for maximum usability with multiple path options for inlet and outlet flow and the vessel is jacketed for additional temperature options. The apparatus is being incorporated into the second lab course in the undergraduate curriculum (second in a sequence of three lab courses). The process of the design and construction, and possible and proposed lab experiments including the materials developed for students and goals of the labs are discussed along with the practical relationship to process control theory and chemical engineering practices. Along with two other process control systems this newest addition have all recently been incorporated formally into the second semester unit operations laboratory course as a month long experimental investigation. The results from both pre and post assessment surveys as well as evaluations of student reporting (written and oral) will be reviewed with the goal of outlining what worked, didn't work and how future improvements might be valuable. Note: The course and this new process control approach will conclude in two weeks. Pre-assessments are complete but this abstract will update to reflect the additional information being compiled and assessed, e.g. specifics will be forthcoming.

See more of this Session: Recent Applications of Active and Cooperative Learning in the Classroom
See more of this Group/Topical: Education

(360e) Triggering Active and Cooperative Learning in a Web-Based Forum

Marcel A. Liauw, Industrial Chemistry, ITMC, RWTH Aachen, Aachen, Germany and Rita Gashi, ITMC, RWTH Aachen, Aachen, Germany

Chemistry students at RWTH Aachen University are introduced to reaction engineering concepts as early as in their second year. While being well-trained in chemical synthesis, the students turn out to be ill-prepared for the math-based engineering approach. As an example, Fick' laws and mass balances are taught in the second year. The exam results are acceptable. However, re-testing the students' knowledge of these concepts two years later reveals that the subject matter was only memorized short-term but not really internalized. In order to avoid this learning-to-the-test, the effective use of problem-based assignments seems to be promising. When posted in a web-based forum, students can work on solving such assignments in an active and cooperative way. The positive influence of interactive discussions on reaction engineering problems has already been found highly motivating for more advanced students.

The students' attitude prior to the launch of the forum was rather positive (average 3.6 with a standard deviation of 0.81 on a Likert scale from 1 to 5, 40 students). This finding was not gender-specific. Passive participation (subscription) was strong from the start. Active participation (solving the problems) was initially started by just a few students. It is remarkable that despite the rather static character of a forum, a detectable degree of immersion was observed: students started to address avatars (virtual characters) in a direct manner, as if these were real characters. Reportedly, students continued discussing the problems in their spare time.

See more of this Session: Recent Applications of Active and Cooperative Learning in the Classroom
See more of this Group/Topical: Education

(360f) Developing the Complete Chemical Engineer: Creative, Innovative and Entrepreneurial

Vinten Diwakar, Jennifer Pascal and Pedro Arce, Chemical Engineering, Tennessee Technological University, Cookeville, TN

The need for innovative and creative engineers has never been greater than it is now. Through a number of sponsored programs, NSF, NAE and ASEE independently promote and indentify the importance of innovative and creative engineers. In addition, many Engineering Research Centers (ERC) sponsored by the NSF have identified that a 21st century engineer should be: Creative, Innovative and with an Entrepreneurial spirit, while the NAE clarifies the need for innovation both in classrooms and the real world. The need for transforming the engineer is therefore a more fundamental one and they should be developed during their years of training in colleges and universities. The traditional classroom is mostly a one-way street with instruction/lecturing centered on the faculty member. Studies show that the large majority of engineering programs in the US still employ this form of instruction. While this may appear to be the preferred method of instruction even in other disciplines, e.g., humanities, arts, etc…, the brazen fact that innovation is stymied by traditional lectures in engineering classrooms cannot be ignored. The traditional classroom needs to be replaced with a model that fosters innovation and creativity. In this contribution, we present an effective environment, i.e., Hi-PeLE, a High Performance Learning Environment that highly drives students to be creative and innovative by building a conducive environment for learning as well as entrepreneurial activities. Details that illustrate the particular characteristics of Hi-PeLE are presented and discussed.

See more of this Session: Recent Applications of Active and Cooperative Learning in the Classroom
See more of this Group/Topical: Education

Teaching Nanobiotechnology

Agnes E. Ostafin, Materials Science and Engineering, University of Utah, Salt Lake City, UT and Donna Ziegenfuss, Center for Teaching and Learning Excellence (CTLE), University of Utah, Salt Lake City, UT

Extended Abstract:
Nanobiotechnology is a subject of great interest for engineering students and is expected to be one of the technology growth areas in the next decade. A course designed around this multidisciplinary topic is an engineering elective in the University of Utah Professional Masters and Nanotechnology Training programs. This course involves a multidisciplinary subject that straddles basic science, materials and chemical engineering principles. Teaching such a course is challenging because students in the class come from a wide range of disciplines and many of the important issues are considered to be specialty topics not generally taught in lower level courses in typical science and engineering programs. To teach these topics at a level, which builds real expertise requires assumption of background knowledge not uniformly held by students in the class. Dealing with the challenges related to teaching this type of course led to a collaboration between an engineering faculty member and a pedagogy expert in the university's Center for Teaching & Learning Excellence.

This session will focus on the process that was used to redesign and implement the nanobiotechnology course and the assessment/evaluation strategies used to collect data from the students to measure the success of the course and identify additional recommendations that could further enhance the course. The course re-design used a "backwards design" process developed by L. Dee Fink (2003) that first identifies the primary student learning goals and objectives and then works backwards to design the teaching/learning activities and assessments that best align to those objectives. This process requires viewing the course structure through a different lens and resulted in a modularly designed course with units that focused on application and integration of nanobiotechnology content with problem solving that utilized group active learning strategies and real world examples. In addition to regular homework problems, the final course design included, team and individual based projects on topics in the class, in class discussions on journal publications in the field, a writing workshop where student practiced making a coherent argument for a proposed problem solving plan, and regular oral presentations by students. The team projects included a team based design project applying nanobiotechnology in water remediation, and a literature review. The findings from a student focus group and future course enhancements about the student course experience will also be discussed.

Fink, L. D. (2003). *Creating significant learning experiences: An integrated approach to designing college courses.* San Francisco: Jossey-Bass

Computing in Chemical Engineering Education: From Mainframes to Main Street
2010 AIChE Annual Meeting, CACHE Symposium, Abstract #193683

Duncan Mellichamp, Professor Emeritus
Department of Chemical Engineering
University of California
Santa Barbara, CA 93106

A paper marking receipt of the CACHE Award for 2010 provides an opportunity to present personal recollections and reflections about computers, computing, and applications to chemical engineering education based on 50+ years of direct participation and observation.

Modern computing technology has successfully changed chemical engineering education and practice so they now derive substantially from the predictive ability of our science-based models and much less from intuition, *ad hoc* methods, and correlations that were taught and used in the past. Employing first principles models requires truly substantial computing capability, far beyond the capabilities of early mainframe computers of the 1950's. No one could have guessed how rapidly mainframe performance "just" 50 years ago would become so widespread and then so much more powerful—first with desktops, then in laptops, and now in "walking-around" cell phones (ergo, the reference in the title to "main street.")

Trying to predict where computing technology will go, and how fast, is full of uncertainty. In the 1960's, control engineers enthusiastically spent much time/effort learning to use and then employ analog computers in the applications areas of that time. These relatively specialized devices were ideally suited (at least in principle) for representing and solving ordinary differential equation models. They had been first employed in the aerospace industry to explore flight characteristics of candidate airframes without having to build them. By 1960, the special capabilities of analog computers, with the ability to solve large ode models at computation rates of 1000 solutions per second, were being coupled to the relatively primitive digital computers of the era. The result, "hybrid computers," provided a way to solve complex ode models using the "repetitive operation" mode of the analog machine and employing the complex logic capability of the digital to provide very rapid model optimization, model parameter fitting, etc.

It is interesting to conjecture that many of the first chemical engineers who worked with hybrid computers were also among the first to jump into the use of the early commercial computers as they became available for data acquisition and control, then to the first laboratory minicomputers for data acquisition and control (at that time referred to as "real-time computing"), then to the first meaningful on-line process control implementations. This paper simply poses that hypothesis in order to outline the career path of the author who was among the earliest chemical engineering digital and analog computer users and who did follow that path.

One characteristic of operating along this trajectory, at least for an academic, is that a tremendous amount of energy was required to get into the field, to stay current, and then to utilize the current technology for productive research. But that investment inevitably was lost as equipment became cheaper, faster, more capable, and completely ubiquitous and as software designed to deal with many previously specialized applications became commercially available.

Specialized computer hardware was replaced by standardized (commercial) analog and digital interfaces. And custom software, often laboriously developed in machine or assembly language, was replaced by off-the-shelf applications packages, such as Labview. These could be used to connect experimental processes to computers for data acquisition/control applications quickly, and allowed a relatively unsophisticated user to implement sophisticated methodology in the laboratory. Suddenly, it was possible to implement an adaptive control algorithm that took the author at least six months in the early 1960's—essentially building a specialized computer out of a laboratory analog computer and a bagful of relays ordered from Allied Radio (the Radio Shack of that period)—a mere several weeks; and that was roughly just 20 years later!

It might be asked: did the urge to design/build applications for research or production purposes force laboratory and process computer users into a mode where much time/energy was expended pushing an envelope that from the user's perspective was "self-receding?" In other words, with technology developing and becoming generally available almost as fast as applications could be implemented/demonstrated, were users chasing a moving target?

Here two quantitative analytical terms from mathematical control theory can be stretched to describe this qualitative situation. From the users' viewpoint, such a changing environment was "uncontrollable" to the extent that, though desirable, these rapid commercial developments in the computer field were completely out of users' hands. The situation also was (at least) partially "unobservable" by virtue of the new/better/cheaper technology that appeared on the market with no advance information. Even if one spent much time attempting to stay current with the field, there was no way to anticipate the market.

The author takes no position on whether "chasing such a will-o'-the-wisp applications culture" was an intelligent way to pursue a career, nor does he have proof that the above characterization is a true representation of what actually happened to anyone besides himself. But whether that is a reasonable characterization of a key development in the early history of process control and in computer data acquisition/control applications, or not, that whole period was still exciting and fun.

In the last 20 years the author has taken a different path, one letting someone else do all the work and leaving the user to reap the benefits. This approach has taken him away from laboratory computing and experimental work, away from anything that requires any specialized hardware/software much beyond what a typical undergraduate can utilize. Beginning well before retirement, it became clear that the future of computing was in the use of standardized software—spreadsheet methods (e.g., Excel), editing packages (Word), presentation packages (Power Point), computation packages (Mathematica, Matlab), and design packages (Aspen, Hysys). Finally! Let someone else do all the work and take the risks!

Since retirement in 2003 and a return to teaching plant design with colleague Mike Doherty, now *pro bono*, I have come to an opinion tolerated perhaps only with older engineers. Computing has become so powerful, so cheap, so available that it has begun to force out the fundamental ways of thinking that made engineers different from scientists, namely the ability to use *ad hoc* and short cut methods. The old way attempted to "get 90% of the desired information and important conclusions from just 10% of the effort." Now, using "10% of the effort" and modern computer technology one can develop an exceptionally detailed process model, run it, optimize it, and "*know* 100% of what is desired" to as many digits of precision as desired.

Well, maybe not! Students today, perhaps even practicing engineers, can easily be led into a trap in which the detail of the computer solution appears to represent reality, but may not because of process modeling errors. Or the results of a carefully constructed optimization study may totally miss the optimum point because of unsuspected parameter errors. Or a straightforward and valuable generalization cannot be deduced from a detailed computer study because no compact mental model exists that might be exploited by a mere human to collapse a large number of numerical/graphical results into clear/concise and generalized conclusions.

An example of the latter situation will be illustrated in an area the author has re-explored in the past few years, namely use of spreadsheets to simplify Discounted Cash Flow (DCF) calculations. These typically determine potential profitability of plant designs at the conceptual design stage, i.e., before going into detailed mechanical design. Most chemical engineering bachelors programs still expect students to analyze their plant designs at that level, i.e., where estimates are available for capital expenditures and for annual revenues/expenses.

Predicting the Future of Computing

Probably the most graphic illustration of just how far off an earnest prediction of the future of computing can be was provided by a projection that came out of the Rand Laboratories in Santa Monica in 1954, was published in one of the popular science magazines of the time, and now is found easily on the Internet. The famous "think tank's" study attempted to provide a mock-up of a home computer in 2004, 50 years after the 1954 study. Just looking at the frightfully large equipment, even by 1960's standards--the old fashioned TV monitor, printer and gauges--and by reading the caption announcing that the computer would be easy to program (in Fortran), one can sense that order of magnitude errors were involved in this projection. Seeing the scientist of the period posing with this equipment, and wondering what he was to do with the two large wheels, makes one all too glad to shy away from predictions, particularly if there is any chance that Moore's Law will continue to apply in the future. Below: "Home Computer of 2004."

Scientists from the RAND Corporation have created this model to illustrate how a "home computer" could look like in the year 2004. However the needed technology will not be economically feasible for the average home. Also the scientists readily admit that the computer will require not yet invented technology to actually work, but 50 years from now scientific progress is expected to solve these problems. With teletype interface and the Fortran language, the computer will be easy to use.

Times haven't changed. Consider Nokia, just a few years ago the dominant force in the cell phone market, now scrambling for survival. Their "mistake" was failing to abandon a successful product line and enter a new market (so-called "smart phones"), which momentarily is dominated by Apple. "Unforeseeable" major changes in technology are difficult to predict.

One feature of the author's career was the opportunity to work with several early mainframe computers in the 1950's. The widespread availability of information on the Internet makes it relatively easy today to document experiences related to computers from the past.

Several mainframes and other interesting equipment from the author's background are discussed in the full talk. There the pleasure of working with compiled languages in a pre-Fortran era is noted. One machine few people will have worked with was the Bendix G-15 computer, often referred to as the first personal computer. Below: G-15 opened to expose its innards.

Relatively cheap ($60,000 in 1958) and small (about the size of a home refrigerator), the Bendix G-15 could be programmed and operated by a single individual. (Two here, for good measure.)

God's Test of Frustration-Shedding Ability – Analog and Hybrid Computers

Anyone who came into the process control and automatic control fields around 1960, as the author did, learned how to utilize analog and, later, hybrid computers. These beautiful assemblages of operational amplifiers, potentiometers, precision resistors and capacitors, required manual dexterity (as the components all had to be wired together via a removable patch panel), basic knowledge of DC circuit theory, and infinite patience to "scale" the equations in terms of the extensive variables (everything had to be expressed in a +10 to -10 VDC range) and time (or length, if the independent variable). But once programmed, all potentiometers set accurately to represent the fixed parameters, and all set to work, they could solve differential equation systems at bandwidths that only became achievable recently using digital simulators.

Unfortunately, the early analog computers used vacuum tube-based amplifiers, which had to be checked often and any non-functioning tubes replaced, or the results obtained were just so much garbage. The commercial "king of the hill," an Electronics Associates 231-R Pace machine, such as used by the author in summer 1960, is pictured below. It cost on the order of $1MM and required a squad of electronic technicians to keep it operational. The man lounging in the chair and the attractively attired female attendant were simple decoys, intended to promote the idea that analog computers were easy and convenient to use. Wonderful devices, but not easy to use!

Plain analog computers lingered for a decade; later versions utilized transistorized amplifiers, were fitted with binary logic modules and interfaced to early digital computers to yield the so-called hybrid computer. Hybrids were useful for iterative optimizations, for example, for optimizing trajectories. But eventually engineers discovered that a slow digital simulator would run all night unattended. Slow was acceptable compared with the frustrations of the analog.

Laboratory and Real-Time Computers—ADCs/DACs, Machine and Assembly Language, #%&@!!

A conjecture left to develop in the presentation concerns early workers in the area of real-time computing (aka data acquisition/control computing). While not provable, it is true that many of the first participants on the CACHE Task Force on Real-Time Computing were people with analog or hybrid computing savvy. These people were willing to work hands-on, had knowledge of electronics, and were willing to program in low-level languages (the assembly- or machine-language required for device drivers not supported by the computer manufacturer), etc.

The earliest commercial laboratory digital machines—DEC PDP-8, later -12, and the Data General NOVA (shown next page), later the Eclipse, minicomputer—cost on the order of $10-15,000 for an entry-level machine and typically required a user to build his[1] own input/output peripherals, assemble hardware, design and program interface drivers in machine or assembly code, and write applications software.

---

[1] The author recalls no women involved in these early activities, so a plausible counter thesis is that early participants self-selected on the basis of willingness to endure much futile head banging, a path non-male people apparently were too smart to go down.

On reflection, many early experimental applications were amazing boot strap efforts. The presentation covers several in the author's own laboratory, including the computer-controlled model railroad (below) used to teach real-time computing to undergraduates beginning in 1974.

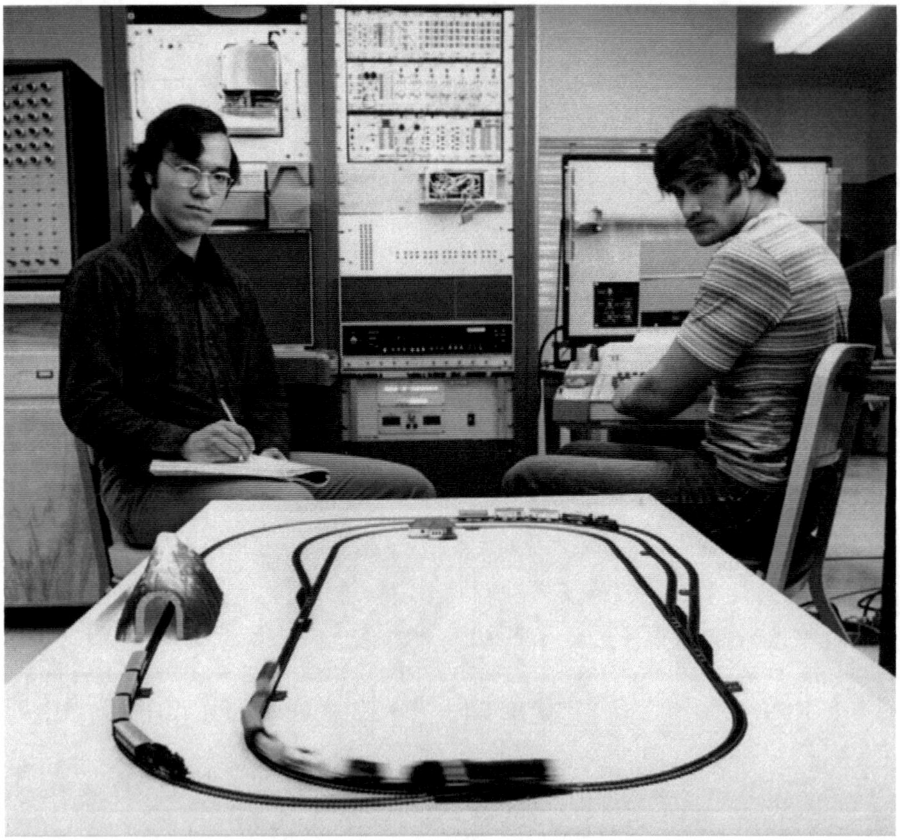

Also elaborated in the presentation are two "firsts" in chemical engineering laboratory applications: (1) of adaptive control of pH (Mellichamp, 1964) and (2) control of an unstable distributed parameter process (Wong, Bonvin etc, 1980). In the former case, an analog computer fitted with binary logic (mechanical relays) was used for the implementation. In the latter, which involved feedback control of an autothermal tubular chemical reactor, a laboratory real-time computing system equipped with standard analog I/O utilized eight axial temperature measurements in the catalyst bed to synthesize a single heat input that shifted the only RHP pole.

Development of Process Control Textbooks

Virtually all of the above work preceded the arrival of Dale Seborg at UCSB in 1976. We quickly established a collaboration that lasted essentially until my retirement. Much research with joint students came out of our partnership, and several additions of newer/commercial control equipment and experiments to our joint laboratory. But clearly the major outcome from our years together was the textbook, *Process Dynamics and Control,* co-authored originally (1989, 2003) with Tom Edgar, another Princeton PhD (associated with a relatively unknown school in Texas). This year, a third edition was published--with Frank Doyle, yet another Princeton PhD, as fourth co-author (Seborg et al, 2010).

During my own PhD work at Purdue, no undergraduate control course in Chemical Engineering, yet existed, so I took a 4 or 5 course minor in EE (automatic control theory). That period coincided with the writing of the text *Process Analysis and Control* (1965) by mentors Don Coughanowr and Lowell Koppel, the first popular undergraduate *process control* text. As eldest graduate student and resident analog guru, I had the pleasure of programming many (most) of the examples in their book using an analog computer at cost to my own research[2]. But when the opportunity came to build the first undergraduate process control laboratory at UCSB (1966-70), the simple liquid level and stirred-tank heating processes used as examples in C&K showed up in physical form. Later, in writing the first edition of "SEM," they naturally moved into our own text.

Spread Sheeting & Plant Profitability Estimates: Overlooking Simplicity Hidden in Complexity

Also left for the presentation is a full discussion of discounted cash flow calculations via spreadsheet and the too-common practice of using such numerical calculations to generate multiple solutions, perhaps even to find the optimum profitability operating point ... but without obtaining fundamental understanding. An example sketched out here shows that we have come to a situation in which computation has become so easy, and computing power so ubiquitous, that key insight and simplifications can be simply overlooked.

The DCF table on the next page was taken from Douglas (1985), Example 4.2. An equivalent analytical expression for NPV_0 [with *Discount Rate* = *CR* = *FR* = *Interest Rate*] at time of start-up, is developed just below the table. This expression appears to be highly non-linear, based on the many exponential terms down to order -10. In the presentation, an alternative way of viewing this result will be described, one that is trivially easy to develop analytically, to manipulate, optimize, etc. But that approach has been overlooked, likely because a spreadsheet is so easy to generate and explore using programs available for any personal computer.

With just one example discussed, it seems a good way now to end a career with a reckless extrapolation from the particular to the general, arguing that we need to emphasize fundamental insight and avoid so many complex approaches that solve problems and yield little understanding of the results, trade-offs, effects of key assumptions, risks associated with the "best solution" chosen, etc.

It is worth recalling Jim Douglas' message: let's go back to being more like real engineers.

---

[2] No reason to feel sorry for the author; the authors of SEM continued the noble tradition of using graduate and undergraduate student forced labor to produce our own books. Apologies to many ...

Table. Discounted Cash Flow Calculations for Example 4-2

	Interest Rate	20.0%		Tax_Rate	48%		
FC	100,000,000		a-3	0		b_1	0.60
WC	20,000,000		a-2	0.15		b_2	0.90
SU	10,000,000		a-1	0.35		b_3	0.95
SV	3,000,000		a0	0.5		Profit_BT	66,770,000
						ROI_BT	51.4%

Year		Cash Flows	Discount Factor	Present Value
-3	Investment Capital	0	1.7280	0
-2		-15,000,000	1.4400	-21,600,000
-1		-35,000,000	1.2000	-42,000,000
0		-50,000,000	1.0000	-50,000,000
	Working Capital	-20,000,000	1.0000	-20,000,000
	Start-Up Capital	-10,000,000	1.0000	-10,000,000
	PV of Investments			-143,600,000
1		26,112,240	0.8333	21,760,200
2		36,528,360	0.6944	25,366,917
3		38,264,380	0.5787	22,143,738
4		40,000,400	0.4823	19,290,316
5		40,000,400	0.4019	16,075,264
6		40,000,400	0.3349	13,396,053
7		40,000,400	0.2791	11,163,378
8		40,000,400	0.2326	9,302,815
9		40,000,400	0.1938	7,752,346
10		40,000,400	0.1615	6,460,288
	Working Capital	20,000,000	0.1615	3,230,112
	Salvage Value	3,000,000	0.1615	484,517
	PV of Operations			156,425,942

NPV(0)=Sum Present Values (Investment Plus Operations) 12,825,942

Eqiivalently, if $TCI = a_{-3}FC(1+CR)^{+3} + a_{-2}FC(1+CR)^{+2} + a_{-1}FC(1+CR)^{+1} + a_0FC + WC + SU$

then $NPV_0 = TR_c\left[(Profit_{BT})\right]\left[b_{+1}(1+FR)^{-1} + b_{+2}(1+FR)^{-2} + b_{+3}(1+FR)^{-3} + \ldots + b_{+10}(1+FR)^{-10}\right]$

$$+0.1TR\ (FC+SU)\left[(1+FR)^{-1} + (1+FR)^{-2} + (1+FR)^{-3} + \ldots + (1+FR)^{-10}\right]$$

$+\left[TR_c(SU+SV) - TCI\right](1+FR)^{-10}$; TR is the tax rate; TR_c is the compl. tax rate, 1 – TR.

References:

Coughanowr, D.R., Koppel, L.B., *Process Systems Analysis and Control*, McGraw-Hill, NY (1965).

Douglas, J.M., *Conceptual Design of Chemical Processes*, McGraw-Hill, NY (1988).

Mellichamp, D.A., *Identification and Control of a Gain-Varying Flow System*, PhD Dissert. Purdue University, Lafayette, IN (1964).

Seborg, D.E., Edgar, T.F., Mellichamp, D.A., *Process Dynamics and Control*, Wiley, NY (ist Ed. 1989, 2nd Ed. 2003).

Seborg, D.E., Edgar, T.F., Mellichamp, D.A., Doyle, F.J., *Process Dynamics and Control*, Wiley, NY (3rd Ed. 2010).

Wong, C., D. Bonvin, D.A. Mellichamp, and R.G. Rinker, "On Controlling an Autothermal Fixed-Bed Reactor at an Unstable Steady State: IV - Model Fitting and Control of the Laboratory Reactor," *Chem. Eng. Sci.*, 38, 619-633 (1983).

(388b) POLYMATH – the Present, the New DIPPR Database Option and the Future of This Popular CACHE Numerical Problem-Solving Package

Michael B. Cutlip, University of Connecticut, Storrs, CT and Mordechai Shacham, Ben Gurion University of the Negev, Beer-Sheva, Israel

This short presentation will highlight current POLYMATH capabilities, the new interface option providing access to the ~2000 component DIPPR database, and the future version that is currently under final development. POLYMATH is currently available under inexpensive CACHE site licenses to departments, engineering colleges and institutions. These licenses allow POLYMATH software to be given to all students, faculty, and staff. Additionally, the software can be installed in computer labs and on other departmental computers. Brief demonstrations will be given. More detailed information and personal demonstrations will be available at the CACHE booth in the Exhibition area.

See more of this Session: CACHE Symposium: Honoring Award Recipients
See more of this Group/Topical: Computing and Systems Technology Division

(388c) A Student Competition to Develop a Chemical Engineering App for the Iphone

Peter T. Cummings and Clare McCabe, Department of Chemical and Biomolecular Engineering, Vanderbilt University, Nashville, TN

Senior students in the Vanderbilt University School of Engineering participate in an interdisciplinary product design course. A group of students took on the challenge of developing an iPhone app for chemical engineering undergraduates. In this talk, we will report our experiences as faculty mentors and on the accomplishments of the students.

See more of this Session: CACHE Symposium: Honoring Award Recipients
See more of this Group/Topical: Computing and Systems Technology Division

(388d) Energy Modules for the ChE Curriculum

Jason M. Keith, Department of Chemical Engineering, Michigan Technological University, Houghton, MI, Thomas F. Edgar, Chemical Engineering, The University of Texas at Austin, Austin, TX, Gavin P. Towler, UOP, Des Plaines, IL, H. Scott Fogler, Chemical Engineering, The University of Michigan, Ann Arbor, MI, David T. Allen, Department of Chemical Engineering, The University of Texas at Austin, Austin, TX and Darlene Schuster, AIChE, New York, NY

The Energy Modules Task Force will present an update on the development of energy modules for use in the Chemical Engineering curriculum.

See more of this Session: CACHE Symposium: Honoring Award Recipients
See more of this Group/Topical: Computing and Systems Technology Division

Modules for High School Engineering Courses

David T. Allen
Department of Chemical Engineering
The University of Texas at Austin

Tens of thousands of students in the United States are now first experiencing engineering through year-long or multi-year high school engineering programs, yet, there is no general consensus regarding the learning objectives for high school engineering courses and few training programs exist for high school engineering teachers. Developing effective high school engineering curricula and teacher training programs is particularly urgent in Texas. A 2006 legislative decision requires all high school students, beginning with those who entered ninth grade in 2007, to complete four years of science to graduate under the state's default degree plan. This fourth year of science must be selected from a group of 18 courses, including a new year-long course in engineering. This represents the first state-wide implementations of a high school engineering course and creates immediate needs for curricula and teacher training.

The University of Texas at Austin is partnering with school districts in Texas to develop and deliver an innovative, design- and challenge-based curriculum for the Texas high school engineering course and to prepare training pathways for secondary teachers of engineering. Four professional development pathways to teacher preparedness have been developed:

1. *UTeach Master of Arts in Science and Engineering Education (MASEE)*, a three summer program for in-service teachers seeking to become leaders in secondary engineering education;
2. *Engineering Summer Institutes for Teachers (ESIT)*, conducted at UT-Austin and 5 dissemination sites, for in-service teachers who want to rapidly prepare to deliver design-based engineering courses;
3. *Engineering Certification Track for Physics, Chemistry and Math Majors*, to prepare the next generation of secondary science and engineering teachers; and
4. *Teacher Preparation Track for Engineering Majors* for engineers who wish to obtain certification as science and engineering teachers while they are completing their engineering degree.

New university level engineering courses have been developed for these teacher professional development pathways and the first cohorts of teachers are now completing these programs.

A high school curriculum has also been developed and is being piloted in 2010-2011 at seven Texas high schools. The course objectives are to introduce students to engineering design methods, the types of challenges addressed by engineers, and the analytical tools that engineers employ in solving problems. The course consists of four components: (i) a six-week unit introducing students to engineering and the world's energy challenges, (ii) a six-week unit on reverse engineering, (iii) a six-week unit on robotics, and (iv) a semester long design project with structured presentation of engineering design methods.

The University of Texas is partnering with CACHE to develop modules, particularly modules making extensive use of visualizations, that can be used in the university-based teacher preparation courses and in the high school engineering course. The framework for this module development will be described in this presentation.

Using Chemsep, COCO and other modeling tools for versatility in custom process modeling

Jasper van Baten – AmsterCHEM (jasper@amsterchem.com)
Ross Taylor – Clarkson University (taylor@clarkson.edu)& Chemsep
Harry Kooijman – Clarkson University (kooijman@clarkson.edu) & Chemsep

Summary

A steady state flowsheeting environment is presented in which out-of-the-box unit operation models can be combined with formula based custom models; the formulas can be presented in Excel, Matlab or Scilab format. The interaction between all simulation software components is based on the CAPE-OPEN standard specifications, which makes that all software components can be reused in other simulation environments. Similarly, software components (such as thermodynamic servers and unit operations) of other simulation environments can be included in the simulation.

The combination of out-of-the-box unit operation models and custom unit operation models makes the presented framework suitable for research as well as teaching purposes.

Introduction

Main stream commercial steady state flowsheet based process simulation applications are available for university courses for attractive prices. The models provided by most simulation environments are often black-box and the model equations may not be readily accessible to the end-user. For many simulation purposes this suffices. However, for the purpose of teaching chemical engineering and for the purpose of research, it may be prudent to allow custom modeling of particular equipment in an easy and transparent way, using generic modeling tools available to students, teachers and researchers.

Although simulation environments generally provide a wide selection of models, it is impossible for any simulation environment to provide all models that the end-user may want to use in a particular context. For example, heat exchangers are often modeled by a counter- or co-current approximation, which does not describe in detail what happens in the most common heat exchanger equipment, the shell-and-tube heat exchanger. In chemical engineering courses plug flow models are often compared to tanks-in-series models. The former is generally provided; the latter is not. Or – the example used in this presentation – a simple model for a membrane separator. Due to the diversity in available data on membrane separations, it is not easy to provide a generic membrane separation model. Other examples of simple equipment that cannot easily be described with generic models are easily provided, all one needs to do is open a text book on chemical engineering and look at the versatility of process equipment provided[1].

Most simulation environments allow for custom models. Often, this involves writing a FORTRAN routine that is suitable only for the simulation environment at hand, i.e. a specialized interface routine only to be used in the simulation environment it was created for. Such an approach requires programming skills on the end-user, which in the case of chemical engineering teaching courses may be a large threshold to use such custom models. In addition, to the use the same custom model in a different simulation environment, the programming exercise needs to be repeated.

Here, we present the COCO[3] simulation environment that is available free-of-charge. For the separation columns, we present ChemSep[4]; the LITE version of which is available free-of-charge (the full version of ChemSep has been available free-of-charge for educational use from the CACHE corporation for more than 15 years now). The combination of COCO and ChemSep has been presented earlier[5-11], and we refer to these earlier publications and presentations for a full introduction to using COCO with ChemSep. In the current presentation, we will focus on the inclusion of custom models using generic modeling tools like Scilab[12], Matlab[13] and Excel[14]. All these software components are interacting via CAPE-OPEN[15] interfaces: an open standard that enables different chemical engineering software libraries and packages to interoperate for the purpose of solving flowsheets; i.e. reactors, distillation columns, and other unit operations with interconnecting material, energy, and information streams. CAPE-OPEN thermodynamic and unit operation interfaces have been well established, validated and demonstrated[8, 16-22].

The COCO steady state flowsheeting suite is entirely based on this open standard; all thermodynamic and unit operations models can be re-used in other simulation environments that are CAPE-OPEN compliant. Similarly, all third party unit operation models and thermodynamic servers that are CAPE-OPEN compliant can be used in COCO. As the COCO suite does not include models for distillation, absorption, and extraction columns, it is bundled with the ChemSep LITE column simulator.

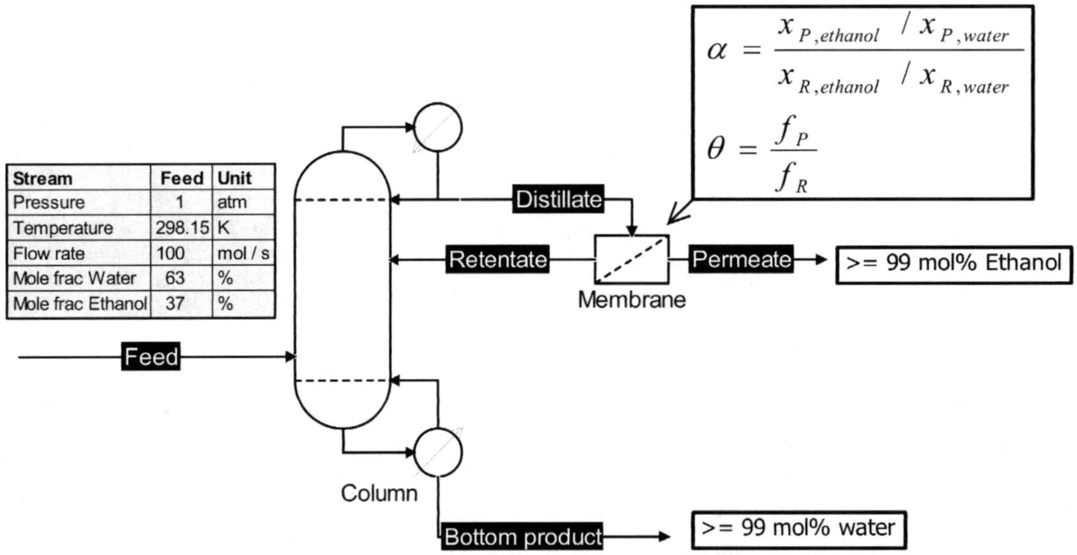

Figure 1: Test case for separation using a distillation column combined with a membrane unit with feed conditions and desired separation targets. The feed goes to a distillation column operating at 1 atmosphere with a partial reboiler and a total condenser.

We will demonstrate how to set up custom models using these modeling tools with a simple test case; this test case has been chosen for its simplicity of demonstration. This simplicity does not reflect on the possibilities of the modeling tools shown here, but is chosen merely for demonstrating the possibilities without getting lost in modeling details. The test case has been taken from a textbook[23]; the goal being to separate water and ethanol. The water-ethanol system has an azeotrope at about 90 mol% ethanol, and direct separation is not possible using a single distillation column. A combined distillation with membrane separation is used as shown in Figure 1. To get started, we start COFE, add a new TEA thermo package called “water-ethanol”, add compounds Water and Ethanol, and select the default NTRL model set. We define and specify the feed stream. We then insert a ChemSep distillation column, and accept all its defaults for “normal distillation” at its creation.

The membrane model is defined by a separation factor α and a permeate cut θ as shown in Figure 1, where $x_{P,ethanol}$ and $x_{R,ethanol}$ are the ethanol mole fractions in permeate and retentate, $x_{P,water}$ and $x_{R,water}$ are the water mole fractions in permeate and retentate and f_P and f_R are the total permeate and retentate flows. Separation factor α and a permeate cut θ are model parameters, with default values of 70 and 0.6 respectively.

The Scilab Unit Operation and the Matlab Unit Operation

To create the simple membrane model using the Scilab scripting language, we start by inserting a Scilab Unit Operation[24]. The first thing we do is define the ports; the feed, the retentate and the permeate products:

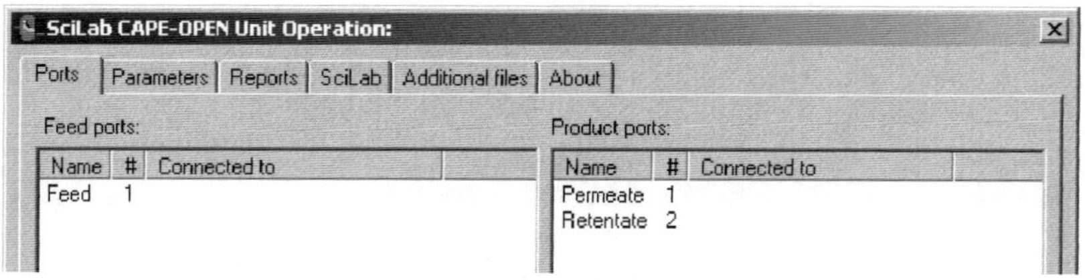

We notice that the ports have received indices; these we will use in the model script to access the values associated with the ports. We then proceed to define the parameters for our model:

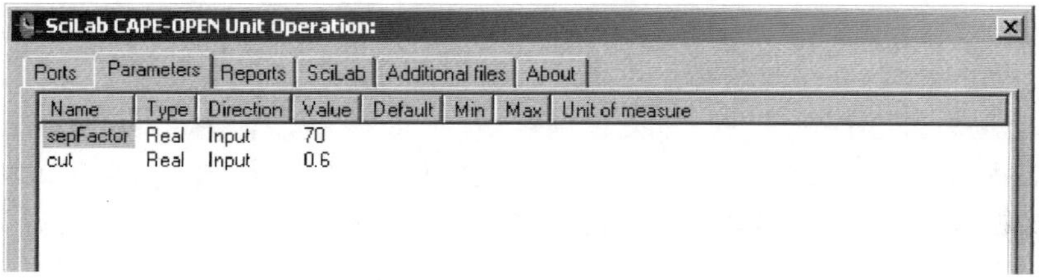

The parameters are in this case dimensionless; dimensions and bounds for parameters are optional. For our simple model, the definition of the parameters, the component balances and the product composition summation equations can be solved explicitly. This is not a requirement however of the modeling tools; all modeling tools that are used come with built-in solvers. To demonstrate their use, we will only use a partly analytic solution. The flows are solved explicitly. To solve the compositions, we will solve a one dimensional problem given by: find $x_{P,ethanol}$ so that the target value of separation factor α is satisfied. Given $x_{P,ethanol}$, the other compositions follow from the component balances and summation as shown by the `GetCompositions` function in the Scilab script for solving the unit:

```
// compound indices:
water=1;
ethanol=2;

//parameter values:
alpha=getParameter('sepFactor');
theta=getParameter('cut');

//feed values
XF=getFeedProp(1,'fraction');
FF=getFeedProp(1,'totalFlow');
TF=getFeedProp(1,'temperature');
PF=getFeedProp(1,'pressure');
```

```
//product flows (explicit solution of theta=FP/FR and FF=FP+FR):
FR=FF/(theta+1);
FP=theta*FR;

//function to calculate compositions from XPE:
function [XP,XR]=GetCompositions(XPE)
 XP=[1-XPE,XPE];
 XRE=(FF*XF(ethanol)-FP*XPE)/FR;
 XR=[1-XRE,XRE];
endfunction

//function to solve compositions
function errorValue=CompositionFunction(XPE)
 //get the compositions given XPE:
 [XP,XR]=GetCompositions(XPE);
 //return the error in alpha
 errorValue= (XP(ethanol)/XP(water))/(XR(ethanol)/XR(water))-alpha;
endfunction

//solve for the compositions to match alpha:
XPE=fsolve(0.999,CompositionFunction);
//the compositions at solutions are then
[XP,XR]=GetCompositions(XPE);

//set the permeate, given FP, XP, TF, PF
setProduct(1,FP,XP,'temperature',TF,'pressure',PF);

//set the retentate, given FR, XR, TF, PF
setProduct(2,FR,XR,'temperature',TF,'pressure',PF);
```

The script obtain the inputs, calculates the product flow rates explicitly, defines a function that gives the product compositions as a function of $x_{P,ethanol}$, defines a function that gives the error in α as a function of $x_{P,ethanol}$, and solves for the error to be zero. The product streams are defined by temperature and pressure of the feed in combination with the calculated flows and compositions. Programming is largely unnecessary; we mostly type the functions and ask the Scilab program to solve them. As pointed out earlier, the iterative solution here is for demonstration purposes, the problem has a simple explicit solution. We could have typed that instead. The `CompositionFunction` has multiple solutions; undesired solutions lie at negative $x_{P,ethanol}$; we choose the initial guess of $x_{P,ethanol}$ at a value of 0.999 to avoid the undesired negative solution.

This example does not use any thermodynamics from the simulation environment; all thermodynamic property calculations and phase equilibrium calculations are however available and can be used in unit operation models. Appendix A provides an example.

The Matlab Unit Operation[25] is based on the same principles as the Scilab Unit Operation and very similar to the Scilab Unit Operation. The example is readily transferred to the Matlab Unit Operation; the Matlab version is not shown here.

To complete the flowsheet, we edit the ChemSep column, set the number of stages to 20, move the feed to stage 17. We set the bottom specification to 99.9 mol% water; this is tighter than required by the problem specification. For the top specification, we use 85% ethanol, to stay below the azeotrope. Running the simulation without the recycle gives us an idea of the retentate composition; we check the column profiles to see where the tray compositions are close to this: stage 6. Here we create a second feed to the column and connect the retentate stream. We solve the problem, and our target specifications have been satisfied.

The Excel Unit Operation

Even though programming is largely unnecessary for the Scilab implementation of the membrane unit as shown above, some users may be more comfortable with entering the formulas in a spread sheet manner. This is possible, using the Excel Unit Operation[26]. We take the flowsheet created above as a starting point, and remove the Scilab unit operation. We insert a new empty Excel Unit Operation. Editing the unit opens Excel; ports and parameters are simply entered on dedicated pre-existing Excel sheets. On the *feeds* sheet, we enter our feed:

	A	B	C	D	E	F	G
1		T	P	H	flow	mole fractions	
2	**Name**	[K]	[Pa]	[J/mol]	[mol/s]		
3	Feed						

We create the Permeate and Retentate on the *products* sheet:

	A	B	C	D	E	F	G
1		T	P	H	flow	mole fractions	
2	**Name**	[K]	[Pa]	[J/mol]	[mol/s]		
3	Permeate						
4	Retentate						

Here, we need to provide values for the flow and compositions for each of the products, and two out of three of T, P and H. The parameters are entered on the *input parameters* sheet:

	A	B	C	D	E	F	G	H	I	J	K	L	M
1	**name**	**value**	**min**	**max**	**m**	**kg**	**s**	**A**	**K**	**mole**	**cd**	**rad**	**relative flag**
2	sepFactor	70											
3	cut	0.6											

Initial values for the parameters are entered; the user can modify these from the simulation environment, and the new values will automatically appear in the sheet. Hence, we can refer to the values in these cells in our formulas; for easy reference, the names *alpha* and *theta* have been assigned to the cells. If we close Excel, connect the ports and re-edit the unit, the feed values are present, which makes entering the formulas easier:

	A	B	C	D	E	F	G
1		T	P	H	flow	mole fractions	
2	**Name**	[K]	[Pa]	[J/mol]	[mol/s]	Water	Ethanol
3	Feed	351.4708	101325	-35678.5	43.4629	0.15	0.85

The formulas we enter on a separate *calculations* sheet:

	A	B	C	D
1	F retentate	=feeds!E3/(theta+1)	mol/s	
2	F permeate	=theta*B1	mol/s	
3				
4	X p ethanol guess	0.999	mol/mol	Initial guess for X p ethanol
5	X p ethanol	0.999	mol/mol	Will be solved to match alpha
6	X p water	=1-B5	mol/mol	
7	X r ethanol	=(feeds!E3*feeds!G3-B2*B5)/B1	mol/mol	
8	X r water	=1-B7	mol/mol	
9				
10	alpha	=(B5/B6)/(B7/B8)		
11	error	=B10-alpha		

We now use the Excel solver to solve `B11` for a value of zero by modifying `B5`. We save the solve scenario, and on the *control* sheet, we indicate that this solver scenario must be used during the calculation, using an initial guess specified in `B4` for cell `B5;` to accomplish this, we assign a name to the saved solver scenario region, the initial guess ad the target cell for the initial guess. These names are specified in the *control* sheet; multiple sequential solver scenarios can be specified. Notice that in this simple problem, the initial guess is a constant; a formula can be entered here as well. On the *products* sheet, we set references to the solution:

	A	B	C	D	E	F	G
1		T	P	H	flow	mole fractions	
2	**Name**	[K]	[Pa]	[J/mol]	[mol/s]	Water	Ethanol
3	Permeate	=feeds!B3	=feeds!C3		=calculations!B2	=calculations!B6	=calculations!B5
4	Retentate	=feeds!B3	=feeds!C3		=calculations!B1	=calculations!B8	=calculations!B7

and our problem specification is complete.

Conclusions

It has been demonstrated how custom models can be incorporated into simulation applications using generic modeling tools such as Scilab, Matlab and Excel, via the CAPE-OPEN interface standards. Together with powerful readily available unit operation implementations such as ChemSep and the unit operations that come with the simulation environment, a highly flexible modeling environment is presented. Formula based custom modeling such as in Scilab, Matlab or Excel, where thermodynamics of the simulation environment are readily available and equation solvers are integrated, allows easy implementation of text-book models and custom models. Each of the units comes with an integrated help and formula reference. Both research and chemical engineering courses may benefit from this setup.

References

1. Coulson, J.M., Richardson, J.F., Sinnot, R.K., (1991), “Chemical Engineering”. Volume 6, Design. Pergamon, Oxford, England.
2. http://www.aspentech.com/
3. COCO is freely available from http://www.cocosimulator.org/
4. http://www.chemsep.org/
5. Kooijman, H., Taylor, R., and van Baten, J.M., 2007. The ChemSep/COCO Casebook: Air Separation Unit, CACHE News, Winter.
6. van Baten, J.M., Kooijman, H. and Taylor, R., 2007. Flowsheeting for Free with COCO, CACHE News, Winter.
7. Taylor, R., Kooijman, H, van Baten, J.M., 2006. CAPE-OPEN Overhead in Distillation Modeling, AIChE presentation by Ross Taylor, San Francisco, Nov

8. Baur, R., van Baten, J., Kooijman, H. and Drewitz, W., 2006, "Cross Platform Chemical Processes". NPT procestechnologie, September
9. van Baten, J.M., 2007. An introduction to COCO. 4th US CAPE-OPEN conference, AIChE annual meeting, Saltlake City, November.
10. Kooijman, H and Taylor, R, 2008. Chapter 3: CAPE-OPEN Flowsheet Simulations with ChemSep, part of The ChemSep book, 2nd edition.
11. Taylor, R. and van Baten, J.M., 2008. Introduction to flowsheeting with COCO and ChemSep. Available from: http://www.cocosimulator.org/down.php?dl=FlowsheetingWithCOCOandChemsep_Notes.pdf
12. Scilab is available free-of-charge from http://www.scilab.org/
13. Matlab is trademarked and licensed by Mathworks: http://www.mathworks.com/
14. Excel is trademarked and licensed by Microsoft: http://www.microsoft.com/
15. The CAPE-OPEN Laboratories Network, CO-LaN, http://www.colan.org/
16. Barrett, William M. and Yang, Jun (2005), "Development of a chemical process modeling environment based on CAPE-OPEN interface standards and the Microsoft .NET framework". *Computers and Chemical Engineering* 30, pp. 191-201
17. Fermeglia, M., Longo, G. and Toma, L., 2009, "Computer Aided Design for Sustainable Industrial Processes: Specific Tools and Applications". AIChE Journal, Vol. 55, No. 4, 1065.
18. Barrett, W., Yang J. and Strunjas, S., 2007, "Final Report For Verification Of The Metal Finishing Facility Pollution Prevention Tool". EPA report, 2007.
19. Lang, Y., Malacina, A., Biegler, L., Munteanu, S., Madsen and J., Zitney, S., 2009, "Reduced Order Model Based on Principal Component Analysis for Process Simulation and Optimization". Energy & Fuels 23, 1695
20. Morales-Rodríguez R., Gani, R., d´Echelotte, S., Vacher A, and Baudouin, O., 2008, "Use of CAPE-OPEN standards in the interoperability between modelling tools (MoT) and process simulators (Simulis® Thermodynamics and ProSimPlus)". Chemical Engineering Research and Design, 86, 823
21. Fermeglia, M., Longo, G., and Toma, L., 2008, "COWAR: A CAPE OPEN Software Module for the Evaluation of Process Sustainability". Environmental Progress 27, No.3
22. Breil, M., Kontogeorgis, G., von Solms, N. and Stenby, E., 2007, "CAPE-OPEN: An International Standard for Process Simulations". Chemical Engineering, Dec., 53
23. Wankat, P.C., 2006. Separation Process Engineering (2nd edition). Prentice Hall, New Jersey, United States.
24. Scilab CAPE-OPEN Unit Operation, available from http://www.amsterchem.com/scilabunitop.html
25. Matlab CAPE-OPEN Unit Operation, available from http://www.amsterchem.com/matlabunitop.html
26. Excel CAPE-OPEN Unit Operation, available from http://www.amsterchem.com/excelunitop.html

Appendix A: using thermodynamics

To illustrate the use of thermodynamic property calculations with a simple example, we take the equations for the membrane separator, and solve for temperature so that the unit operates adiabatically. This means we have to solve the additional equation:

$$f_F h(T,P,X_F) = f_R h(T,P,X_R) + f_P h(T,P,X_P)$$

where the enthalpy terms [J/mol] are obtained from the simulation environment. A small modification to the presented Scilab script is required:

```
// compound indices:
water=1;
ethanol=2;

//parameter values:
alpha=getParameter('sepFactor');
theta=getParameter('cut');

//feed values
XF=getFeedProp(1,'fraction');
FF=getFeedProp(1,'totalFlow');
TF=getFeedProp(1,'temperature');
PF=getFeedProp(1,'pressure');

//product flows:
FR=FF/(theta+1);
FP=theta*FR;

//function to calculate compositions from XPE:
function [XP,XR]=GetCompositions(XPE)
 XP=[1-XPE,XPE];
 XRE=(FF*XF(ethanol)-FP*XPE)/FR;
 XR=[1-XRE,XRE];
endfunction

//function to solve compositions
function errorValue=CompositionFunction(XPE)
 //get the compositions given XPE:
 [XP,XR]=GetCompositions(XPE);
 //return the error in alpha
 errorValue= (XP(ethanol)/XP(water))/(XR(ethanol)/XR(water))-alpha;
endfunction

//solve for the compositions to match alpha:
XPE=fsolve(0.999,CompositionFunction);
//the compositions at solutions are then
[XP,XR]=GetCompositions(XPE);

//now that we have the composition, solve the heat balance
HF=getEquilibriumProperty('enthalpy',XF,'temperature',TF,'pressure',PF);

//function to solve the heat balance at constant P
function errorValue=HeatBalance(T)
 HR=getEquilibriumProperty('enthalpy',XR,'temperature',T,'pressure',PF);
 HP=getEquilibriumProperty('enthalpy',XP,'temperature',T,'pressure',PF);
 errorValue=HF*FF-HR*FR-HP*FP;
endfunction

//solve temperature to close the heat balance
TOUT=fsolve(TF,HeatBalance);

//set the permeate, given FP, XP, TOUT, PF
setProduct(1,FP,XP,'temperature',TOUT,'pressure',PF);

//set the retentate, given FR, XR, TOUT, PF
setProduct(2,FR,XR,'temperature',TOUT,'pressure',PF);
```

For the enthalpy calculations, we cannot get away with cheaper single phase calculations as we cannot be sure what phase(s) the feed and products are in at given temperature. In fact, we know that the feed stream is at its dew point, so we likely undergo phase changes during the solution of the heat balance. The `getEquilibriumProperty` function combines a phase equilibrium calculation at specified conditions with a property calculation.

(396a) Best Practices by Department Chairs

Said AbuBakr, Paper Engineering, Chemical Engineering and Imaging, Western Michigan University, Kalamazoo, MI, **Martin C. Hawley**, Department of Chemical Engineering and Materials Science, Michigan State University, East Lansing, MI, Jeffrey A. Reimer, Chemical Engineering, University of California - Berkeley, Berkeley, CA, Richard B. Dickinson, Chemical Engineering, University of Florida, Gainesville, FL, Valerie L. Young, Chemical and Biomolecular Engineering, Ohio University, Athens, OH and Michael A. Matthews, University of South Carolina, Columbia, SC

The position of Department Chair varies from person to person, from department to department and from university to university. This session is designed to have an open discussion on the department head position description and experience. Department chairs will share their best practices in areas such as mentoring new faculty, dealing with tenure promotion issues, ABET accreditation and assessment, fostering collegiality among faculty, managing students complaints, dealing with industrial advisory boards, alumni and employers, interacting with their deans and upper administration, fund raising and recruiting activities, responding to new and emerging initiatives, teaching, conducting research and advising graduate students, and much more. The recruitment, mentoring, evaluation and integration of teaching-only faculty will be also discussed.

See more of this Session: Department Heads Forum
See more of this Group/Topical: Education

(396b) Continuous Improvement of ABET

Michael Milligan, ABET, ABET, Inc., Baltimore, MD

In June 2009, Dr. Michael K.J. Milligan became the Executive Director of ABET. In addition to his leadership experience, he has five years of experience as an ABET program evaluator and has led the accreditation process at his former institution. Dr. Milligan brings multiple perspectives to the leadership of ABET, and in the first few months, he has reached out to multiple constituents and continues those conversations as part of his ongoing strategic activities. In his discussion with the chemical engineering department heads, Dr. Milligan will share what he has learned from his conversations with ABET constituents and the priorities that he has established to address concerns about ABET processes. Among the topics to be discussed are: continued training of ABET volunteers, expanding communication channels with ABET programs, ABET's role in promoting innovation of technological education, and ABET's vision for international accreditation.

See more of this Session: Department Heads Forum
See more of this Group/Topical: Education

(396c) ABET Update

Jeffrey J. Siirola, Eastman Research Division, Eastman Chemical Company, Kingsport, TN

This session will give an update from ABET regarding lessons learned from 2010-2011 visits and what to look for in 2011-21012 visits

See more of this Session: Department Heads Forum
See more of this Group/Topical: Education

(396d) NSF Update

Thomas W. Peterson, Engineering, National Science Foundation, Arlington, VA

The National Science Foundation promotes the progress of engineering in the United States in order to enable the Nation's capacity to perform. Its investments in engineering research and education aim to build and strengthen a national capacity for innovation that can lead over time to the creation of new shared wealth and a better quality of life. Most NSF programs in engineering are funded through the Directorate for Engineering, which also sponsors the NSF's Small Business Innovation Research (SBIR) program.

See more of this Session: Department Heads Forum
See more of this Group/Topical: Education

AIChE Academic Salary Survey

Geoffrey L. Price
Department of Chemical Engineering
University of Tulsa
Tulsa, OK

Abstract

Each US Chemical Engineering department listed at www.che.utexas.edu/che-faculty/ was contacted via e-mail through the chair/head/director of the department with a request to return a salary survey for their department. Responses have been averaged nationally and by region and presented at the Department Head's Forum. Results will also be compared to results from previous surveys.

Incorporating the Design of Sustainable Systems Into Engineering Curricula

David T. Allen
Department of Chemical Engineering
The University of Texas at Austin

Cynthia Folsom Murphy
Center for Energy and Environmental Resources
University of Texas at Austin

David R. Shonnard
Department of Chemical Engineering
Michigan Technological University

Sharon Austin and Nhan T. Nguyen
Chemical Engineering Branch
Office of Pollution Prevention and Toxics
USEPA, Washington, DC

Sustainability is a powerful, yet abstract concept. The most commonly employed definition is that of the Brundtland Commission report – meeting the needs of the present generation without compromising the ability of future generations to meet their needs (World Commission on Environment and Development, 1987) – but there are many variations on this basic definition.

The work of engineers is central to the development of sustainable systems for meeting human needs. Incorporating sustainability into engineered products, processes, and systems generally means evaluating environmental and social performance of designs. While this may seem straightforward, reducing this concept to the types of quantitative design tools and performance metrics that can be applied in engineering design is a challenge. Yet, it is a challenge that many engineering educators are embracing. This presentation reports on how engineering educators are incorporating sustainability into engineering courses. The analysis is based on a recent benchmarking of the incorporation of sustainability concepts into engineering education in the United States (Murphy, et al., 2009).

The results are presented in two sections. First, the types of courses in which sustainability concepts are being addressed are described, along with the diverse subject areas covered in these courses. Then, common elements in the courses are identified. Based on these common elements projected structures for incorporating sustainabilty into engineering education are presented.

Sustainability Courses and Content

Based on data provided by hundreds of engineering educators, defined here as Sustainable Engineering champions, there are four primary means by which sustainable engineering concepts are being incorporated into engineering courses: dedicated sustainable engineering courses (48% of courses identified in a 2009 benchmarking); integrating sustainable engineering concepts into traditional engineering courses (23%); courses focused on the technologies predicted to be important in developing sustainable engineering designs, such as photo-voltaic solar cells and fuel cells (14%); and interdisciplinary courses done in conjunction with a non-engineering department (15%). These data are summarized in Figure 1. The data presented in Figure 1 also show that most of the courses are stand-alone electives. Roughly a quarter of the courses (23%) are part of a formal major or minor requirement.

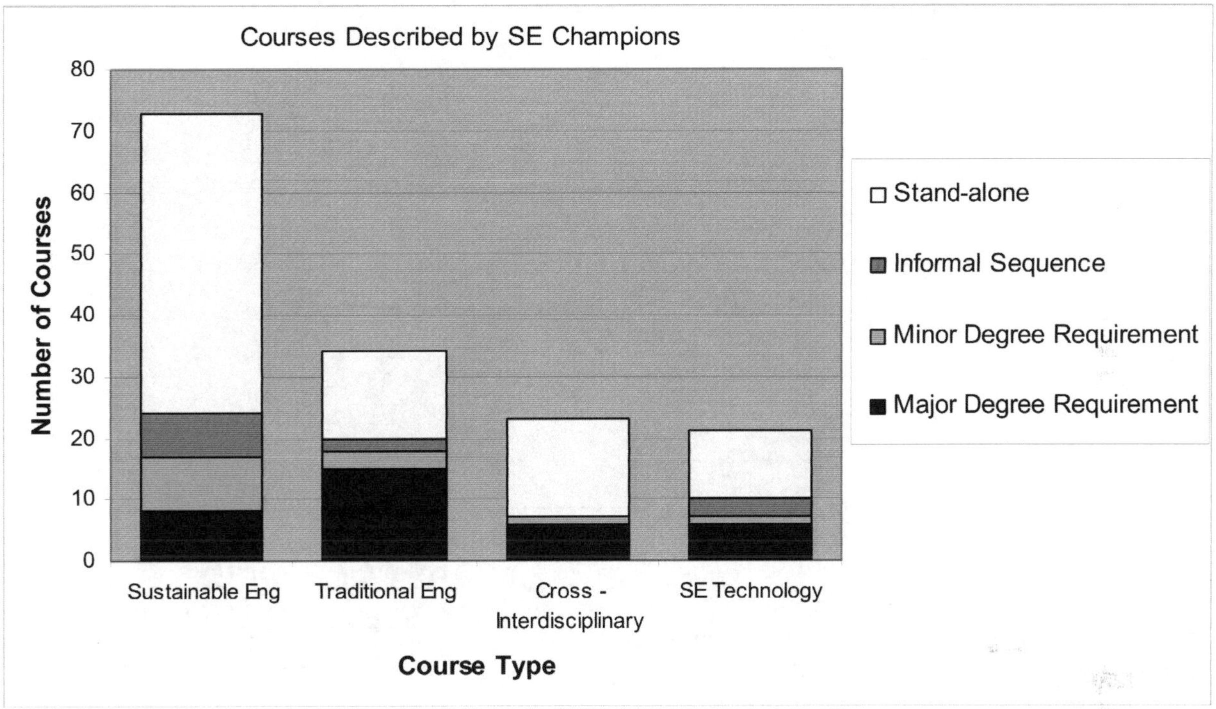

Figure 1. Courses in sustainable engineering are grouped into four categories, Sustainable Engineering (dedicated sustainable engineering courses), Traditional Engineering courses with sustainable engineering content, Cross-disciplinary courses offered jointly with a non-engineering department, and Sustainable Engineering Technology courses which addressed technologies viewed as enabling for sustainability. (Murphy, et al., 2009; Allen, et al., 2009)

The content of the courses can be categorized based on the scale of the systems being analyzed. Figure 2 illustrates the types of scales that are frequently considered in sustainability courses, using mobility systems as a case study. As shown in Figure 2, the personal device that is used to provide mobility in North America is the automobile. One method of incorporating sustainability into engineering design is to assess environmental and social impacts of decisions affecting the design of a new automobile (e.g., choice of paint type, or chassis and engine materials). At a larger scale, sustainability concepts can be incorporated into automotive design decisions involving the recyclability of the vehicle. At even larger scales, the impact of automobile design (e.g., gasoline or electric power) on fuel industries and road construction can be considered. Finally, the overall sustainability of mobility transportation systems is also influenced by and influences the design of homes, communities and workplaces. In this presentation, these scales of design will be referred to as gate-to-gate, cradle to grave, inter-industry interactions and extra-industry interactions. More complete definitions of each of these scales and the topical areas covered at each of these scales are provided in Table 1.

Table 1. System Scales and Topics (Murphy, et al., 2009)

System Size	Description	Topics
Gate to Gate	Decisions made within a single facility or corporation by engineering and/or business units (i.e., site or industry sector specific activities).	Process design, including material and/or energy reduction
		Material or chemical selection
		Product design for a single phase of a product's life (e.g., design for recycling)
		Pollution prevention
		Media-based (i.e., air, water, solid waste) regulations
Cradle to Grave	Decisions made by different entities over the life of a product or sector activity. Activities are typically analyzed as sequential events (i.e., life cycle analysis).	Resource availability and economics
		Consumer behavior
		Product utility
		Reuse and recycling options
		Product based legislation (e.g., WEEE) and standards (e.g., ISO 14000)
		Life cycle inventory development
Inter-Industry (Industrial Symbiosis)	Decisions made by two or more entities (corporations or other stakeholders), often involving multiple sectors. The analysis typically captures spatial as well as temporal effects and scales, although temporal scales may be compressed such that activities are presumed to occur in parallel (i.e., industrial ecology)	Material flow analysis
		By-product synergy
		Eco-industrial development
		Multiple/nested LCA analysis
		Input-output analysis (either physical or economic)
Extra-Industry	Decisions made by multiple stakeholders, including industry, non-governmental organizations (NGOs), policy makers, consumers, etc.	Policy development (current and historical)
		Consumption patterns and preferences
		Eco-industrial development
		Multiple/nested LCA analysis
		Input-output analysis (either physical or economic)

Most engineering courses address gate-to-gate and cradle-to-grave issues. Far fewer courses cover topics at inter-industry and extra-industry scales. The topics covered at gate-to-gate and cradle-to-grave scales are summarized in Figure 3. Most engineering courses cover

i.) Analysis of energy and material use
ii.) Recycling and reuse, and
iii.) Life cycle assessment

The courses cover these topics primarily in the context of product design, process design, and materials selection.

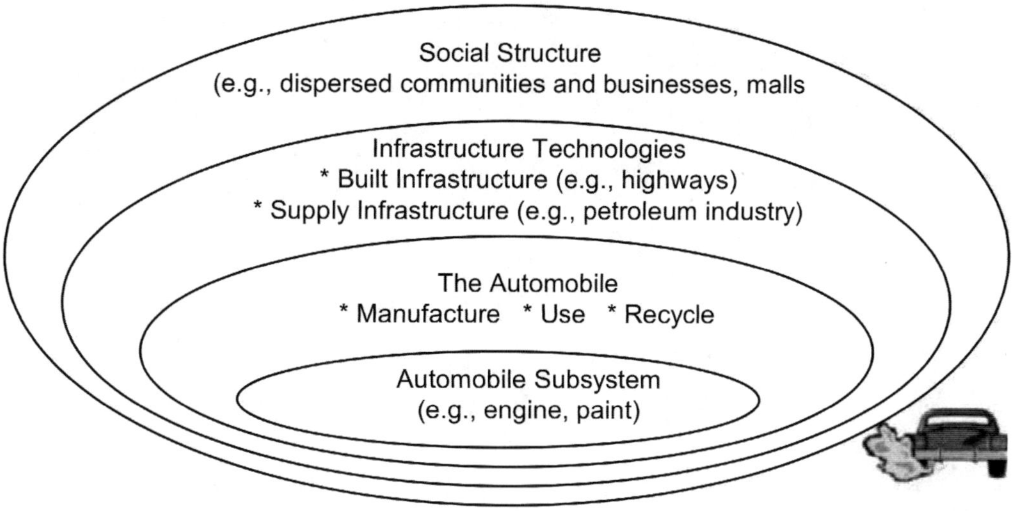

Figure 2. The technological-social system of the automobile exists in multiple layers; design decisions made in any of the layers shown influence decisions in all other layers. (from Graedel and Allenby, 1998)

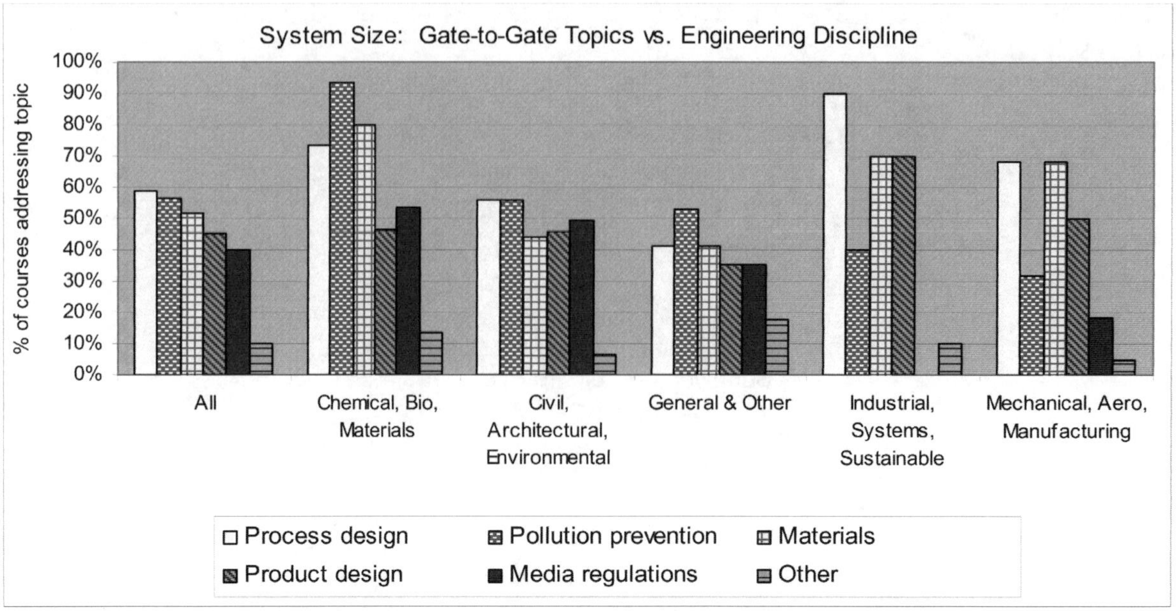

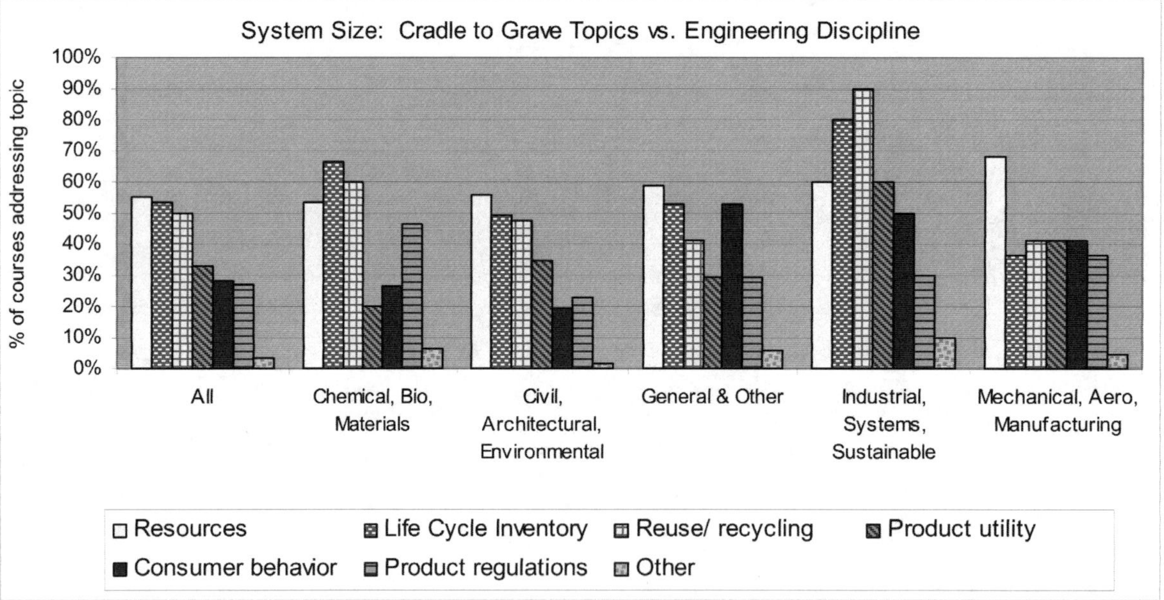

Figure 3. Coverage of topics at gate-to-gate and cradle-to-grave scales in sustainable engineering courses (Murphy, et al., 2009)

To summarize, engineering courses incorporating sustainability will most likely have a strong focus on minimizing material and energy resources required for the designs. The courses also are likely to cover life cycle impacts of the designs and the materials used in the designs. Engineering courses covering sustainability are less likely to include an assessment of the interaction of engineering designs with other industries and societal structures.

Systematically Incorporating Sustainability into Engineering Courses

Roughly 80% of the engineering programs rated among the top 100 programs by US News & World Report have some type of sustainability content in their curricula (Murphy, et al., 2009). As shown in Figure 1, however, this content is most commonly found in elective courses. If this content were to transition into required curricula, the topics covered would need to match educational objectives of the required courses. Based on the content currently covered in courses addressing sustainability, adding sustainability content to required courses would be most logically done in freshman engineering and senior design courses. Outlined below are suggested materials.

Freshman Engineering Required freshman engineering courses are beginning to emerge in engineering curricula. The most common educational goals in freshman engineering courses are to expose students to the nature of the design process, the creativity inherent in design, the trade-offs associated with meeting design objectives, and the iterative nature of the engineering design process. Courses necessarily have no pre-requisites and often involve engineers from multiple disciplines. Because these courses frequently seek to expose students to the contributions that engineers can make in solving the grand challenges facing human societies, the design problems used in these courses frequently incorporate environmental constraints and objectives. For these types of freshman design courses, the commonly taught principles of green or sustainable engineering that could be incorporated are life cycle assessments, environmental cost accounting, energy and material use footprints of designs, and deign for recyclability. All of these topics can be covered with a limited amount of mathematics, physics and chemistry.

Senior Design Senior capstone design courses are ubiquitous in engineering curricula. As capstone courses, they seek to synthesize material students have been exposed to throughout their curriculum through a design challenge. The educational goals parallel freshman design courses. The goals are to expose students to the nature of the design process, the creativity inherent in design, the trade-offs associated with meeting design objectives, and the iterative nature of the engineering design process. Just as in freshman courses, the design problems frequently incorporate environmental constraints and objectives. So for senior design courses, like freshman design courses, the commonly taught principles of green or sustainable engineering that could be incorporated are life cycle assessments, environmental cost accounting, energy and material use profiling of designs, and deign for recyclability. In senior design, case studies can draw on more sophisticated engineering analyses, but the underlying elements remain the same.

References

Allen, D.T., C.F. Murphy, B. Allenby and C. Davidson, "Incorporating Sustainability in Chemical Engineering Education, *Chemical Engineering Progress*, 105 (1) 47-53 (2009).

Murphy, C., D.T. Allen, B. Allenby, J. Crittenden, C. Davidson, C. Hendrickson and S. Matthews "Sustainability in Engineering Education and Research at U.S. Universities" *Environmental Science and Technology (Feature and cover article)*, 43, 5558-5564 (2009).

World Commission on Environment and Development, "Our Common Future," Oxford University Press, Oxford, U.K. (1987).

(405b) Lifelong Learning Training through Sustainability-Focused Problems Using Information Literacy

Paul Blowers and Dan Faetz, Chemical and Environmental Engineering, The University of Arizona, Tucson, AZ

Lifelong learning is a difficult topic to convey to students as it involves training them in a process and not in a specific and well defined topical knowledge content area. One way to think of lifelong learning is to break it up into the context of activities that need to occur in order for self-learning to be undertaken. First, learners must identify a topic that they are interested in that is relevant to their needs. They must then identify the limits of their current knowledge within the framework of what needs to be learned. Once this groundwork has been established, the learners must then identify the information gaps they have and the methods they will use through information literacy to seek out and validate the available information. The found data will then need to be used to address the original topical needs before the learner can reach understanding.

In this pedagogical work, information literacy has been combined with sustainability topics in order to train students how to independently learn information in parallel with topics in a required class. Sustainability issues were selected due to the high student interest in these topics and their immediacy in the public media. Stories were collected from prominent news sources and were discussed briefly at the beginning of each class meeting for the 100 person sophomore class. These stories were then reintroduced in the context of homework problems that were specifically constructed around the topical area, often in a thematic way where the same newspaper article may be used for three different chapters of material. The construction of the problems guided students to use information literacy techniques that were introduced in one lecture to obtain and evaluate information as they used it in order to solve the problems. As the semester progressed, students were required to obtain more peer reviewed information and to compare public media data with peer reviewed data in addition to comparing to their computed results.

An algorithm for creating information literacy-based problems to foster lifelong learning is presented. Additionally, all prior materials constructed for a materials and energy balances course, with complete solutions, is available by request so that faculty do not need to construct their own problems. Key to this algorithm is an emphasis on learning styles. Specific students seem to learn most efficiently from either visual or textual or audio sources, perhaps in a unique mixture different from other students. For this reason, an attempt was made to present focused problems in a multitude of forms so the students may learn in different congruent contexts.

Availability of course materials is of paramount concern to both student and teacher. All materials presented to students at any point during the semester should be easily accessible to them, independent of location and time. Explanations, visual aids, problems and solutions were placed in an online database accessible to both students and teacher. This expanded the students' abilities to learn by allowing them to watch and listen during lectures instead of taking notes, and to stay current in spite of absences. Any references to outside data sources were carefully cited and compiled, with concern to the future availability of archived web resources.

Students responded positively to the materials over a three year span, both quantitatively and with anecdotal evidence. Student scores on exams rose about 10% after the homework problems were introduced while homework scores decreased slightly. It was not unusual, even for non-major students who had taken the class to contact the professor in subsequent semesters to suggest problems that they had thought of in the context of the class. For example, one non-major student emailed a newspaper article from a media story where a doctor commented that people living in deserts had to drink a certain amount of water per day due to that lost during respiration. The student went on to suggest that a problem be constructed around the ideal gas law, Raoult's Law, and mole balances. Students also anecdotally commented that while the homework was very challenging, they appreciated the skills they obtained during the semester.

See more of this Session: Incorporating Green Engineering and Sustainability Into the Curriculum
See more of this Group/Topical: Education

(405c) Engineering Solutions for Sustainable Communities

Kyriacos Zygourakis, Chemical and Biomolecular Engineering, Rice University, Houston, TX and Richard Johnson, Sociology, Rice University, Houston, TX

This presentation will outline our experiences from a design course focusing on energy and sustainability. The course emphasizes the integration of engineering fundamentals with societal issues, environmental and safety considerations, sustainability and professional communications. We introduced this course in the spring of 2007 and taught it every year since then to highly interdisciplinary classes that included students from various engineering, natural science and social science majors.

The course begins with the introduction of a general framework for assessing the impact of humans on the environment. Several examples are used to present and critically analyze the IPAT equation and its variants. This introduction is followed by a series of lectures designed to help our students develop a broad understanding of the current and future availability of various primary energy sources, along with the environmental consequences of consuming these resources. Students are also introduced to the topic of global warming, and over the course of the semester, are exposed to linkages between global warming and energy technologies. We compare mature energy technologies to emerging technologies that are touted as "green" solutions, and challenge our students to think critically about these solutions.

In parallel to these lectures, students collect data from the operation of various parts of the university using an expanding array of instruments that include: pyranometers for measuring solar insolation or the reflectivity (albedo) of building roofs or other surfaces, temperature sensors, devices that measure the consumption of electricity by lab instruments or computers, and light and motion sensors. Through a series of assignments, in-class discussions, and a final report, students learn how to process data in order to assess environmental performance and how to evaluate potential solutions for improving the environmental performance of the university enterprise.

To highlight the successes and underscore the challenges encountered in this course, we will present representative data and action plans from three projects considered over the past four years: (1) an assessment of potential energy savings for Houston buildings from "cool roofs" (flat roofs covered with special paints to increase their reflectivity); (2) a study of the energy and water sustainability of Rice; and (3) the development of a detailed plan that will help Rice University become "carbon-neutral" by a yet-to-be-determined future date through a combination of energy efficiency gains, changes in the fuel mix used to power the university and off-campus carbon sinks.

See more of this Session: Incorporating Green Engineering and Sustainability Into the Curriculum
See more of this Group/Topical: Education

(405d) Project-Based Learning for Sustainability and Life-Cycle Assessment

Benjamin John Davis, Chemical Engineering, The Cooper Union, New York, NY

Sustainability is a very hard concept to define, defying the traditional disciplines of ecology, economics, sociology, and engineering. Even harder is the task of explaining to undergraduate chemical engineering students what sustainability means and how it relates to their life experiences to date. These students, familiar with rigid concepts from chemistry, physics, mathematics, and other hard sciences often have a very difficult time approaching a concept which is inherently "fuzzy" and does not have right or wrong answers.

How do we engage and teach these students? How do we get a student, who at this point in his or her intellectual development believes that there are only correct and incorrect answers, to consider the question of whether the paper and plastic bags they are trained to make are good or bad for the environment? (The answer to that question, of course, is "It depends.")

In my talk, I will outline the motivations and goals of my course "Sustainability and Pollution Prevention" at The Cooper Union, which is based on a similar course at The University of California, Los Angeles. The course is primarily inquiry-based, with two month-long open-ended assessment projects forming the bulk of the work product.

In the first project, the students learn a fuzzy-logic based assessment model (grounding them in their comfort zone of rigid concepts) and apply that model to assess the sustainability of a corporation or industry of their choosing. They explore their chosen entity's impacts on land use, water resources, air emissions, and biodiversity, while also considering things like employee health and wellness, revenue and economic viability, good corporate policies, and research and development work.

In the second project, the students (hopefully now more comfortable with open-ended questions) perform a life-cycle assessment on a product or process of their choosing. This can be related to their work on their capstone design project or on a completely different product that the student has been able to collect data about.

My hope is that this talk will inspire others to offer a similar course at their institution to engage and challenge chemical engineering students to think about social, political, and economic issues. These topics, related to their traditional coursework but not always formally addressed in class, are essential to their development as future engineers and citizens.

See more of this Session: Incorporating Green Engineering and Sustainability Into the Curriculum

(405e) Developing Sustainable Engineering Modules for the Chemical Engineering Curriculum

Jeffrey Seay, Richard Rezek and Luke Richardson, Chemical and Materials Engineering, University of Kentucky, Paducah, KY

In order to incorporate the principles of sustainability into the Chemical Engineering curriculum, educational resources are needed for course instructors. Over the last few years, the AIChE Sustainable Engineering Forum has worked to develop educational modules for 5 core courses in the chemical engineering curriculum. The current Chair of the Education Committee of the Sustainable Engineering Forum will provide an overview of the modules developed, and provide information on how course instructors can access these modules through the Sustainable Engineering Forum website.

Modules have been developed for the following undergraduate courses: Separations, Thermodynamics, Engineering Economics, Transport Processes (Heat Transfer, Mass Transfer and Fluid Flow) and Reactor Design. Each of the modules consists of an instructor overview of the engineering topic, focused on how sustainability can be integrated into the course. Additionally, each module includes 10 – 15 sample sustainable engineering problems with solutions. Modules will be made available free of charge to course instructors.

See more of this Session: Incorporating Green Engineering and Sustainability Into the Curriculum
See more of this Group/Topical: Education

(405f) Hydrogen and Fuel Cell Workbook for Material and Energy Balances

Daniel Lopez Gaxiola and Jason M. Keith, Department of Chemical Engineering, Michigan Technological University, Houghton, MI

Studies have shown that performance in the first chemical engineering course has a large impact on student retention in the discipline. One successful approach to teaching the course is to engage the students in repetitive problem solving. In this project, we describe our efforts to create a workbook on material and energy balance problems that relate to the generation of hydrogen and their resulting use in fuel cells. These problems can be used in parallel with the classic text by Felder, Rousseau, and Newell or as a component of a separate course on hydrogen energy.

See more of this Session: Incorporating Green Engineering and Sustainability Into the Curriculum
See more of this Group/Topical: Education

(435a) Insights Into a Successful Academic Career: An Educational Perspective

Margot Vigeant, Chemical Engineering, Bucknell University, Lewisburg, PA

This presentation will provide an educational perspective on developing a successful academic career for young faculty.

See more of this Session: Young Faculty Forum
See more of this Group/Topical: Education

(435b) Insights Into a Productive Academic Career: A Research Perspective

Christopher W. Jones, Chemical & Biomolecular Engineering, Georgia Institute of Technology, Atlanta, GA

This presentation will provide a research perspective on developing a successful academic career for young faculty.

See more of this Session: Young Faculty Forum
See more of this Group/Topical: Education

(435c) Presentations by NSF CBET Program Managers

CBET National Science Foundation, National Science Foundation, Arlington, VA

Presentations will be made by NSF CBET program managers regarding the National Science Foundation.

See more of this Session: Young Faculty Forum
See more of this Group/Topical: Education

(435d) Break-out Session with NSF Program Managers

CBET National Science Foundation, National Science Foundation, Arlington, VA

Project Summaries have been collected from young faculty and given to CBET Program Managers from NSF for review. Program Managers from select CBET programs will critique the summaries and meet with the PIs during the break-out session. Program Managers will not discuss any likelihood of funding for the described research nor is this any guarantee of funding. The purpose of this exercise is to provide constructive feedback to the young faculty for proposal writing.

See more of this Session: Young Faculty Forum
See more of this Group/Topical: Education

(459a) Group Mini Design Projects for Freshman Intro to Engineering

Noelle K. Comolli, Chemical Engineering, Villanova Univesity, Villanova, PA and Randy D. Weinstein, Department of Chemical Engineering, Villanova University, Villanova, PA

Freshman engineering programs vary greatly at different universities and are constantly a topic of debate for educators seeking to inspire, excite and educate the next generation of engineers. The demand from industry is to provide engineers that are not only technically competent, but that can thrive in a group environment and lead a team to design a better product. The challenge for educators then, is to introduce more design within socially relevant contexts into their curriculum starting at the freshman level without decreasing student retention. The dilemma is in introducing design problems, which are open-ended, with more than one solution, to freshman[1]. Most freshman still learn by acquiring knowledge from the teacher, assuming that all knowledge is a collection of facts that are right or wrong[2]. Therefore asking them to achieve the next leaving of learning that is required of engineering design is not an easy task. The goal of this paper is to evaluate using the artificial kidney design, as well as some smaller chemical engineering specific design projects, as a stepping stone for the introduction of multiple disciplines of engineering to freshman at Villanova University in a seven week mini project based course. The effectiveness of this course at introducing freshman students to engineering concepts and retaining their enthusiasm for engineering was evaluated using a pre/post course survey, student quiz performance and final student design reports (both oral and written). Overall, the instructors found the course to be effective at improving student understanding of engineering fundamentals and increasing their enthusiasm for engineering.

See more of this Session: Communication and Collaboration
See more of this Group/Topical: Education

(459b) Integrating Communication Skills and Fostering Collaboration in the Mass and Energy Balances Course with Team Activities

Susan Daniel, Chemical and Biomolecular Engineering, Cornell University, Ithaca, NY and Kathryn Dimiduk, Teaching Excellence Institute, Cornell University, Ithaca, NY

This year we changed our course to implement a collaborative learning environment as opposed to a competitive model. An important part of this change was to integrate communication and teamwork skills throughout the semester to facilitate collaborative learning in group projects and activities. We designed the course to progressively increase communication skill level from initially writing simple memos to writing formal project reports, from small team presentation activities in recitations to giving a final oral presentation on a team project. Communication and teamwork tasks were embedded within many different assignments including homework problems, recitation activities, lecture "think-pair-share" activities, and three multi-week team projects. We chose the team project topics to 1) explore a breadth of topics within chemical engineering (environment, human health, and energy) to engage the interests of students from information that we collected in a beginning of the semester survey, 2) connect the projects to societal impact, as many students said on the survey that this was an important motivating factor for exploring chemical engineering as a career choice, 3) provide students with a sense of the size scales over which chemical engineers solve problems, and 4) provide a forum in which students could learn important communication and collaboration skills in teams as a foundation for the rest of their career. We carefully selected groups based on a variety of data to create groups that facilitated collaboration. We were sensitive to selecting groups to positively integrate women and underrepresented groups and foster team work to promote communication and critical thinking skills of all students through collaboration. Student feedback on this approach and statistics will be provided.

See more of this Session: Communication and Collaboration
See more of this Group/Topical: Education

(459c) Improving Communication Between Sophomores and Juniors through Chem-E-Car Project

Sundararan. V. Madihally, School of Chemical Engineering, Oklahoma State University, Stillwater, OK and **Karen A. High**, Chemical Engineering, Oklahoma State University, Stillwater, OK

Chem-E-Car Competition has become one of the signature events of the Annual AIChE student conference gaining attention from various student as well as professional attendees. Despite the increasing participation by a wide number of AIChE Student Chapters, opportunities to use chem-E-Car as a teaching tool have been overlooked. For example, the Chem-E-Car competition is a powerful tool to enhance technical writing skills, provide engineering analysis opportunities, and apply team management skills. The Chem-E-Car competition has been utilized for the last few years as part of multiple courses in chemical engineering curriculum at Oklahoma State University. Typically, a number of teams comprising of three to five juniors were formed in the Fall semester for the competition to be held in the subsequent Spring semester. Three to six sophomores were included to enhance cross-class participation and to provide application-oriented examples. A folder containing compartments for log sheets, pictures/sketch, reaction/safety, analysis, and calibration was given to each team. The sophomore chemical engineering students were engaged in many components of technical communication. They kept lab books on their experiences, developed posters for the competition, reflected on meetings with the junior students, and wrote team reports and individual reflection. Lessons learned from the point of instructors will be presented.

See more of this Session: Communication and Collaboration
See more of this Group/Topical: Education

(459d) Teaching the Value of Communication in the Engineering Design Cycle

Taryn M. Bayles, University of Maryland Baltimore County, Baltimore, MD

For the last ten years, the Introduction to Engineering Design course at the University of Maryland Baltimore County (UMBC) has evolved from a traditional lecture and design-on-paper course, to an active learning lecture and project-based learning engineering design course. The importance of learning teamwork skills and communication skills are emphasized in the course. Every effort is made to ensure that the design teams are diverse, interdisciplinary, yet academically balanced. The design teams are required not only to research, design, construct, evaluate, test and present (through oral presentation and written reports) their product, but also to develop a mathematical model to predict their product's performance. It is important that the students have a fun yet inexpensive product to design and build, but they must also develop a mathematical understanding of the fundamental engineering principles that make their design work. Through this mathematical modeling the students cultivate the connection between mathematics, science and engineering. Successful engineering design projects have included human powered pumps, hot air balloons, hemodialysis systems, chemically powered vehicles, renewable energy systems, contaminated water purification systems and simulated heart-lung machine.

Over the years it has been observed that the students often overlook the importance of communication during the design process. Therefore, the teams were also evaluated on their ability to effectively communicate their design with their team members and with other teams in their discussion session. The class consists of 210 students; broken into seven discussion sessions, with five teams each – for a total of 35 design teams. The communication criteria was assessed in addition to the design project criteria. Each team was required to complete their design, mathematical model and testing plans well in advance to their scheduled testing date. Each team gave their design project construction materials and design plans (which include CAD drawings and instructions) to another team in the class which constructed the design. The construction team had one week to construct the design and then passed the constructed design to an evaluation team. The evaluation team tested and evaluated the design based on the testing / evaluation documentation provided by the original design team (the original design team constructs and evaluates two different designs while their design is worked on by other teams). Three days later the evaluation team returned the design, with the construction and evaluation reports to the original design team. The design team then had time to make any necessary modifications / improvements before the final demonstration / testing / evaluation of their design with the instructor. It was hoped that this rotation of the designs within the discussion session will facilitate better communication (both written and oral) between the teams and more successful designs; since each team is required to work on three different designs. The success (or failure or – logistical nightmare) of the communication criteria of the project is assessed with student surveys, written project report scores and oral presentations. Comparisons are made with previous semester surveys and scores when the designs did not rotate between the teams in each discussion session.

(459e) Student Assessment Results of Incorporating Multidisciplinary Industrial Design Problems in the Chemical Engineering Curriculum

Jeffrey R. Seay[1], Jimmy L. Smart[2], Stephen Hutcheson[3] and David L. Silverstein[3], (1)Department of Chemical and Materials Engineering, University of Kentucky, Paducah, KY, (2)Chemical & Materials Engineering, University of Kentucky, Paducah, KY, (3)University of Kentucky, Paducah, KY

Recently the students and faculty from the Chemical and Materials Engineering Department at the University of Kentucky Paducah Extended Campus have entered into a partnership with a local specialty chemical company to develop student design problems that not only provide a benefit to the company at minimal cost but are multidisciplinary in nature and can be taken on by student groups in the sophomore, junior and senior classes. Although this partnership would appear to be an obvious benefit to students, a detailed assessment survey has been completed by student participants to gauge their perceived benefits from participation. Additionally, the survey has been used to allow the students to self assess their knowledge of industrial projects before participating and how their perceptions and knowledge have changed after participating. This contribution will present an overview of the organization of and the deliverables from these projects as well as the results of the student assessment survey.

A typical road block to incorporating industrial design projects outside the capstone course is that prior to this course, students simply don't have the experience to contribute to a real world industrial design problem. The assessment included with these projects will provide data on the students' own perception of their readiness to participate in such projects, both prior to participating and after participating in one or more project cycles. In addition to the assessment results, anecdotal reports from the students participating in the project will be included. Furthermore, the specific challenges of organizing multidisciplinary teams involving students at different academic levels as well a different faculty members will be addressed based on recent experience with these projects.

See more of this Session: Communication and Collaboration
See more of this Group/Topical: Education

USE OF INTERNET TO DO EXPERIMENTS IN DYNAMICS AND CONTROL FROM ZACATECAS MEXICO IN THE LABORATORY OF THE UNIVERSITY OF TENNESSEE AT CHATANOOGAA.

Jim Henry[1*], José Alberto González Guerrero[2], Benito Serrano Rosales[2].

1. The University of Tennessee at Chattanooga, 615 McCallie Ave., Chattanooga, TN 37403

2. Universidad Autonoma de Zacatecas, Unidad Academica de Ciencias Quimicas, Campus Siglo XXI, Edificio 6, Carr. a Guadalajara Km. 6, Ejido la Escondida, Zacatecas Zac., 98160, Mexico, beniserra@prodigy.net.mx Phone 52-492-925-6690 Ext 6137

Abstract

The internet was used to do remote experiments in Universidad Autonoma de Zacatecas Mexico, using the laboratories in the University of Tennessee at Chattanooga, specifically, the system of two non-interacting tanks. In the first part, step and pulse perturbations were applied, and in the second part, a proportional controller was used and step perturbation on the set point was applied. The mathematical models were deducted in all the cases, linearizing the expression for the flow between tank 1 and 2.

In all the cases, the agreement between experimental and predicted results was very good, indicating the models were conceptually correct, and the most important, the use of remote laboratory was possible and allowed a relationship between two universities.

Introduction

Currently, the use of internet to do experiments in a remote laboratory is a reality. The experiments were performed in the Program of Chemical Engineering, Faculty of Chemical Sciences, Autonomous University of Zacatecas Mexico, at any day, any time, using the two non-interacting tanks in the laboratory of the University of Tennessee at Chattanooga. Each tank was 6 inches in diameter and 12 inches in height. The feed from the first tank to the second tank is by gravity (See figure 1). A mathematical model was constructed and it was necessary to linearize it.

Experimentation

The first perturbation was to turn on the equipment of the two non-interacting tanks and to let it to reach the steady state, which means a step perturbation. The implementation of an interface through internet http://flow.engr.utc.edu:443/ allows access to this piece of equipment from anywhere in the world. See figure 2.

The first part of the experimental section was to run some experiments for different values of the pump. In the second part, one step perturbation was applied, with magnitude A, and in other experiment, after the start of the equipment, a pulse perturbation was applied. For both cases, the response of the height of the level of the second tank was registered, and the experimental results were compared with the predictions of the model. In the third part, the experiments were performed with step perturbation, regulated by a proportional controller.

Figure 1.- System of two non-interacting tanks.

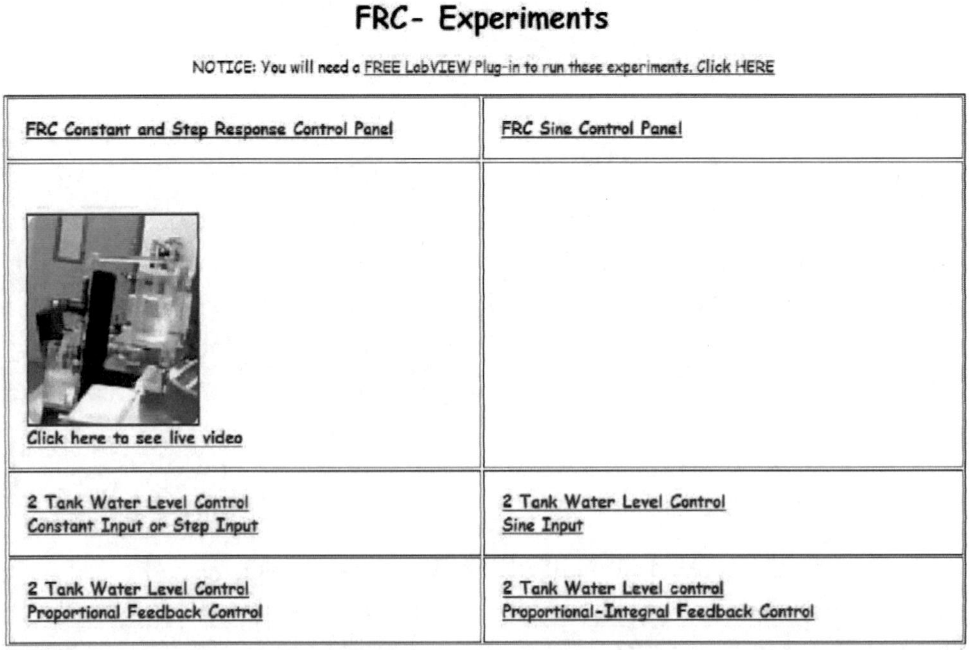

Figure 2.- Screen to select the piece of equipment.

Click the link “2 Tank Water Level Control Constant Input or Step Input.”

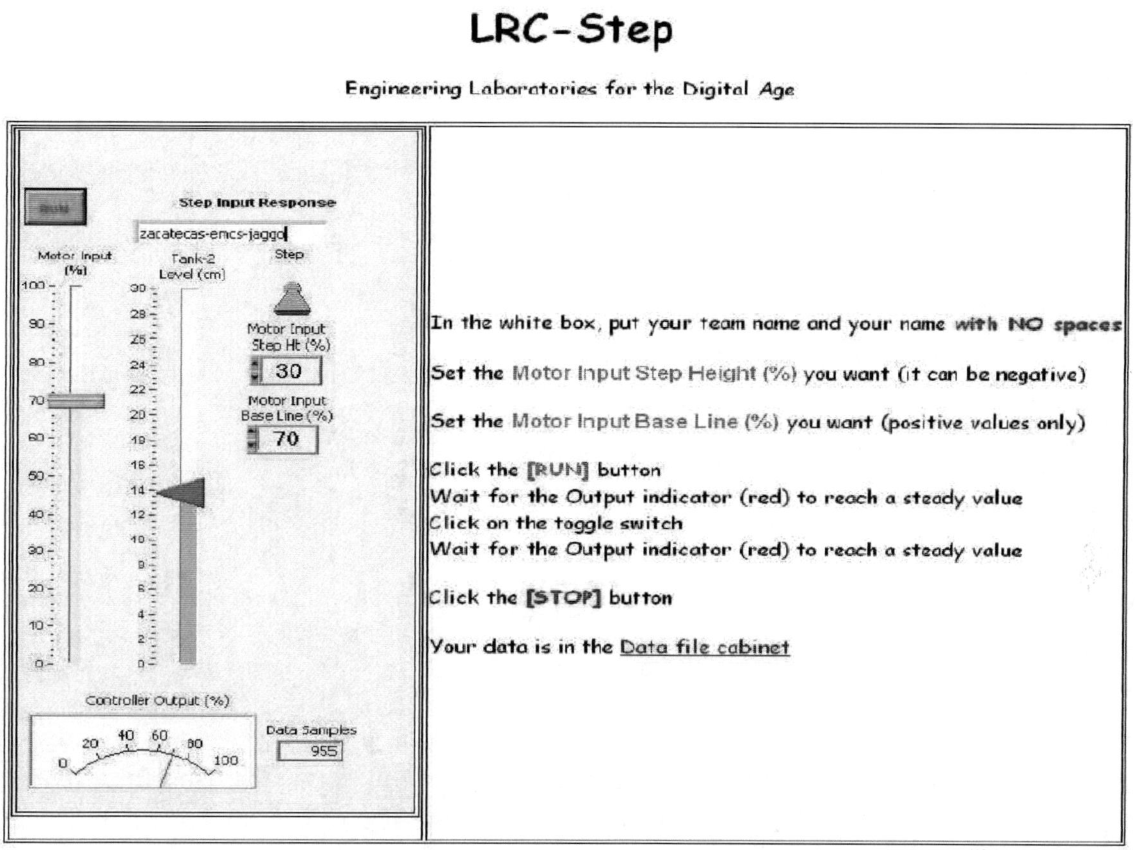

Figure 3.- Screen to start the experiments with the system of two non interacting tanks.

A typical experiment for the start of the experiments is shown in figure 3, with 100% of the capacity of the pump. It is observed that the steady state height of the level in tank 2, 22.5 cm, is reached in 150 seconds.

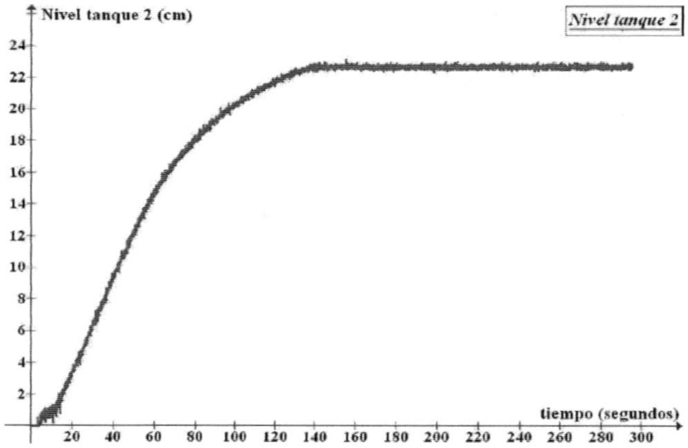

Figure 4.- Level of tank 2, with 100 % capacity of the pump.

The next experiment was to work with the pump at its 80 % of capacity, and to apply a step perturbation of 20 %. Figure 5 shows the results of this experiment.

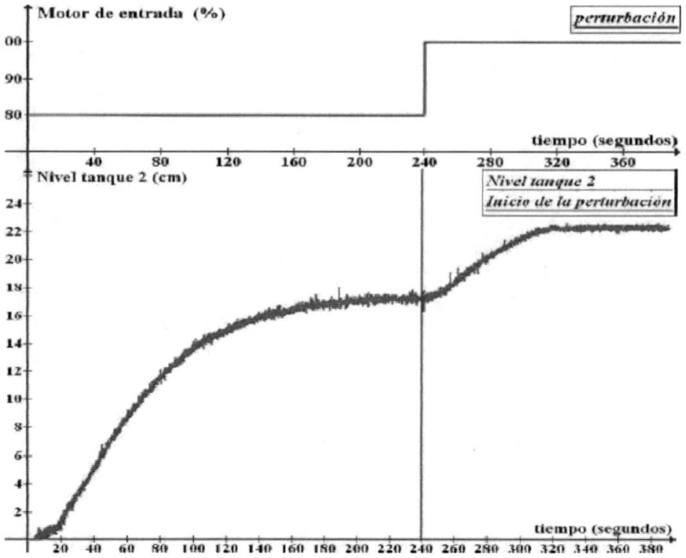

Figure 5.- Response of the level of tank 2 to a step perturbation.

The next experiment was carried out with the capacity of the pump of 70 %. Once the stady state was reached, a pulse perturbation was applied increasing the pump capacity in other 30%, for 100 seconds. After this period, the pump was returned to 70 %. The results are shown in figure 6.

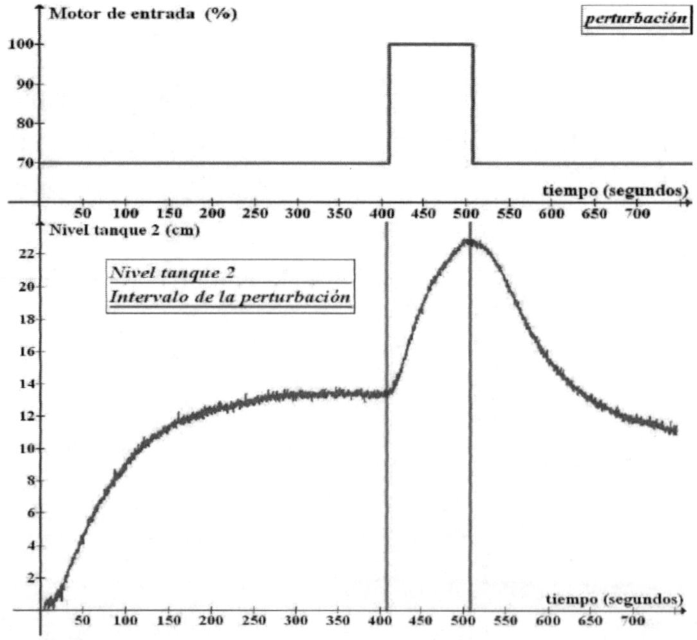

Figure 6.- Response of the level of tank 2 under a pulse perturbation of 100 seconds.

In the next experiment, the pump was operated at 60 % of its capacity, using a proportional controller with gain 0. After the steady state was reached, a step perturbation was applied increasing another 40 %. Several values of the gain were used. The experimental results are shown in Figure 7. It is observed that increasing the value of the gain, an unstability is generated, as it is predicted by theory.

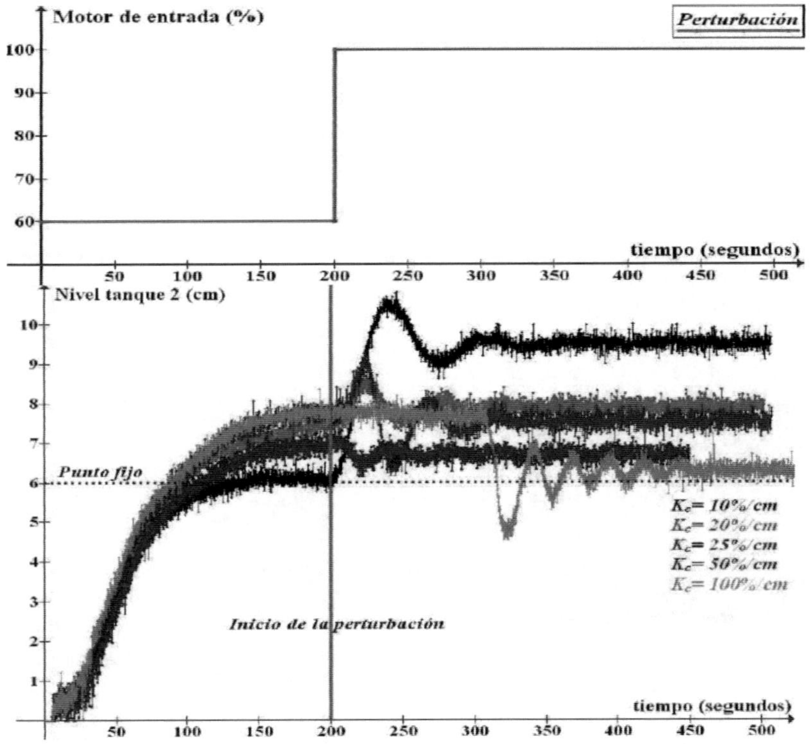

Figure 7.- Response of the level of tank 2 under a step perturbation for several values of the gain of the proportional controller.

It is very important to note that all these experiments were done using internet, in the laboratory of Unviersity of Tennessee from the Autonomous University of Zacatecas, Mexico.

Mathematical modeling.

The mass balances for tanks 1 and 2 are:

$$q_i(t) - q_1(t) = A_{T1}\frac{dh_1(t)}{dt}$$

$$q_1(t) - q_2(t) = A_{T2}\frac{dh_2(t)}{dt},$$

Using the Taylor series, the linearized expression for q1(t), the flow leaving tank 1 is obtained:

$$q_1(t) \cong q_{1s} + \frac{(h_1(s) - h_{1s})}{R_1}$$

The transfer function for this system of non interacting tanks for the liquid level of the second tank is:

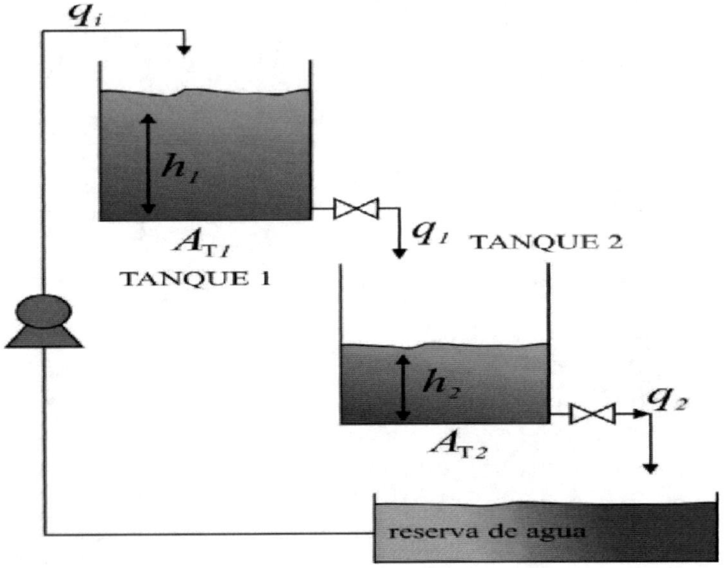

$$H_2(s) = \frac{Q_i(s)}{\tau_1 A_{T2}\left(s + \frac{1}{\tau_1}\right)\left(s + \frac{1}{\tau_2}\right)}$$

Where:

$D_1 = D_2 = 6in = 15.24cm$, $h_{2s} = 17.2cm$, $H = 30\frac{cm^3}{s}$ $A_{T1} = A_{T2} = 182.415\ cm^2$

$\tau_1 = R_1 A_{T1} = 1.6317s$, $\tau_2 = R_2 A_{T2} = 31.1929s$, $R_1 = \frac{2\sqrt{h_{1s}}}{C_{v1}} = 0.008945\frac{s}{cm^2}$

$R_2 = \frac{2\sqrt{h_{2s}}}{C_{v2}} = 0.171\frac{s}{cm^2}$ $C_{v2} = C_{vT2}[p_{v2}(t)]\sqrt{\rho r\frac{g}{g_c}} = 48.5\frac{cm^{5/2}}{s}$

Cv_1 is the secondary coefficient of valve in tank 1. Cv_2 is the secondary coefficient of valve in tank 2.

CvT_1 coefficient of valve in tank 1

CvT_2 coefficient of valve in tank 2

$Pv_1(t)$ = position of the valve opening in tank 1.

$Pv_2(t)$ = position of the valve opening in tank 2.

If we apply a step perturbation, with magnitude H, and with transfer function H/s. The response of the second tank is:

$$h_2(t) = h_{2s} + HR_2\left[1 + \frac{\tau_1}{(\tau_2 - \tau_1)}e^{(-t/\tau_1)} + \frac{\tau_2}{(\tau_1 - \tau_2)}e^{(-t/\tau_2)}\right]$$

Substituting the specific values, we obtain:

$$h_2(t) = 22.33 + 0.283e^{(-t/1.6317)} - 5.413e^{(-t/31.197)}$$

The comparison of the predictions of this equation with the experimental results obtained after using the remote laboratory of the University of Tennessee campus Chattanooga, are shown in Figure 1. In this case, the agreement is very good.

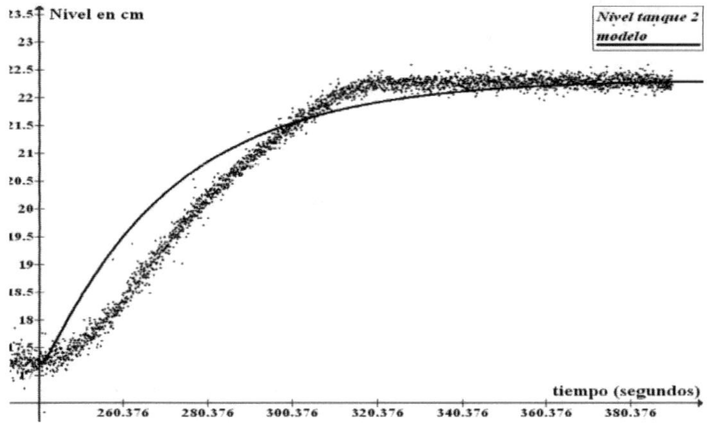

Figure 8.- Comparison between the mathematical model and experimental results for a step perturbation.

Now, applying a rectangular pulse of magnitude H and duration T, The response is:

$$h_2(t) = h_{2s} + HR_2 + \frac{H\tau_1 R_2}{\tau_2 - \tau_1} e^{(-t/\tau_1)} + \frac{H\tau_2 R_2}{\tau_1 - \tau_2} e^{(-t/\tau_2)}, \quad Para\ t < T$$

$$h_2(t) = h_{2s} + \frac{H\tau_1 R_2}{\tau_2 - \tau_1}\left[e^{\left(-\frac{t}{\tau_1}\right)} - e^{\frac{(T-t)}{\tau_1}}\right] + \frac{H\tau_2 R_2}{\tau_1 - \tau_2}\left[e^{\left(-\frac{t}{\tau_2}\right)} - e^{\frac{(T-t)}{\tau_2}}\right], \quad Para\ t > T$$

The numerical values are:

$$h_{2s} = 11\,cm, \quad H = 45\frac{cm^3}{s}, \quad T = 100s, \quad R_1 = 0.008954\frac{s}{cm^2}, \quad C_{v2} = 21\frac{cm^{5/2}}{s}$$

Substituting these values in the pulse response:

$$h_2(t) \cong 25.214 + 0.4146e^{(-t/1.6317)} - 14.6283e^{(-t/57.6192)}$$

$$Para\ t < T$$

$$h_2(t) \cong 11 + 0.4146\left[e^{(-t/1.6317)} - e^{\frac{(100-t)}{1.6317}}\right] - 14.6283\left[e^{(-t/57.6192)} - e^{\frac{(100-t)}{57.6192}}\right]$$

$$Para\ t > T$$

A comparison between the expermiental and calculated results is shown in figure 2. The agreement during the first stage are excellent, however, some diference is observed after the perturbation.

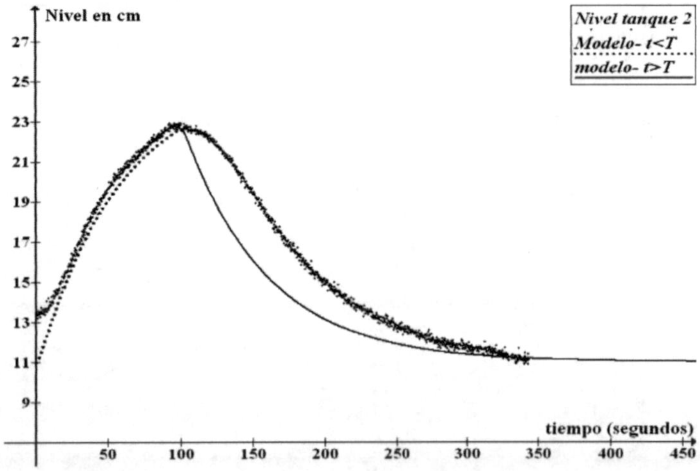

Figure 9.- Comparison between the mathematical model and experimental results for a pulse perturbation.

Experiments were also performed using a proportional controller. The block diagram is shown in the following figure:

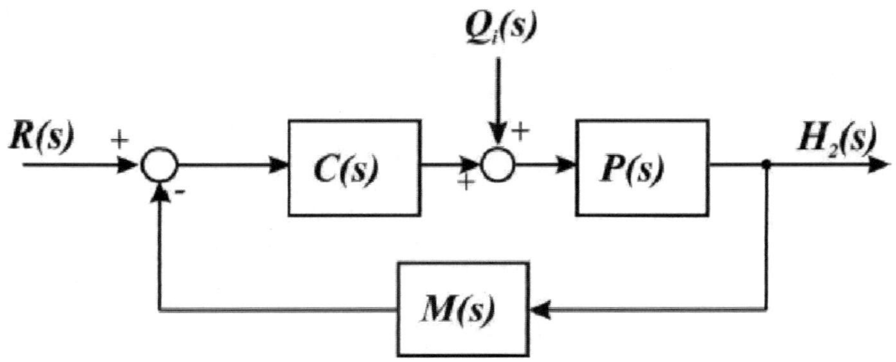

Where:

R(s) = Set point.

$Q_i(s)$ = Flow entering the first tank, or load variable.

C(s) = Controller transfer function.

P(s) = Transfer function for the process of two non-interacting tanks.

M(s) = Transfer function of the measurement device.

$H_2(s)$ = Transfer function of the level of the second tank, or exit variable.

The following transfer function is obtained:

$$\frac{H_2(s)}{Q_i(s)} = \frac{P(s)}{[1 + M(s)C(s)P(s)]}$$

$$C(s) = K_p, \quad P(s) = \frac{1}{\tau_1 A_{T2}\left(s + \frac{1}{\tau_1}\right)\left(s + \frac{1}{\tau_2}\right)}, \quad M(s) = 1$$

Performing algebra:

$$\frac{H_2(s)}{Q_i(s)} \cong \frac{1}{\tau_1 A_{T2}\left(s + \frac{1}{\tau_1}\right)\left(s + \frac{1}{\tau_2}\right) + K_p}$$

$$H_2(s) = \frac{H}{\tau_1 A_{T2} s_1 s_2 s} + \frac{H}{\tau_1 A_{T2}\left(s_1^2 - s_1 s_2\right)\left(s - s_1\right)} + \frac{H}{\tau_1 A_{T2}\left(s_2^2 - s_1 s_2\right)\left(s - s_2\right)}$$

Performing the inverse transform:

$$H_2(t) = \frac{H}{\tau_1 A_{T2} s_1 s_2} + \frac{H}{\tau_1 A_{T2}\left(s_1^2 - s_1 s_2\right)} e^{s_1 t} + \frac{H}{\tau_1 A_{T2}\left(s_2^2 - s_1 s_2\right)} e^{s_2 t}$$

Where: H = 30 cm, $\tau_1 A_{T2} = 297.6465$, $s_1 = -0.3224 + \sqrt{0.0844 - 3.3597x10^{-3} K_p}$

$s_1 = -0.3224 - \sqrt{0.0844 - 3.3597x10^{-3} K_p}$, $s_1 s_2 = 19.54x10^{-3} + 3.3597x10^{-3} K_p$

$$s_1^2 = 0.18844 - 0.6448\sqrt{0.0844 - 3.3597x10^{-3} K_p} - 3.3597x10^{-3} K_p$$

$$s_1^2 = 0.18844 + 0.6448\sqrt{0.0844 - 3.3597x10^{-3} K_p} - 3.3597x10^{-3} K_p$$

$$\left(s_1^2 - s_1 s_2\right) = 0.1688 - 0.6448\sqrt{0.0844 - 3.3597x10^{-3} K_p} - 6.7194x10^{-3} K_p$$

$$\left(s_2^2 - s_1 s_2\right) = 0.1688 + 0.6448\sqrt{0.0844 - 3.3597x10^{-3} K_p} - 6.7194x10^{-3} K_p$$

The comparison between the experimental and predicted results is done in figure 10.

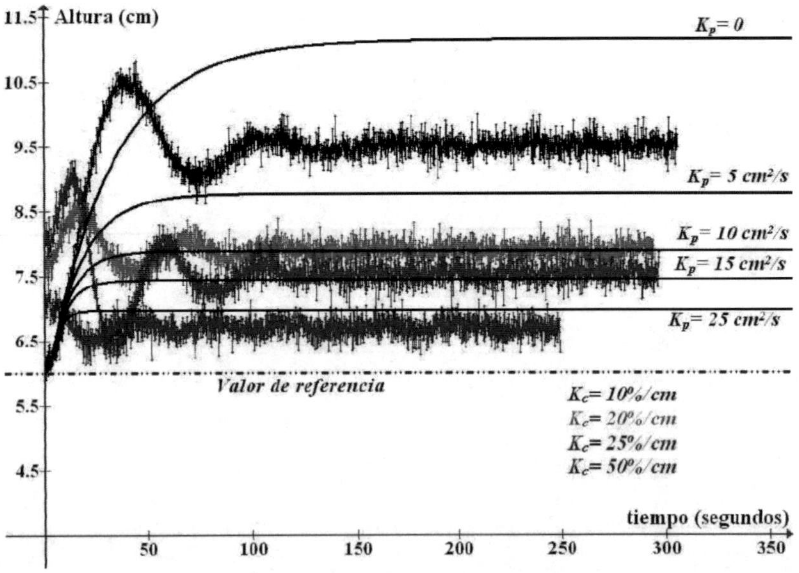

Figure 10.- Comparison of experimental and predicted results using a step perturbation and a proportional controller.

Certainly, the agreement is not very good but the model indicats very well the tendency of the experimental results.

Conclusions.

1.- The agreement between the experimental and modelled results is good, and this validates the mathematical model, which is able to predict the response of the system to any step perturbation, whether at the start or after a steady state is reached.

2.- In general, the model using a pulse perturbation is conceptually correct and reproduces the experimental data, in spite of after the perturbation, some diference between data is observed.

3.- The mathematical model using proportional controller predicts the response tendency when the value of K_c is increased, indicating that the model is conceptually correct.

References.

1.- Donald R. Coughanowr, "Process system analysis and control", Mc. Graw Hill, second edition, 1991.

2.- Wodek K. Gawronski, "Advanced Structural Dynamics and Active Control of Structures", Spronger, 2004.

(498a) How We Teach Kinetics and Reactor Design

David L. Silverstein, University of Kentucky, Paducah, KY, Margot Vigeant, Chemical Engineering, Bucknell University, Lewisburg, PA and Donald R. Woods, McMaster University, Hamilton, ON, Canada

The authors will present the results of the 2010 AIChE Education Division survey on how chemical engineering courses are taught. This year's survey focuses on the kinetics and reactor design sequence. Both statistical type of information (hours taught, class sizes, chapters covered) and innovative teaching approaches and labs will be presented.

See more of this Session: Survey Results and Best Practices in Kinetics
See more of this Group/Topical: Education

(534a) Plantwide Dynamic Simulation in the Lehigh Design Course

William L. Luyben, Chemical Engineering, Lehigh University, Bethlehem, PA

A two-semester senior design course has been taught at Lehigh for a decade. The course expands the traditional steady-state aspects of process design (flowsheet topology, engineering economics, separation sequences, reactor/recycle trade-offs, etc) with dynamic plantwide control. Student groups work on individual design projects provided by industrial consultants who serve as mentors. Effective plantwide control structures are developed for the multi-unit processes with recycle and tested for realistically large disturbances using dynamic simulation. Considerations of process safety is incorporated in the projects. The paper will discuss the objectives, content and format of the course and provide examples of typical student design projects.

See more of this Session: Effective Use of Simulation and Numerical Methods in Various Courses
See more of this Group/Topical: Education

THE PROCESS SIMULATION COURSE – THE CULMINATION OF CORE UNDERGRADUATE COURSEWORK IN CHEMICAL ENGINEERING

Mordechai Shacham, Ben-Gurion University, Beer-Sheva, Israel

Michael B. Cutlip, University of Connecticut, Storrs, CT 06269, USA

Introduction

The status of the incorporation of computational tools in the undergraduate Chemical Engineering curriculum has been recently reviewed by Shacham et al. (2009). This review, as well as some additional references (Dahm et al., 2002, Rockstraw, 2005, Dias et al., 2010), reveal that process simulation is being used to some extent in various courses. Most often commercial simulators (such as HYSIS, AspenPlus and PRO II) are used to model the steady state or dynamic operations of processes. The benefit of such use of a process simulator (as stated by Dahm et al., 2002) is that it "provides a time-efficient and effective way for students to examine cause-effect relationships" among various parameters of the process. A pedagogical drawback to the use of such packages is that "it is possible for students to successfully construct and use models without really understanding the physical phenomena within each unit operation" (Dahm et al., 2002). Furthermore, Streicher et al. (2005) have found that "the majority of students see simulations merely as sophisticated calculators that save time."

In order to promote and emphasize the full educational benefits of process simulation, we have developed a process simulation course where the students need to prepare process models that, in most cases, are ultimately simulated with MATLAB. The course content begins with the steady state operation of simple recycle processes that contain mixers, simple splitters and conversion type reactors. Material balances on such systems can be represented by linear models. For such systems, a direct solution (of a system of linear equations) can be obtained and the performance of various iterative algorithms can be examined. The students learn to establish the computational sequence in the flow-sheet using various partitioning and tearing algorithms. Then, students solve the problems directly and iteratively using the successive substitution, dominant eigenvalue, Wegstein's and Broyden's methods. After understanding the basic principles of the operation of the steady state simulators, models of unit operations with increasing level of complexity (such as isothermal, adiabatic and three-phase flash) are prepared. The models are implemented as MATLAB functions where the input parameters are the vectors of flow-rates and enthalpy of the inlet streams and the design parameters of the process unit. The output parameters are the vectors of flow-rates and enthalpies of the outlet streams. Physical property data and correlations that required for the modeling the unit operation are taken from the DIPPR thermo physical database (Rowley et al., 2006, http://dippr.byu.edu/). Binary equilibrium data, needed for calculations of activity coefficients, are retrieved from the Dortmund Data Bank (http://www.ddbst.com/en/ddbst/index.php) and regressed with the Polymath program. The systems of nonlinear algebraic equations that represent typically the steady state models of the various units operations are solved using the constrained version of the Newton-Raphson method (Shacham, 1986).

In dynamic simulation, the emphasis is on the solution of Multiple-Model, Multiple-Algorithm (MMMA, see Cutlip et al., 2009) problems. An example is a process unit that can operate in different modes (such as a reactor with heating and cooling periods) and requires different integration algorithms (stiff and non-stiff) in each period. For example, the students can model the operation of a runaway chemical reactor which is described in detail by Eisenberg et al., 2006.

The process simulation course is a challenging course to teach, from the technical standpoint. Derivation of complex equations require the use of a chalkboard, complex diagrams (like flow-sheets) require the use of pre-prepared PowerPoint slides, and demonstrations of computer programs require live interaction with the program in class. In addition to the presentations, note taking by the students is also demanding as they have to add notes to a hardcopies of the PowerPoint slides during the presentations. To assist the students, we have been using a Tablet-PC (which can serve both as chalkboard and computer for PowerPoint presentations and program demonstrations) and can also record the classroom activities and lectures that are later made available from a class site on the internet.

This paper will emphasize in some detail the preparation of complex process models, the use of the DIPPR database as the source of physical and thermodynamic properties, and the use of the "Self-recording with a Tablet-PC" technique.

Non-Isothermal Multi-component Flash Calculation - The Structure of a Unit Operations Module

The general structure of a unit operations module is shown in Figure 1. The purpose of this model is to calculate the flow rate temperature and pressure of the outlet streams using the information on the input streams, design parameters and the physical/thermodynamic properties of the compounds involved.

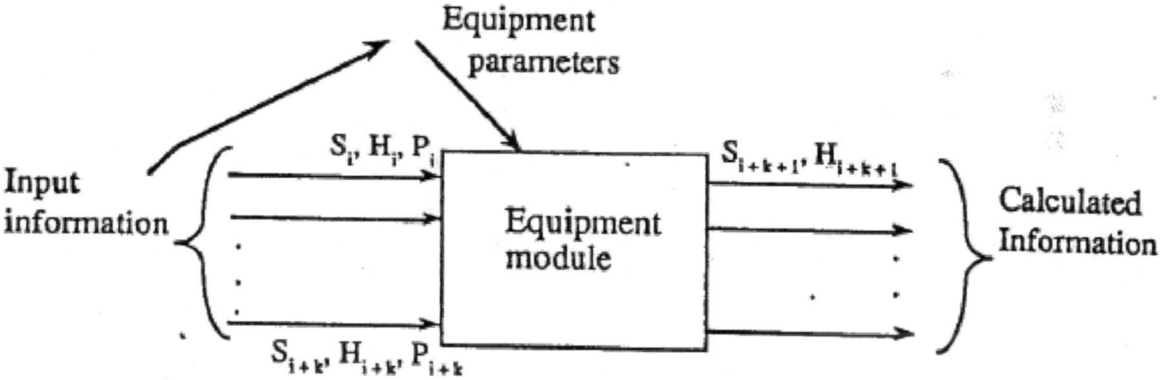

Figure 1. The General Structure of a Unit Operations Module

The notation used in Fig. 1: $\mathbf{S}_i$ is a vector of flow-rates in stream i, H_i is the total enthalpy of stream i, P_i is the pressure of stream i, and i and k – are stream indices.

When process simulators used for modeling a unit operation (or a complete process), the user is responsible for specifying the type of the equipment item, the connections between the various items, input stream flow-rates, compositions, enthalpies and pressure as well as the parameters of the processing units.

Constructing a unit operations module requires much more involvement on the student's behalf. The required knowledge and activities will be demonstrated for the case of a non-isothermal multi-component flash unit. In addition to the information concerning the input stream(s), there is a need to specify the parameters of the unit which include the heat added or removed from the flash drum (Q) and the pressure. Based on this information, the module has to determine whether under the stated conditions the outlet will be all gas, all liquid, or two streams of different phases. Then the calculations can be carried out accordingly. To determine the state of the outlet stream(s), the bubble point and the dew point temperatures of the feed and the enthalpies at these temperatures need to be calculated. These enthalpies have to be compared with the total enthalpy available (feed stream and heat input) in order to determine the state of the output stream(s). Thus the non-isothermal flash module must contain the model equations for calculating the following: 1. Bubble point temperature of a multi-component mixture, 2. Dew point temperature of a multi-component mixture, 3. Non-isothermal flash of a multi-component mixture, and 4. Mixture enthalpy at a specified temperature.

In addition to the mathematical model, physical property data related to phase equilibrium calculations (vapor pressure as function of temperature, activity coefficients) and to vapor and liquid enthalpy calculations (ideal gas enthalpy and heat of vaporization as function of temperature) are needed.

The calculation of the bubble point and dew point temperatures, and the non-isothermal flash calculations require solution of systems of nonlinear algebraic equations (NLE). The multidimensional Newton-Raphson method can be used for solving the NLE systems. Because of the complex temperature dependence of the vapor pressure (Riedel's equation) and the composition dependence of the activity coefficients, no analytical expression for the partial derivatives can be obtained but they must be estimated by finite differences. The many orders of magnitude differences between the terms of the material balance equations and the enthalpy balance equations require proper scaling of the equations in order to prevent ill-conditioning of the NLE system.

Proper safeguards should be included in the solution algorithm to enable providing sensible output stream flow-rates even in abnormal situations in the early stages of the solution process (zero input to the unit, failure to converge of one of the computations, etc.).

```
% function [SV,HV,SL,HL,err]=adbf(SF,HF,P,Q)
% Calculates the composition and the enthalpy of a flash unit with
% specified pressure, feed enthalpy and heat input
% Description of the input parameters
% SF - vector of feed flow rates (kg-mole/hr)
% HF - feeds enthalpy ( J/hr)
% P - pressure (Pa)
% Q - heat input (J/hr)
% Description of the output parameters
% SV - Outlet vapor stream flow rates vector (kg-mole/hr)
% HV -  outlet vapor enthalpy (J/hr)
% SL - outlet liquid stream flow rates vectot (kg-mole/hr)
% HL - outlet liquid enthalpy (J/hr)
% err - Eucledian norm of the final corrections in alpha and TF
% subprograms used bubble.m, dew.m, enthalpy.m, kvalue.m and adbfun.m
```

Figure 2. MATLAB function interface for the non-isothermal flash module

The MATLAB function interface for the non-isothermal flash module is shown in Figure 2. The interface provides details regarding the input and output parameters of the function and the subprograms used by this module (function). The function parameters coincide with the input and output information of the modeled unit operation (Figure 1).

One or more assignments are given to the students which involve non-isothermal flash calculations. In all the assignments the students receive from the instructor some of the MATLAB functions that are needed for these calculations the others they need to prepare on their own. The functions used for physical property calculations (kvalue.m and enthalpy.m) are always prepared by the students. The use of the Polymath 7.0 interface for the DIPPR database (Rowley et al. 2006) which is used to prepare property-related functions is discussed in the next section.

The use of the Polymath 7.0 interface for the DIPPR database in physical property calculations

Constant physical properties (such as critical properties, normal boiling temperature, etc.) and equations of temperature-dependent properties can be copied from the DIPPR database to Excel spreadsheets, Polymath, and MATLAB using the newly developed Polymath 7.0 interface for the DIPPR database. The process starts with user identification of the compounds and the properties involved. Figure 3 shows the "Temperature-Dependent Properties" communication box where the properties needed for non-isothermal flash calculations (vapor pressure, ideal gas heat capacity and heat of vaporization) are selected for the seven compounds from the database that were identified earlier using the interface.

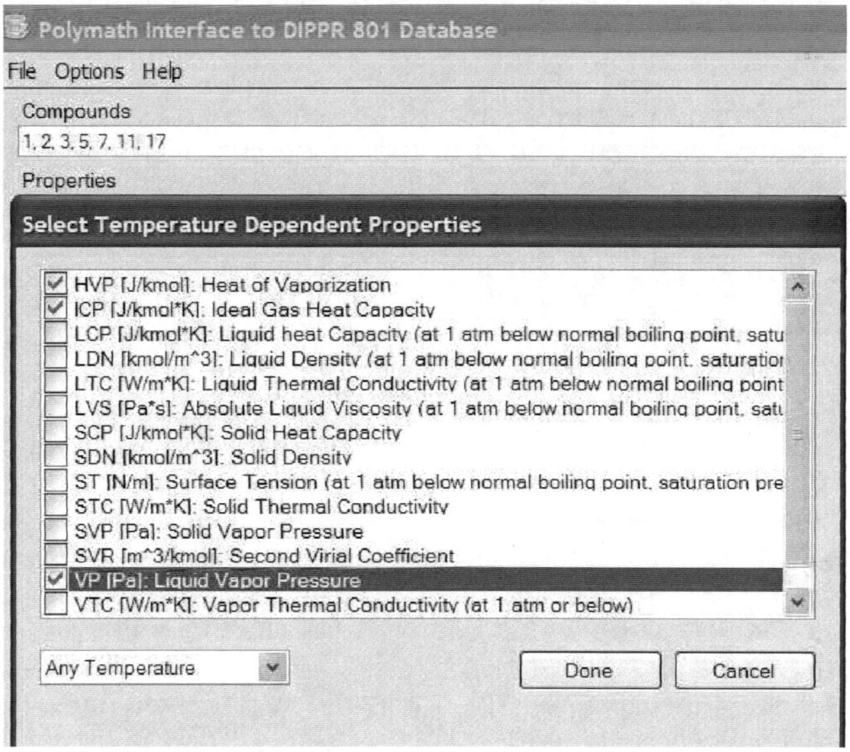

Figure 3. Polymath 7.0 Interface for the DIPPR database showing desired selection from among the available temperature-dependent properties.

After identifying the compounds and desired properties, a report is generated. The report contains the relevant data, equations and additional information available in the DIPPR database for each property. The data, equations and supporting information can then be copied into an Excel spreadsheet (in tabular form) and into a Polymath or MATLAB program. The relevant property equations for n-Pentane, as copied into a MATLAB m-file are shown in Figure 4.

```
1  % Heat of Vaporization of C5H12
2  % n-PENTANE
3  % Uncertainty: < 1%; Acceptance: A; Data Type: Experimental; Source Type: Unevaluated;
4  % Tc=469.7
5  % Min_T=143.42, Max_T=469.7, Min_Val=33968000, Max_Val=0 [K ; J/kmol]
6  HVP_C5H12 = 39109000 * (1 - (T / 469.7)) ^ 0.38681; % J/kmol
7  % Ideal Gas Heat Capacity of C5H12
8  % n-PENTANE
9  % Uncertainty: < 1%; Acceptance: A; Data Type: Smoothed; Source Type: Evaluated; Sourc
10 % Min_T=200, Max_T=1500, Min_Val=94039, Max_Val=329270 [K ; J/kmol*K]
11 % HIG_C5H12 = 88050*T+301100*1650.2*coth(1650.2/T)-189200*747.6*tanh(747.6/T)+HCON %
12 ICP_C5H12 = 88050 + 301100 * (1650.2 / T / sinh(1650.2 / T)) ^ 2 + 189200 * (747.6 / T
13 % Liquid Vapor Pressure of C5H12
14 % n-PENTANE
15 % Uncertainty: < 3%; Acceptance: A; Data Type: Experimental; Source Type: Unevaluated;
16 % Min_T=143.42, Max_T=469.7, Min_Val=0.068642, Max_Val=3364200 [K ; Pa]
17 VP_C5H12 = exp(78.741 - (5420.3 / T) - (8.8253 * log(T)) + 9.6171E-06 * T ^ 2); % Pa
18 
```

Figure 4. MATLAB formatted physical property equations and supporting information as generated by the Polymath 7.0 DIPPR interface

The property equations are shown in lines 6, 12 and 17, respectively. The names of the variables which store the numerical values of the properties at a specified temperature are consist of the symbol of the property (HVP-heat of vaporization, ICP-ideal gas heat capacity, VP-vapor pressure) and the chemical formula of the compound involved. The units of the property are shown (as a comment) in the same line as the equation.

The supporting information is provided as "comments". There are three user-selectable levels for this information. The report in Figure 4 was generated using the "basic" (intermediate) level. The first line of supporting information includes: the full name of the property and the formula of the compound, and the 2nd line contains the full name of the compound. In the 3rd line, information concerning the reliability of the property equation, as determined by the DIPPR staff, is shown. This includes the uncertainty in the property value (an upper limit on the experimental or prediction error), information whether the equation is based on experimental data or prediction, and whether the equation was evaluated and accepted by the DIPPR staff. An additional comment line contains the range of validity of the equation in terms of temperature and property limits. In the same line, the units of the temperature and the property are also shown.

The ideal gas heat capacity (line 12 in Figure 4) deserves special discussion. The transcendental equation of Aly and Lee (1981) is used by DIPPR to model this property. The integrated form of this equation for calculation of the ideal gas enthalpy is considerably different than the heat capacity equation. For the benefit of the users, whenever the ideal gas heat capacity is requested, the equation for ideal gas enthalpy is also included as a comment (see line 11). The enthalpy equation includes an integration constant H_{con}. This constant depends on the standard state selected for the enthalpy calculation. For example, selection of a pure liquid component at 0 °C (273.15 K) as a standard state yields H_{con} as ideal gas enthalpy at 273.15 K subtracted from the heat of vaporization of the component at the same temperature.

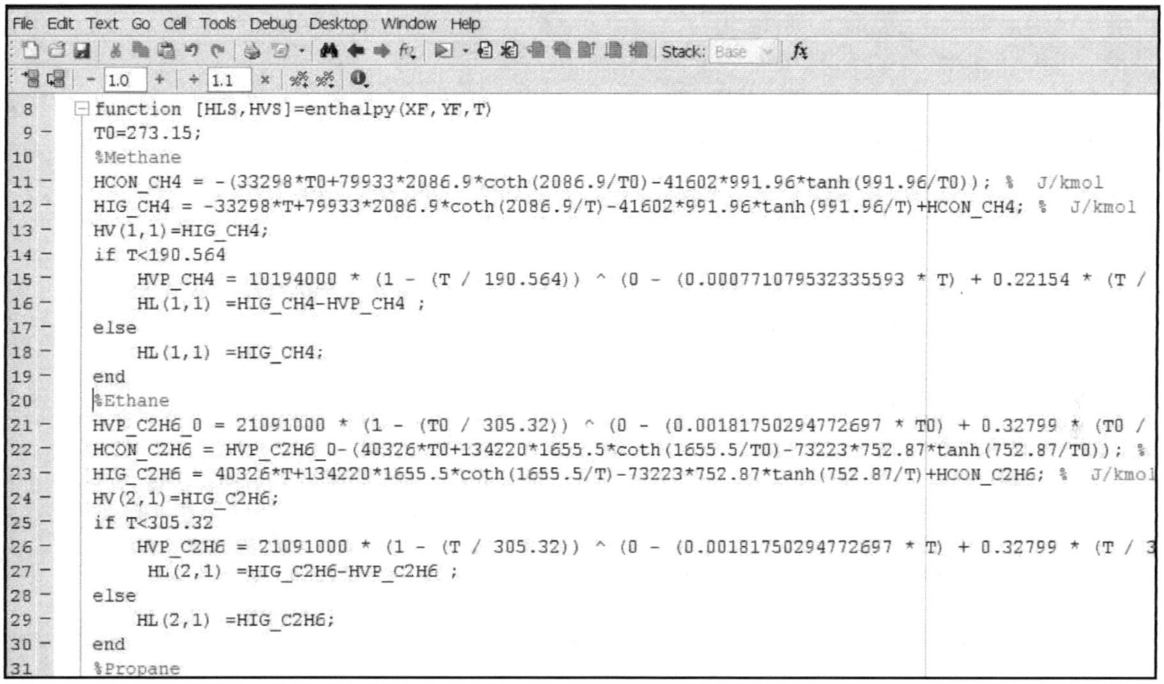

Figure 5. Partial display of the MATLAB enthalpy.m function

Part of the MATLAB function: enthalpy.m for calculating the molar enthalpy of liquid with composition of $\mathbf{x}_F$ and molar enthalpy of vapor with composition of $\mathbf{y}_F$ at temperature T, is shown in Figure 5. For brevity, most supporting information is not shown. The ideal gas enthalpy (HIG) and heat of vaporization (HVP) equations are used. The integration constants H_{con} are calculated for all the compounds using the standard state of liquid at 273.15 K. Additional programming is required to separate cases where a compound above or below the critical temperature, and the enthalpies must be introduced into arrays.

Thus the Polymath 7.0 DIPPR interface enables very rapid program development for calculating properties of mixtures using state-of-the-art industrial standard correlation equations.

Self-Recording of the Lectures Using a Tablet-PC and a Camtasia Recorder

This new technology allows the lectures to still be delivered face-to-face; however, instead of writing on the chalkboard the instructor writes on the computer screen. At the same time, the instructor uses the computer to record voice and the visual material that is displayed on the PC screen (and also projected to a large screen in the classroom). The Camtasia recorder (a software package loaded on the instructor's computer, http://www.softwarecasa.com) is used for recording.

The visual material of the lecture can be presented with the Tablet PC either using the Windows Journal or Power Point slides. When Windows Journal is used, the computer screen essentially serves as a chalkboard. When power point slides are used, it is preferable to put on them only part of the lecture material, while the rest is added by writing during the presentation. The students are encouraged to bring hard copies of the incomplete slides to the class and to add missing parts during the class presentations. The instructor writes on the Tablet PC screen while explaining the material, and the students complete the missing information on the hard copies. This method gives students sufficient time to absorb the more difficult material and to ask questions of the instructor.

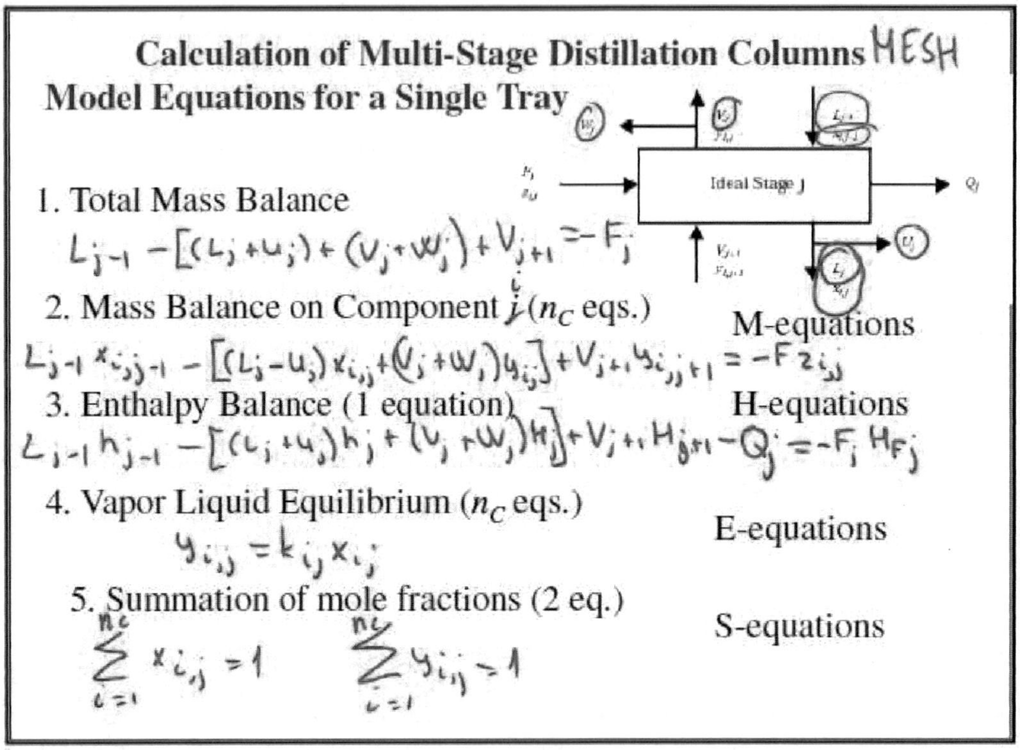

Figure 6. Introduction of the MESH equation for multistage distillation calculations

Consider, for example, the derivation of the MESH equations for multistage distillation calculation, as shown in Figure 6. The sketch of an equilibrium stage and the text characterizing the equations needed are placed in advance on the PowerPoint slide. The equations are derived in class with considerable student involvement.

Figure 7 demonstrates another option in the effective use of the Tablet PC. In this figure, a part of a computer program for solving a recycle problem using Broydens' method is shown. The students bring a hard copy of the program. The connection between the mathematical formulas used, the solution algorithm, and the coding of the method is explained in class by putting markings on the Tablet PC screen while the students write on the hard copy.

The instructor starts the recording at the beginning of the lecture and stops it toward the end. The recorded lecture becomes available on-line within a few hours after some modest editing is done. The main purpose of the editing is to add subtitles that enable accessing separate sections of the lecture rapidly. The students can watch the recorded lectures wherever they have access to the Internet.

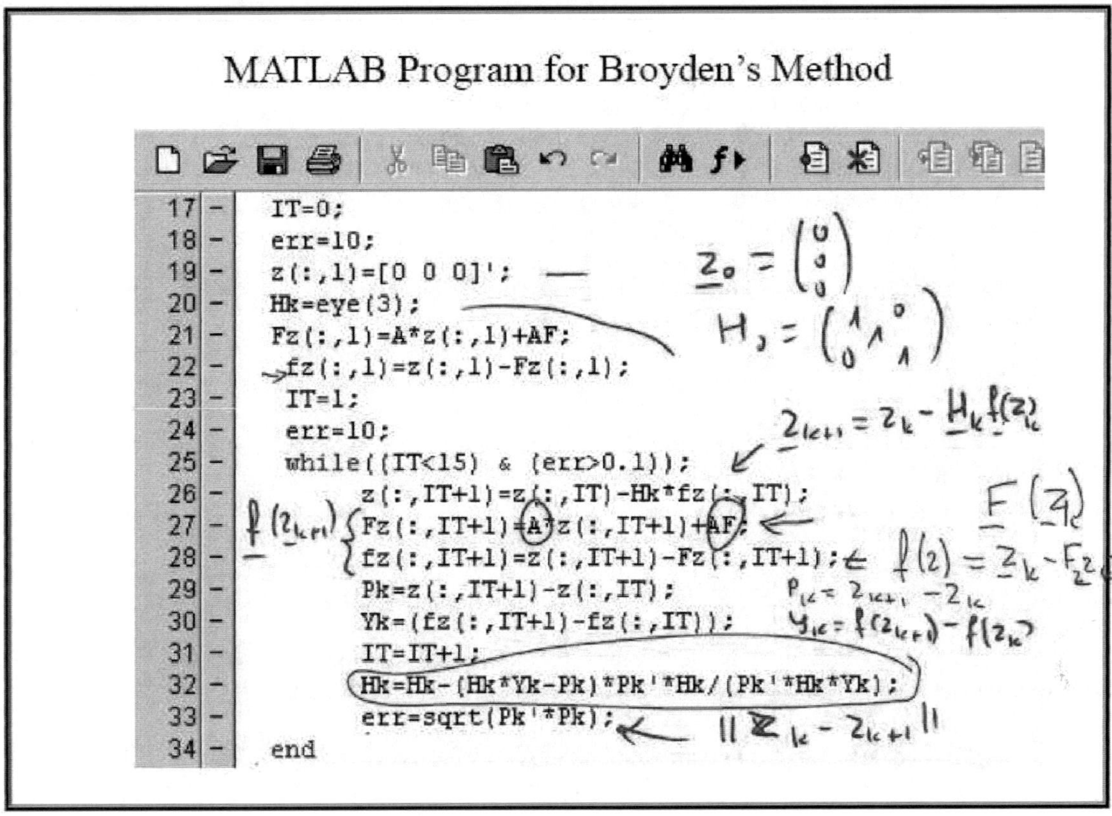

Figure 7. Use of Broyden's Quasi-Newton method for recycle convergence

Conclusions

The process simulation course is given as an elective course for fourth year undergraduate students and new graduate students. Success in this course requires that students must review, enhance, update and make practical use of their knowledge of programming, material and energy balances, thermodynamics, numerical methods and reaction engineering. With the integration of such broad content, the course can be considered as a culmination of core chemical engineering coursework.

Use of the Polymath 7.0 – DIPPR database interface enables the students to use state of the art, industrial-standard physical and thermodynamic properties as well as property correlation expressions. They become aware of the uncertainty associated with the property data as well as with the need to respect the range of validity of the correlation equations.

Student evaluation of the new instructional technique, enabled by the Table PC, yielded very positive results. In particular, the students pointed out the following advantages: 1. The lectures are better structured and more interesting than the traditional methods of teaching (PowerPoint or chalkboard) 2. The records kept in their notes are clear and understandable 3. The recorded lectures are beneficial when someone misses a lecture or wishes to review the lecture material when completing homework assignments or preparing for an exam.

References

1. Aly, F. A. and L. L. Lee, "Self-Consistent Equations for Calculating the Ideal Gas Heat Capacity Enthalpy, and Entropy". Fluid Phase Equilib. 6, 169(1981)
2. Cutlip, M. B., N. Brauner and M. Shacham, "Biokinetic Modeling of Imperfect Mixing in a Chemostat – an Example of Multiscale Modeling", Chem. Eng. Ed., 43, 243 (2009)
3. Dahm, K. D., R. P. Hesketh and M. J. Savelski, "Is process simulation used effectively in ChE courses?" Chem. Eng. Ed., 36, 192 (2002)
4. Dias, R. S., L. C. Silva, L and A. J. De Assis, Plant wide simulation using the free chemical process simulator Sim42: Natural gas separation and reforming, Published online in Wiley InterScience; DOI 10.1002/cae.20200
5. Eisenberg, S., M. Shacham and N. Brauner, "Combining HAZOP with Dynamic Simulation - Applications for Safety Education", Journal of Loss Prevention in the Process Industries 19, 754–761 (2006)
6. Rockstraw, D. A., "ASPEN Plus in the ChE curriculum. Suitable course content and teaching methodology," Chem. Eng. Ed., 39, 68 (2005)
7. Rowley R. L., W. V. Wilding, J. L. Oscarson, Y. Yang and N. A. Zundel. DIPPR Data Compilation of Pure Chemical Properties Design Institute for Physical Properties. http//dippr.byu.edu, Brigham Young University Provo Utah (2006)
8. Shacham, M., "Numerical Solution of Constrained Non-Linear Algebraic Equations", International Journal of Numerical Methods in Engineering, 23, 1455-1481 (1986).
9. Shacham, M., M. B. Cutlip and N. Brauner, "From Numerical Problem Solving to Model Based Experimentation – Incorporating Computer Based Tools of Various Scales into the ChE Curriculum", Chem. Eng. Ed., 43, 299 (2009)
10. Streicher, S. J., K. West, D. M. Fraser, J. M. Case and C. Linder, "Learning through simulation, student engagement," Chem. Eng. Ed., 39, 288 (2005)

(534c) Numerical Methods and Simulation in the Minnesota ChEn Curriculum

Jeffrey J. Derby[1], Satish Kumar[2], Prodromos Daoutidis[1], Kevin D. Dorfman[3] and Alon V. McCormick[4], (1)Chemical Engineering and Materials Science, University of Minnesota, Minneapolis, MN, (2)Department of Chemical Engineering and Materials Science, University of Minnesota, Minneapolis, MN, (3)Department of Chemical Engineering and Materials Science, University of Minnesota - Twin Cities, Minneapolis, MN, (4)University of Minnesota, Minneapolis, MN

The current approach to computing in our curriculum is to address different skills at different class levels. In the Sophomore year, we have the students use Matlab as a tool to address problems of significant size and complexity, but postponing full understanding of the numerical method to a later course (and allowing students to experience the risk of numerical difficulty). In the Junior year, the students take a numerical methods class that shows how the algorithm makes a difference when approaching large or challenging problems; at this stage the difficulties encountered earlier are now appreciated and understood. In the Senior year, commercial simulation software is used in Design and Control, with some attention to and awareness of the advantages that come with understanding the numerics.

See more of this Session: Effective Use of Simulation and Numerical Methods in Various Courses
See more of this Group/Topical: Education

(534d) Online Simulations for the Demonstrations of Proper Experimental Design and Data Analysis

Anthony Butterfield, Chemical Engineering, University of Utah, Salt Lake City, UT

Undergraduate instruction in experimental design and data analysis is often left to elective courses or is given as a fraction of lab courses. Furthermore, some engineering students find concepts of statistics difficult to grasp when these concepts are divorced from hands-on experiments. However, real world experiments can be time consuming and costly, and with such limitations a thorough demonstration of statistical concepts in a classroom setting is problematic.

To address these issues and promote efficient and statistically sound experimental design throughout our department, a system of online experiment simulations and associated demonstrations in experimental design and data analysis has been developed. This online curriculum was constructed using a combination of Javascript and php coding languages and may be accessible by students in every course throughout the department. Graphical interfaces, and visual displays of data and statistical plots are emphasized in order to help student develop greater insight into and comfort with the topic. Simulations of experiments are geared towards the experience of a Chemical Engineer, in order to address student's concern that their training in statistics felt as though it had little bearing on real world problems. Topics covered include basic statistics, errors in physical measurements, probability distributions, hypothesis testing, error propagation, and factorial experimental design.

These web tools allow students to develop an insight and intuition into statistically sound methods and results, and at minimal investment of student's time and department resources. Students using these online tools report significantly more satisfaction with their training, and confidence in their abilities to properly apply statistics to chemical engineering problems.

See more of this Session: Effective Use of Simulation and Numerical Methods in Various Courses
See more of this Group/Topical: Education

(534e) Modeling and Simulation of a Two Phase Flow as An Example of Effective Research Computation for a Graduate Seminar Class

Richard L. Long, New Mexico State University, Las Cruces, NM

Graduate level research can emphasize either laboratory results or modeling results or both. Effective modeling results often have an important computational component. An important goal of modeling is often to match experimental resultswith model computations. Such a match can give confidence that the model, as a scientific explanation of experimental observations, is correct. If computations do not match experimental observations, then either the computations are in error or the model is not a good explanation of the observed phenomena. The most confidence results when the model and the experiments are done by different researchers. New graduate students do not always grasp some of these basic points. One approach to teaching them the significance of modeling in research is to derive a model based on coursework they have completed for a particular research topic and show them a set of computational results obtained from the model and compared to a set of experimental observations for that topic. This compels them to realize that not all problems have already been solved and that effective computation applied to good models can lead to an accurate representation of observations of the phenomena. A graduate seminar course is a good place to do this.

See more of this Session: Effective Use of Simulation and Numerical Methods in Various Courses
See more of this Group/Topical: Education

(534f) Dynamic Simulation Troubleshooting and Control Exercises On the DCS

Peyton C. Richmond, **M.A.K. Rasel** and Daniel H. Chen, Dan F. Smith Department of Chemical Engineering, Lamar University, Beaumont, TX

Several troubleshooting cases involving operating issues with a de-propanizer de-butanizer twin tower process (Woods, 2006) were simulated in HYSYS and implemented on an industrial DCS system. The towers, equipment and instrumentation were modeled dynamically and connected to an industrial DCS system using the OPC protocol. A commercial multivariable controller was installed to determine the optimal operation and to expose the students to the modern manufacturing environment. An evaluation (test) and student satisfaction surveys were conducted following the class exercises.

Two DCS-based cases were provided as exercises to an Advanced Analysis class following application of similar paper-based "triad" troubleshooting exercises as described by Woods. The DCS-based troubleshooting cases included operating problems caused by (1) high overhead cooling water temperature, (2) a reflux block valve left partially closed following maintenance activities. The first DCS-based troubleshooting case was also used as the basis of a "DCS troubleshooting competition" held at the Southwest Region student AIChE conference at Lamar University, March 26-28, 2010.

A feed forward controller was configured and tuned for the de-propanizer tower without a multivariable controller. The feedforward controller was demonstrated in a Process Control II class where students made a disturbance move (10,000 lb/hr feed change) to observe the controller in action in real time. A performance comparison between the feedforward plus feedback controller and the feedback only controller was shown to the students. During the DCS demonstration, historical closed-loop tuning records were displayed to illustrate the online feedforward controller tuning procedure. Students were exposed to the feedforward tuning procedure and the effect of feedforward parameter changes on the controller performance.

Woods, D.R., Successful Trouble Shooting for Process Engineers, WILEY-VCH Verlag GmbH & Co., Weinheim (2006).

See more of this Session: Effective Use of Simulation and Numerical Methods in Various Courses
See more of this Group/Topical: Education

(534g) Student Learning in Hands-On, Remote, and Virtual Laboratory Experiences: What Works (or doesn't)

David DiBiasio, Chemical Engineering Department, Worcester Polytechnic Institute, Worcester, MA, Jim Henry, Chemical Engineering, University of Tennessee at Chattanooga, Chattanooga, TN, William M. Clark, ChE, Worcester Polytechnic Institute, Worcester, MA and Marina Miletic, Department of Chemical and Biomolecular Engineering, University of Illinois Urbana-Champaign, Urbana, IL

Background Student learning outside the classroom may be complemented by a variety of structures. Laboratory and project work are typical modes. However, lab work can be hands-on, remote, or based upon simulations (virtual). Ideally, remote experiments can be conducted at any time from any place. They are particularly useful for students at universities where resources are severely limited and there is no access to any significant experimental equipment. Simulations are also accessible 24/7, do not require substantive resources (unless the software is expensive) and can provide efficient and easy ways to study multiple variables. Multiple runs are frequently impossible or too costly with real lab equipment whether remotely operated or not.

In all contexts, understanding student learning is critical. Each mode has unique advantages and disadvantages that contribute to learning. Basically, knowing what works and what doesn't is necessary to structure course material and inform implementation of any or all modes in a given course.

At WPI distillation is taught in the sophomore year throughout a project-based spiral curriculum. All projects are team-based. We introduce basic concepts early in the sophomore year then revisit distillation throughout the year with successively more complex assignments and projects. This includes at least two lab experiences. The first experiment uses a batch column operated at total reflux to introduce students to multistage distillation including efficiency and energy balances at total reflux. A follow-on course explores pressure swing distillation without a lab component. Near the end of the year, the batch column operated at a constant external reflux ratio is used again in a lab project. This project engages students in process dynamics and challenges them to compare differences between theory and reality using the Rayleigh analysis. Large enrollments forced us to "outsource" lab work by taking advantage of the available but remotely-operated column at UTC. The column is the same as WPI's but is accessible and controllable through a web-based interface. Methodology At the last AIChE meeting we presented a study that showed student learning in remote and hands-on environments was essentially equivalent. We have supplemented that work by comparing learning in the same course described above in three cohorts: hands-on, remote, and virtual. The virtual column was constructed using Aspen BatchSep®.

Teams were four students each (teams were self-selected), and they were randomly assigned to cohorts.

Seven teams ran the local column (hands-on), six teams ran the UTC column (remote) and seven teams used the simulation (virtual). Remote and hands-on teams conducted identical experiments with identical assignments. Virtual teams had an identical base case assignment but made additional simulation runs. Analysis and project reporting requirements were the same for all cohorts.

Evaluation had three components with direct and indirect assessments, giving us some degree of triangulation. All cohorts completed surveys managed by a third partner: UIUC. The online survey included closed-end, Likert-scaled responses and open-ended questions. The open-ended questions included student attitudes but also probed team processing since we wanted to learn more about how each team handled roles, logistics, scheduling, data collection, and analysis.

Course instructors compared final reports from all teams. A pre/post in-class quiz compared individual learning for students in the remote cohort compared to hands-on cohort in a quantitative manner. Qualitative measures include evaluating student attitudes about the experience and assessing students' improvement in ABET outcomes such as: (b) – design, analysis and interpretation of data; (d) – functioning on multidisciplinary teams; (g) – effective remote communication; (h) - have broad education necessary to understand the impact of engineering solutions in a global, economic, and environmental societal context; and (k) - using techniques, skills, and modern engineering tools (such as computers and web interfaces) necessary for engineering practice. Results Logistics As previously reported, having completed two cycles of this pilot we have a fairly good understanding of what it takes to put 20 teams through this lab project. It would have been essentially impossible to run all teams through the local WPI lab or the UTC lab, during the time allotted for the course. However, splitting effort between WPI, UTC, and Aspen makes logistics workable. We learned that running either the remote or virtual column presents different but no more burdensome scheduling issues than running the local column. Student Attitudes Survey results will not be available until sometime after May, 2010. Previous results indicated that there are few, if any, serious issues. Students don't mind being assigned to cohorts nor was there any indication of teams feeling they lost out on something because they weren't in a different cohort. New results involving virtual team experiences will be available over the summer and will be analyzed in time for the conference.

Student Learning All teams produced a significant final report that was graded by the WPI instructor using the same rubrics. We evaluated presentation quality, analysis of results, and demonstration of appropriate concepts. We also examined learning issues as they specifically related to two important concepts: distillation column dynamics and comparison of theory and experiment using the Rayleigh analysis. We also gave a pre/post quiz regarding simple concepts in these areas to compare differences among cohorts.

Our previous content analysis of the reports showed there were common deficiencies that were independent of remote, or hands-on cohorts. These included the typical mistakes with energy balance calculations; and another typical confusion over quantitative comparison of experiment to Rayleigh analysis. Students expect exact comparisons yet reality is never that predictable. In too many cases, the report presentation quality was marginal. This result is not unusual given the end-of-semester time demands.

Our report analysis is ongoing and details will not be available as of the deadline abstract. However, a preliminary read showed that the virtual cohort may have shown some interesting positive learning gains when compared to remote or hands-on. This, at first, seems surprising but virtual students had the advantage of being able to make several runs and may show learning gains because of that. The authors note that this is very preliminary and will only be confirmed (or not) when the content analysis is done. Previously we've used the psychology of "presence" (Ma and Nickerson, ACM Computing Surveys, v. 68 no. 3, article 7, 2006) as a framework for interpretation. Student learning in lab environments is connected to their perception of presence and the process of collaboration. Presence can be physical, telepresence, or virtual (for example, a simulation). In our case the remote teams experienced telepresence, local teams experienced physical presence, and simulation teams experience virtual presence. Hence a hands-on lab that has minimal physical interaction may not be significantly different than a remote one where the interaction is mediated via a computer. In fact, the UTC column has three cameras that can be manipulated by students bringing an element of physical presence into the mix. However, the Aspen simulation is viewable only through the standard Aspen interface. By meeting time we will also know more about students' collaboration process, and how that dynamic affected interactions, and hence learning, within each cohort.

See more of this Session: Effective Use of Simulation and Numerical Methods in Various Courses
See more of this Group/Topical: Education

(534h) The Paradigm Shift From 'in Vitro' to 'in Silico' Experimentation, Offers Important New Benefits for Education

David A. Gallagher, CAChe Research, Beaverton, OR

The latest generation of software modeling tools enables students to quickly and easily perform virtual experiments in thermodynamics and chemistry on their own laptop computers (in-silico). Software tools that incorporate quantum chemistry and statistical thermodynamics methods do not require empirical data and so, can release the students from the constraints of their own limited experience and available chemicals. They are able to predict a wide range of properties of compounds and mixtures, including vapor pressure, boiling points, activity coefficients, azeotropes, partition coefficients, liquid-liquid equilibria, chemical reactivity, etc., for even hypothetical compounds.

However, more than just numbers, graphical visualization of the computational models offers an intuitive understanding of the fundamental phenomena that influence physical properties, fluid thermodynamics, or chemical reactivity. For example, visualization of electrostatic potentials on the solvent-accessible surface and comparison of sigma profiles, illustrate the potential for hydrogen bonding, dipole-dipole, and Van der Waals interactions, providing clear insight into the molecular interactions that determine the physical properties.

Additionally, these new tools enable the student to explore their own creativity and experiment with off-the-wall ideas, in complete safety on a laptop computer, in the classroom, at home, or virtually anywhere.

This talk will be illustrated with examples of experiments performed with commercial software that is available free to academics and students for education (e.g. COSMOtherm), along with a brief overview of its underlying computational methodologies.

See more of this Session: Effective Use of Simulation and Numerical Methods in Various Courses
See more of this Group/Topical: Education

(542a) STEM Educational Outreach Using An Inquiry-Based Radio Broadcast

Peter J. Ludovice[1], William Hunt[2], Adam Wathen[2], Donna Llewellyn[3] and Marion Usselman[4], (1) School of Chemical & Biomolecular Engineering, Georgia Institute of Technology, Atlanta, GA, (2) School of Electrical & Computer Engineering, Georgia Institute of Technology, Atlanta, GA, (3)Center for the Enhancement of Teaching and Learning, Georgia Institute of Technology, Atlanta, GA, (4)Center for Education Integrating Science, Mathematics, and Computing, Georgia Institute of Technology, Atlanta, GA

A recent report from the National Academy of Engineering, suggested an updated approach to engaging students in a dialogue about engineering and technical careers.[1] This report suggested stressing the aspects of engineering with broad appeal such as its creative-side and positive impact on society, in contrast to its liberal use of math. While this new message is necessary, it is not sufficient to engage young people in a conversation that will recruit the future engineers required in the United States. We contend that the delivery method, as well as the message must be changed to sufficiently engage people in a conversation about technical careers such as engineering. The proposed updated delivery method includes two important elements: (i) the modern infotainment elements used successfully in many educational television programs, and (ii) inquiry based learning.

This updated informal delivery method has been used in a weekly, live radio broadcast in Atlanta since April 2006. The weekly program entitled "Inside the Black Box," uses humor to engage the public in a conversation that attempts to demystify science and technology.[2] Humor has been used previously to engage audiences regarding topics that they might otherwise find uninteresting. Examples include everything from the humorous approach to teaching auto mechanics on the radio (NPRs Car Talk), to the successful use of humor in teaching driver safety courses in the State of Texas. Other infotainment techniques such as drama have also been used successfully (the Crime Scene Investigators Television Series). This radio program is hosted by two Georgia Tech Engineering Professors and produced on the Georgia Tech Campus. Technical experts from the Atlanta area and beyond discuss various topics in science and engineering live, as the listening audience calls in with specific questions. Recently, the program has added live teleconferences with High School science classes to its weekly show. The program is then driven by the student questions to the expert guest and the hosts, rather than the technical expert and hosts. A study of one such student inquiry-based interaction produced an increase in the technical level of student inquiry following the radio program.[3]

This program represents a potentially successful formula that can be transplanted easily to any college or university with a radio station. Transplanting such a simple program nation-wide could result in a significant outreach program in a number of technical fields. We will characterize the basic elements of this program and illustrate with examples the best practices that have been learned over the past several years. The specifics of inquiry-based component will also be described.

[1]Committee on Public Understanding of Engineering Messages, National Academy of Engineering, Changing the Conversation: Messages for Improving Public Understanding of Engineering, NAE Press: Washington, D.C., 2008.

[2]www.insidetheblackbox.org, WREK-Atlanta, 91.1FM

[3]Wathen, A.; Hunt, W.; Llewellyn, D.; Ludovice, P.; Usselman "A Study of Inquiry-Based Informal Science Education in an Urban High School Physics Class," Proc. ASEE National Meeting, Louisville, KY, Paper AC-2010-674, 2010.

See more of this Session: Interactions with K-12
See more of this Group/Topical: Education

SWEET-CREAM: Combining Teacher and Student Centered Activities to Maximize Impact on K-12 Interactions

Richard Zollars
Gene and Linda Voiland School of Chemical Engineering and Bioengineering
Washington State University

Over the past two years both an RET activity (Summer at WSU – Engineering Experiences for Teachers – SWEET) and a GK-12 activity (Culturally Relevant Engineering Applications in Mathematics – CREAM) have been underway at Washington Sate University. The GK-12 program places STEM graduate students into classrooms to provide a real world application for the mathematics that students are studying. The RET program brings teachers to the WSU campus during six weeks of the summer to work on research projects in engineering in order to gain a familiarity with engineering. Both programs seek to increase their impact in underrepresented groups by focusing participation on selected high schools in the state. Our biggest impact has been on the high schools in the Sunnyside/Mabton area. In both of these schools the percentage of Hispanic students is much higher than the state average. To date eight of the mathematics/science teachers have participated in one or both of the programs; four graduate students have participated as GK-12 fellows at Sunnyside High School. The impact of the interactions on the effectiveness of the two programs will be highlighted.

Background

For the past two years the Voiland School of Chemical Engineering and Bioengineering at Washington State University has been running two significant activities aimed at improving the backgrounds and numbers of high school students interested in pursuing engineering degrees. One of these is a Research Experiences for Teachers (RET) site activity funded by the National Science Foundation (grant # 0808716). The activity was entitled Summer at WSU – Engineering Experiences for Teachers (SWEET). In this activity science and/or mathematics teachers at the high school and middle school level come to the Washington State University campus for six weeks to work on ongoing research projects. During their six-week stay the teachers are directly involved with a research project and are expected to develop a teaching module, which incorporates some aspect of their research effort, which they would use during the following school year in their classes. The focus of this program is on the teachers as an efficient means of getting engineering concepts to middle school and high school students. Developing a program aimed at students would mean that only a relatively small number of students could be contacted each year and, since they graduate, this would have to be repeated to a new set of students every year. By directing the programs towards teachers, and their needs, we develop a person who can contact many students each year and who will hopefully remain in that position for many years. The six week duration of the program is also felt to be necessary. Since engineering is new to almost all of the teachers they need to have enough experience so that they will feel comfortable in using engineering concepts in their classrooms. Thus the two major goals of this activity are (1) to introduce the math/science teachers to engineering so that they can act as knowledgeable resources in their classrooms and (2) to develop classroom materials involving engineering that can be used at the middle school and high school level. The classroom materials are made available to all teachers via a website (http://eerc.wsu.edu/SWEET/) and are tested by an average of 30 middle school and high school students before being made publically available.

The second activity is a GK-12 activity funded by a second National Science Foundation grant (# 0538652). The activity was entitled Culturally Relevant Engineering Applications in Mathematics (CREAM). The GK-12 activities involved placing graduate students in STEM fields into high school classrooms. The graduate students selected for this program first went through a summer-long course introducing them to teaching philosophies and techniques. During the summer they also met with the teacher in whose classroom they would be working during the following year. One of the goals of these summer sessions is to acquaint the GK-12 fellow with the high school and community. In this way the GK-12 fellow is introduced to the culture of the area in preparation for assisting in or developing activities in mathematics that relate to that culture. During the following academic year the GK-12 fellows would spend one to two days per week assisting with mathematics courses in their respective high schools. They also continued to take classes on teaching methodologies. During the first year of this activity a set of predetermined mathematics activities were used by all of the GK-12 fellows. During the last two years of this activity, while these prior activities were still available, the GK-12 fellows were encouraged to develop classroom activities having some connection to their thesis research.

In both of these activities a focus was placed on interactions with students from underrepresented groups in engineering. In the CREAM program this was accomplished by placing the GK-12 fellows into schools with a higher than usual proportion of students from underrepresented groups. During the CREAM program these schools have been Lake Roosevelt High School (58% Native American), Omak High School (26% Native American), and Stevens Middle School (89% Hispanic). In the SWEET program teachers were recruited from schools with a larger proportion of underrepresented groups or who were themselves members of underrepresented groups. Two of the targeted schools in the SWEET program have been Sunnyside High School (82% Hispanic) and Mabton High School (94% Hispanic). In addition, three of the teachers who had participated in the SWEET program from either Sunnyside or Mabton were Hispanic.

In the summer of 2009 the activities in these two programs were combined. Using the contacts that had been established during the SWEET program three more teachers from either Sunnyside or Mabton High were recruited for participation in the SWEET program. In addition, three GK-12 fellows (Courtney Buonucelli, Elvie Brown, and Nathan Hamlin) were placed into classrooms in Sunnyside or Mabton High. During the summer of 2010 these activities were continued with another two teachers participating in the SWEET program and the three GK-12 fellows continuing at the two high schools. Thus, between the two programs, eight science or mathematics teachers from these two high schools have participated in the SWEET program. GK-12 fellows have assisted in the classrooms of five of these eight teachers for the past two years. This combination, having a classroom teacher with the RET experience and an engineering/science graduate student as an assistant, provides even greater impact on the STEM education for the students.

Results

At present the measures of the impact that these joint activities have had on the students, teachers and fellows are mainly obtained via reflective assessments of each group. For example, one teacher reported difficulty helping students understand and appreciate the value of matrix algebra. The GK-12 student in this classroom then presented a project in which matrix algebra was an integral part. Following this project the teacher reported an improvement in performance by the students.

The influence of engineering in the classroom was highlighted by the events at one of the schools. The students in the low performing math class performed at a level above that expected by the end of this activity. While there are many factors that may have contributed to this improvement the teacher and GK-12 fellow involved noted that the concepts presented were done so in a "project" oriented manner. If a person in the low performing group were there because their interests were more in the application of mathematics rather than mathematics for mathematics sake the project orientation would be more appealing thus promoting greater performance. Because engineering uses mathematics in applications, students with this more utilitarian viewpoint would likely show improved performance.

Teachers also report a change as a result of their experiences. A number felt that they have learned new applications and examples of how mathematics is applied in an engineering setting. Among the comments from the teachers were:

> *"This program will have an impact on both my teaching & student learning."*
>
> *"I learned a lot about engineering and enjoyed applying it to math and science."*
>
> *"My project was very applicable to the classroom and I know that my students will benefit from my participation."*
>
> *"Now that I have a better idea of what engineering is & what engineers really do, I'll be talking about engineering to my students .... a lot!"*
>
> *"The program was a great opportunity to develop my knowledge of engineering and I feel well prepared to talk with students about what engineering is and why it is important."*

"I personally put a lot of effort into it and took a great deal from the program. My students will gain a lot from it. Thanks."

In response to a question on whether their students would benefit from their (the teacher's) participation in this project the teachers response was a 4.8 ± 0.45 on a 0 – 5 Likert scale (0 = no benefit, 5 = greatly benefit).

In response to the question "This kind of science lesson makes me interested in engineering" the students using the modules developed by teachers who had participated in both the SWEET and CREAM programs reported an evaluation of 1.8 ± 0.8 on a 1 – 5 Likert scale (1 = strongly agree, 5 = strongly disagree) compared to an evaluation of 2.53 ± 0.97 on the same question for all teachers who had participated in the SWEET program alone.

Conclusion

Thus the combination of the SWEET and CREAM programs seems to be having an impact beyond that of either program by itself. There is evidence that by introducing the application side of mathematics via engineering situations a broader range of students succeed in their mathematics courses. The students themselves also report greater interest in engineering after exposure to the modules presented by teachers who have participated in the combined activity. Future follow-up will attempt to measure whether these effects carry-over to increased participation and success in engineering for these students.

(542c) Integrating Engineering to K-12 by Training Teachers Using REU Concept

Sundararajan. V. Madihally and Karen High, School of Chemical Engineering, Oklahoma State University, Stillwater, OK

Providing K-12 students an earlier exposure to engineering is recognized as a way to prepare them for postsecondary courses. There has been a significant increase in academies and workshops addressing this concept. Since there is a total turnover of students each year, training K-12 teachers has been thought as an efficient approach. Further, teachers can integrate engineering concepts during the academic year which provides a natural transition. Based on this concept, we started a Research Experience for Teachers (RET) project, where teachers were introduced to research dealing with tissue engineering. Teachers worked for six weeks (four days a week) within the research laboratory on formation of porous structures using biodegradable polymers. One unique feature of RET project was pairing teachers with undergraduate students (referred as REU students in the subsequent discussion) to bridge gaps between the knowledge and experience of faculty and teachers. REU students were selected for this project from the engineering programs, so that they are comfortable with instrumentation and understand measurement procedures that may not be familiar to teachers. REU students were also required to continue to work with teachers during the fall semester to help implement engineering research knowledge in the classroom. Teachers were exposed to the technique of forming porous structures using biocompatible and biodegradable polymers into various shapes using the apparatus available in the laboratory. The concept of design cycle was utilized with the initial starting location of researching existing polymers. Next they performed modeling of the cost followed by implementation of the freeze-drying technique. Students are also asked to measure the number of pores using light microscopy. They also had to communicate their experience to the teachers. Presence of REU student was integral to continuous input and integration during the academic year. Extended presence of the REU student benefits the project since the REU student can continue to work with the teacher to solve technical issues that arise once the teacher begins to implement the research problem at their home school. Further, REU student also had significant exposure to research. In 2010, we have expanded this experience to four teachers with one teacher returning from the previous year. We anticipate that the entire hands-on experience will stimulate their interest towards engineering or STEM fields.

See more of this Session: Interactions with K-12
See more of this Group/Topical: Education

(542d) Chemical Engineering TIGERs – Practices at Auburn University

William E. Josephson[1], David Mills[1], Christopher Roberts[1], Bonnie Wilson[2] and Tiffany Ostertag[3], (1) Chemical Engineering, Auburn University, Auburn, AL, (2)WISE Institute, Auburn University, Auburn, AL, (3)Samuel Ginn College of Engineering, Auburn University, Auburn, AL

Auburn University's Samuel Ginn College of Engineering conducts a number of outreach programs designed to expose K-12 students to various aspects of engineering. Among these programs are the annual Teams and Individuals Guided by Engineering Resources (TIGERs) Camps. The TIGERs Camps are weeklong residential summer camps that immerse rising 8th - 11th graders in a number of different activities pertaining to the engineering disciplines. The typical structure of the camps includes campers spending half-day sessions with individual engineering departments. The Chemical Engineering Department has been involved with the TIGERs Camps since their inception. Over the years we have refined our sessions and attempted to continuously improve our offering to the TIGERs. We have succeeded in operating a program that is highly rated by the participants and stimulates their interest in engineering. In this work we describe the activities and examine what makes for a successful program. During the course of a given summer two camps are conducted: one for rising 8th and 9th graders and one for rising 10th and 11th graders. During the Chemical Engineering Department's session the campers are split up into teams and shuttle among 5 or so different activity stations. From the departmental perspective these activities involve chemical engineering topics such as transport phenomena and reaction engineering as well as more general engineering matters such as design of experiment methodologies and technical communications. The TIGERs Camper perspective of these activities, with names such as "Water Challenge" and "The Perfect Cookie", is obviously a bit different. Many of the activities have some sort of inherent scoring element (e.g., make a cookie that's rated by outside tasters) and this naturally leads to a competitive aspect to the entire program. We have found that this competition enhances the TIGERs experience, stimulates the TIGERs Campers and helps them understand engineering principles.

See more of this Session: Interactions with K-12
See more of this Group/Topical: Education

(542e) Self- Assembly and Nanotechnology: Real-Time, Hands- On, and Safe Experiments for K-12 Students

Hitesh G. Bagaria, Dept of Chemical and Biomolecular Engineering, Rice University, Houston, TX, Michelle R. Dean, Bellaire High School, Houston, TX, Carolyn A. Nichol, Center for Biological and Environmental Nanotechnology, Rice University, Houston, TX and Michael S. Wong, Chemical and Biomolecular Engineering and Chemistry, Rice University, Houston, TX

One question that students and teachers often ask is how are nano-sized things made when they are so small? One answer is through the process of self assembly, in which molecules, polymers, and nanoparticles can connect to form larger objects of a defined structure and shape. Two hands-on experiment procedures are presented, in which students can prepare capsules in real time using simple and safe ingredients, and can study them for encapsulation and release of food coloring dye. These experiments provide a visual and interactive manner for the concepts of self-assembly (as a synthesis tool) and nanotechnology (in which nanomaterial can be made to perform useful functions) to be learned.

See more of this Session: Interactions with K-12
See more of this Group/Topical: Education

(542f) Assessing Fundamental Conceptual Understanding with a High School Outreach Project

Taryn M. Bayles, University of Maryland Baltimore County, Baltimore, MD

In order to create an interest among high school students in STEM fields a high school outreach program was developed. Students from a core Chemical & Biochemical Engineering class (Transport Phenomena II: Heat and Mass Transfer) are required to go to a local high school or middle school and make a presentation to increase the high school students' awareness of the importance of mathematics, physics, chemistry, and biology to the field of engineering. Teams of 3-5 college students deliver a presentation, provide hands-on activities, and provide an evaluation to be completed by the class and the teacher. This project provides the college students with an opportunity to be creative, to share their experiences with high school students, and to introduce the high school students to technical areas and careers that they might not have considered. During the presentation, the high school students are made aware of the various paths and diverse coursework that college students have taken in order to study engineering and what they plan to do upon graduation. The college students also discuss what skills they learned in high school that have been helpful in their college education and during their summer research experiences and internships.

Hands-on activities tie concepts that the high school students have learned to what the college students are learning in their advanced university courses. The college students are then required to explain how the activities relate to practical industrial applications. The college students are allowed to form their own teams, select the school, and the topic that they will teach. Each team must practice their presentation with the instructor before they are allowed to present to the high school or middle schools students; and in addition the instructor attends each of the school presentations. The college students must also critically assess the efficacy of their outreach project – and determine how this can be measured. Through this project the college students have the opportunity to practice their communication skills with an audience which has little technical background – which has resulted in being able to assess fundamental conceptual understanding for each of the individual college students.

See more of this Session: Interactions with K-12
See more of this Group/Topical: Education

(542g) Research Experience for Teachers in Manufacturing for Competitiveness in the United States

Joseph J. Biernacki[1], Donald P. Visco[2], Mario Oyanader[1], Holly Stretz[1] and Mohamed Abdelrahman[3], (1)Chemical Engineering, Tennessee Technological University, Cookeville, TN, (2)Chemical Enigneering, Tennessee Technological University, Cookeville, TN, (3)Electrical Engineering, Tennessee Technological University, Cookeville, TN

Research experiences for teachers are one way to interject engineering content into existing science and mathematics programs across the K-12 curriculum. The National Science Foundation (NSF) Research Experience for Teachers (RET) program is one initiative that offers funding to organize and implement interactions between university faculty and K-12 teachers. Tennessee Technological University (TTU) presently has an RET site program called Research Experience for Teachers in Manufacturing for Competitiveness in the United States (RETainUS). The TTU program endeavors to immerse teachers from local high schools in research activities, mentored by TTU engineering faculty. The faculty mentors represent an interdisciplinary group representing electrical, mechanical and chemical engineering. TTU's Department of Chemical Engineering has four faculty participants working with teachers to develop a broad spectrum of applications for classroom and laboratory use. Through the training or retraining of teachers RET is one way to reach large numbers of students. A brief overview of the RETainUS program will be given along with descriptions of teacher research, course/laboratory materials developed and outcomes realized after implementation.

See more of this Session: Interactions with K-12
See more of this Group/Topical: Education

AUTHOR INDEX

AUTHOR INDEX

AUTHOR INDEX

AUTHOR INDEX

American Institute of Chemical Engineers
3 Park Avenue
New York, New York 10016-5991